Universitext

Bernt Øksendal · Agnès Sulem

Applied Stochastic Control of Jump Diffusions

2nd Edition

With 27 Figures

 Springer

Bernt Øksendal
University of Oslo
Department of Mathematics
0316 Oslo
Norway
e-mail: oksendal@math.uio.no

Angnès Sulem
INRIA Rocquencourt
Domaine de Voluceau
78153 Le Chesnay CX
France
e-mail: agnes.sulem@inria.fr

Mathematics Subject Classification (2000): 93E20, 60G40, 60G51, 49L25, 65MXX, 47J20, 49J40, 91B28

Library of Congress Control Number: 2007921874

ISBN-10: 3-540-69825-6 Springer Berlin Heidelberg New York
ISBN-13: 978-3-540-69825-8 Springer Berlin Heidelberg New York
ISBN-13: 978-3-540-14023-8 1st ed. Springer Berlin Heidelberg New York

Springer is a part of Springer Science+Business Media
springer.com
© Springer-Verlag Berlin Heidelberg 2004, 2007

Cover design: WMXDesign, Heidelberg
Typesetting by the authors and SPi using a Springer LaTeX macro package

Printed on acid-free paper   SPIN: 11872405   41/2141/SPi   5 4 3 2 1 0

To my family

Eva, Elise, Anders, and Karina

B. Ø.

A tous ceux qui m'accompagnent

A. S.

Preface to the Second Edition

In this second edition, we have added a chapter on optimal control of random jump fields (solutions of stochastic *partial* differential equations) and partial information control (Chap. 10). We have also added a section on optimal stopping with delayed information (Sect. 2.3). It has always been our intention to give a contemporary presentation of applied stochastic control, and we hope that the addition of these recent developments will contribute in this direction.

We have also made a number of corrections and other improvements, many of them based on helpful comments from our readers. In particular, we would like to thank Andreas Kyprianou for his valuable communications. We are also grateful to (in alphabetical order) Knut Aase, Jean-Philippe Chancelier, Inga Eide, Emil Framnes, Arne-Christian Lund, Jose-Luis Menaldi, Tamás K. Papp, Atle Seierstad, and Jens Arne Sukkestad for pointing out errors and suggesting improvements. Our special thanks go to Martine Verneuille for her skillful typing.

Oslo and Paris, November 2006   Bernt Øksendal and Agnès Sulem

Preface of the First Edition

Jump diffusions are solutions of stochastic differential equations driven by Lévy processes. Since a Lévy process $\eta(t)$ can be written as a linear combination of t, a Brownian motion $B(t)$ and a pure jump process, jump diffusions represent a natural and useful generalization of Itô diffusions. They have received a lot of attention in the last years because of their many applications, particularly in economics.

There exist today several excellent monographs on Lévy processes. However, very few of them – if any – discuss the optimal control, optimal stopping, and impulse control of the corresponding jump diffusions, which is the subject of this book. Moreover, our presentation differs from these books in that it emphasizes the applied aspect of the theory. Therefore, we focus mostly on useful verification theorems and we illustrate the use of the theory by giving examples and exercises throughout the text. Detailed solutions of some of the exercises are given at the end of the book. The exercises to which a solution is provided, are marked with an asterix *. It is our hope that this book will fill a gap in the literature and that it will be a useful text for students, researchers, and practitioners in stochastic analysis and its many applications. Although most of our results are motivated by examples in economics and finance, the results are general and can be applied in a wide variety of situations. To emphasize this, we have also included examples in biology and physics/engineering.

This book is partially based on courses given at the Norwegian School of Economics and Business Administration (NHH) in Bergen, Norway, during the Spring semesters 2000 and 2002, at INSEA in Rabat, Morocco in September 2000, at Odense University in August 2001 and at ENSAE in Paris in February 2002.

Oslo and Paris, August 2004   Bernt Øksendal and Agnès Sulem

Acknowledgments   We are grateful to many people who in various ways have contributed to these lecture notes. In particular, we thank Knut Aase, Fred Espen Benth, Jean-Philippe Chancelier, Rama Cont, Hans Marius Eikseth, Nils Christian Framstad, Jørgen Haug, Monique Jeanblanc, Kenneth Karlsen, Arne-Christian Lund, Thilo Meyer-Brandis, Cloud Makasu, Sure Mataramvura, Peter Tankov, and Jan Ubøe for their valuable help. We also thank Francesca Biagini for useful comments and suggestions to the text and her detailed solutions of some of the exercises. We are grateful to Dina Haraldsson and Martine Verneuille for proficient typing and Eivind Brodal for his kind assistance. We acknowledge with gratitude the support by the French–Norwegian cooperation project Stochastic Control and Applications, Aur 99-050.

Oslo and Paris, August 2004   Bernt Øksendal and Agnès Sulem

Contents

Stochastic Calculus with Jump Diffusions

1.1 Basic Definitions and Results on Lévy Processes

In this chapter we present the basic concepts and results needed for the applied calculus of jump diffusions. Since there are several excellent books which give a detailed account of this basic theory, we will just briefly review it here and refer the reader to these books for more information.

Definition 1.1. *Let $(\Omega, \mathcal{F}, \{\mathcal{F}_t\}_{t \geq 0}, P)$ be a filtered probability space. An $\mathcal{F}_t$-adapted process $\{\eta(t)\}_{t \geq 0} = \{\eta_t\}_{t \geq 0} \subset \mathbb{R}$ with $\eta_0 = 0$ a.s. is called a* Lévy *process if η_t is continuous in probability and has stationary and independent increments.*

Theorem 1.2. *Let $\{\eta_t\}$ be a Lévy process. Then η_t has a càdlàg version (right continuous with left limits) which is also a Lévy process.*

Proof. See, e.g., [P, S]. □

In view of this result we will from now on assume that the Lévy processes we work with are càdlàg.

The *jump* of η_t at $t \geq 0$ is defined by

$$\Delta \eta_t = \eta_t - \eta_{t-}. \tag{1.1.1}$$

Let $\mathbf{B}_0$ be the family of Borel sets $U \subset \mathbb{R}$ whose closure $\bar{U}$ does not contain 0. For $U \in \mathbf{B}_0$ we define

$$N(t, U) = N(t, U, \omega) = \sum_{s:0 < s \leq t} \mathcal{X}_U(\Delta \eta_s). \tag{1.1.2}$$

In other words, $N(t, U)$ is the number of jumps of size $\Delta \eta_s \in U$ which occur before or at time t. $N(t, U)$ is called the *Poisson random measure (or jump measure)* of $\eta(\cdot)$.

Remark 1.3. Note that $N(t, U)$ is *finite* for all $U \in \mathbf{B}_0$. To see this we proceed as follows: Define

$$T_1(\omega) = \inf\{t > 0; \eta_t \in U\}.$$

We claim that $T_1(\omega) > 0$ a.s. To prove this note that by right continuity of paths we have

$$\lim_{t \to 0^+} \eta(t) = \eta(0) = 0 \quad \text{a.s.}$$

Therefore, for all $\varepsilon > 0$ there exists $t(\varepsilon) > 0$ such that $|\eta(t)| < \varepsilon$ for all $t < t(\varepsilon)$. This implies that $\eta(t) \notin U$ for all $t < t(\varepsilon)$, if $\varepsilon < \mathrm{dist}(0, U)$.

Next define inductively

$$T_{n+1}(\omega) = \inf\{t > T_n(\omega); \Delta\eta_t \in U\}.$$

Then by the above argument $T_{n+1} > T_n$ a.s. We claim that

$$T_n \to \infty \quad \text{as} \quad n \to \infty, \text{a.s.}$$

Assume not. Then $T_n \to T < \infty$. But then

$$\lim_{s \to T^-} \eta(s) \quad \text{cannot exist,}$$

contradicting the existence of left limits of the paths.

It is well known that Brownian motion $\{B(t)\}_{t \geq 0}$ has stationary and independent increments. Thus $B(t)$ is a Lévy process. Another important example is the following.

Example 1.4 (The Poisson Process). The Poisson process $\pi(t)$ of intensity $\lambda > 0$ is a Lévy process taking values in $\mathbb{N} \cup \{0\}$ and such that

$$P[\pi(t) = n] = \frac{(\lambda t)^n}{n!} \mathrm{e}^{-\lambda t}; \quad n = 0, 1, 2, \dots$$

Theorem 1.5. *[P, Theorem 1.35].*

1. *The set function $U \to N(t, U, \omega)$ defines a σ-finite measure on $\mathbf{B}_0$ for each fixed t, ω. The differential form of this measure is written $N(t, \mathrm{d}z)$.*
2. *The set function $[a, b) \times U \to N(b, U, \omega) - N(a, U, \omega); [a, b) \subset [0, \infty), U \in \mathbf{B}_0$ defines a σ-finite measure for each fixed ω. The differential form of this measure is written $N(\mathrm{d}t, \mathrm{d}z)$.*
3. *The set function*

$$\nu(U) = E[N(1, U)], \tag{1.1.3}$$

   where $E = E_P$ denotes expectation with respect to P, also defines a σ-finite measure on $\mathbf{B}_0$, called the Lévy measure of $\{\eta_t\}$.
4. *Fix $U \in \mathbf{B}_0$. Then the process*

$$\pi_U(t) := \pi_U(t, \omega) := N(t, U, \omega)$$

   is a Poisson process of intensity $\lambda = \nu(U)$.

Example 1.6 (The Compound Poisson Process). Let $X(n)$, $n \in \mathbb{N}$ be a sequence of i.i.d. random variables taking values in $\mathbb{R}$ with common distribution $\mu_{X(1)} = \mu_X$ and let $\pi(t)$ be a Poisson process of intensity λ, independent of all the $X(n)$s.

The *compound Poisson process* $Y(t)$ is defined by

$$Y(t) = X(1) + \cdots + X(\pi(t)), \quad t \geq 0. \tag{1.1.4}$$

An increment of this process is given by

$$Y(s) - Y(t) = \sum_{k=\pi(t)+1}^{\pi(s)} X(k), \quad s > t.$$

This is independent of $X(1), \ldots, X(\pi(t))$, and its distribution depends only on the difference $(s - t)$ and on the distribution of $X(1)$. Thus $Y(t)$ is a Lévy process.

To find the Lévy measure ν of $Y(t)$ note that if $U \in \mathbf{B}_0$ then

$$\nu(U) = E[N(1, U)] = E\left[\sum_{s; 0 < s \leq 1} \mathcal{X}_U(\Delta Y(s)) \right]$$

$$= E[(\text{number of jumps}) \times X_U(\text{jump})] = E[\pi(1)\mathcal{X}_U(X)] = \lambda\mu_X(U),$$

by independence. We conclude that

$$\nu = \lambda\mu_X. \tag{1.1.5}$$

This shows that a Lévy process can be represented by a compound Poisson process if and only if its Lévy measure is finite. Note, however, that there are many interesting Lévy processes η_t with infinite Lévy measure ν, in fact even with $\int_{|z|<1} |z|\nu(\mathrm{d}z) = \infty$. See, e.g., [B]. In general, one can prove that for any fixed $R > 0$ the processes

$$M_t^{(k)} := \int_{1/k \leq |z| \leq R} z(N(t, \mathrm{d}z) - t\nu(\mathrm{d}z)), \quad k = 1, 2, \ldots$$

are $L^2(P)$-martingales and they converge in $L^2(P)$ to a martingale M_t denoted by

$$M_t = \int_{|z| \leq R} z(N(t, \mathrm{d}z) - t\nu(\mathrm{d}z)).$$

In fact we have

Theorem 1.7 (Itô–Lévy Decomposition [JS]). *Let $\{\eta_t\}$ be a Lévy process. Then η_t has the decomposition*

$$\eta_t = \alpha t + \sigma B(t) + \int_{|z|<R} z\tilde{N}(t, \mathrm{d}z) + \int_{|z|\geq R} zN(t, \mathrm{d}z), \tag{1.1.6}$$

for some constants $\alpha \in \mathbb{R}$, $\sigma \in \mathbb{R}$, $R \in [0, \infty]$. *Here*

$$\widetilde{N}(\mathrm{d}t, \mathrm{d}z) = N(\mathrm{d}t, \mathrm{d}z) - \nu(\mathrm{d}z)\mathrm{d}t \qquad (1.1.7)$$

is the compensated Poisson random measure *of* $\eta(\cdot)$, *and* $B(t)$ *is a Brownian motion independent of* $\widetilde{N}(\mathrm{d}t, \mathrm{d}z)$. *For each* $A \in \mathbf{B}_0$ *the process*

$$M_t := \widetilde{N}(t, A) \quad \text{is a martingale.} \qquad (1.1.8)$$

If $\alpha = 0$ *and* $R = \infty$, *we call* η_t *a Lévy martingale.*

Theorem 1.8. *We can always choose* $R = 1$. *If*

$$E|\eta_t| < \infty \quad \text{for all } t \geq 0,$$

then

$$\int_{|z| \geq 1} |z|\nu(\mathrm{d}z) < \infty$$

and hence we may choose $R = \infty$ *and write*

$$\eta_t = \alpha t + \sigma B(t) + \int_{\mathbb{R}} z\widetilde{N}(t, \mathrm{d}z).$$

(See [S, Theorem 25.3].)

Theorem 1.9 ([P]). *A Lévy process is a strong Markov process.*

Theorem 1.10 (The Lévy–Khintchine formula [P]). *Let* $\{\eta_t\}$ *be a Lévy process with Lévy measure* ν. *Then* $\int_{\mathbb{R}} \min(1, z^2)\nu(\mathrm{d}z) < \infty$ *and*

$$E[\mathrm{e}^{\mathrm{i}u\eta_t}] = \mathrm{e}^{t\psi(u)}, \quad u \in \mathbb{R}, \qquad (1.1.9)$$

where

$$\psi(u) = -\frac{1}{2}\sigma^2 u^2 + \mathrm{i}\alpha u + \int_{|z| < R} \{\mathrm{e}^{\mathrm{i}uz} - 1 - \mathrm{i}uz\}\nu(\mathrm{d}z) + \int_{|z| \geq R} (\mathrm{e}^{\mathrm{i}uz} - 1)\nu(\mathrm{d}z).$$

$$(1.1.10)$$

Conversely, given constants α, σ^2, *and a measure* ν *on* $\mathbf{B}_0$ *s.t.*

$$\int_{\mathbb{R}} \min(1, z^2)\nu(\mathrm{d}z) < \infty,$$

there exists a Lévy process $\eta(t)$ *(unique in law) such that (1.1.9–1.1.10) hold.*

Note. It is possible that $\int_{|z| \leq R} |z|\nu(\mathrm{d}z) = \infty$.

Theorem 1.11. *[P, Corollary p. 48]. A Lévy process is a* semimartingale.

Definition 1.12. *[P]. Let* $\mathbf{D}_{\mathrm{ucp}}$ *denote the space of càdlàg adapted processes, equipped with the topology of uniform convergence on compacts in probability* (ucp) : $H_n \to H$ ucp *if for all* $t > 0 \sup_{0 \le s \le t} |H_n(s) - H(s)| \to 0$ *in probability* ($A_n \to A$ *in probability if for all* $\varepsilon > 0$ *there exists* $n_\varepsilon \in \mathbb{N}$ *such that* $n \ge n_\varepsilon \Rightarrow$ Prob.$(|A_n - A| > \varepsilon) < \varepsilon$).

Let $\mathbf{L}_{\mathrm{ucp}}$ *denote the space of adapted* càglàd *processes (left continuous with right limits), equipped with the* ucp *topology. If* $H(t)$ *is a step function of the form*

$$H(t) = H_0 \mathcal{X}_{\{0\}}(t) + \sum_i H_i \mathcal{X}_{(T_i, T_{i+1}]}(t),$$

where $H_i \in \mathcal{F}_{T_i}$ *and* $0 = T_0 \le T_1 \le \cdots \le T_{n+1} < \infty$ *are* $\mathcal{F}_t$*-stopping times and* X *is càdlàg, we define*

$$J_X H(t) := \int_0^t H_s \mathrm{d}X_s := H_0 X_0 + \sum_i H_i (X_{T_{i+1} \wedge t} - X_{T_i \wedge t}), \quad t \ge 0.$$

Theorem 1.13. *[P, p. 51]. Let* X *be a semimartingale. Then the mapping* J_X *can be extended to a continuous linear map*

$$J_X : \mathbf{L}_{\mathrm{ucp}} \to \mathbf{D}_{\mathrm{ucp}}.$$

This construction allows us to define stochastic integrals of the form

$$\int_0^t H(s) \mathrm{d}\eta_s$$

for all $H \in \mathbf{L}_{\mathrm{ucp}}$. (See also Remark 1.18.) In view of the decomposition (1.1.6) this integral can be split into integrals with respect to $\mathrm{d}s$, $\mathrm{d}B(s)$, $\tilde{N}(\mathrm{d}s, \mathrm{d}z)$, and $N(\mathrm{d}s, \mathrm{d}z)$. This makes it natural to consider the more general stochastic integrals of the form

$$X(t) = X(0) + \int_0^t \alpha(s, \omega)\mathrm{d}s + \int_0^t \beta(s, \omega)\mathrm{d}B(s) + \int_0^t \int_{\mathbb{R}} \gamma(s, z, \omega)\bar{N}(\mathrm{d}s, \mathrm{d}z),$$
$$(1.1.11)$$

where the integrands are satisfying the appropriate conditions for the integrals to exist and we for simplicity have put

$$\bar{N}(\mathrm{d}s, \mathrm{d}z) = \begin{cases} N(\mathrm{d}s, \mathrm{d}z) - \nu(\mathrm{d}z)\mathrm{d}s & \text{if } |z| < R \\ N(\mathrm{d}s, \mathrm{d}z) & \text{if } |z| \ge R, \end{cases}$$

with R as in Theorem 1.7. As is customary we will use the following shorthand differential notation for processes $X(t)$ satisfying (1.1.11):

$$\mathrm{d}X(t) = \alpha(t)\mathrm{d}t + \beta(t)\mathrm{d}B(t) + \int_{\mathbb{R}} \gamma(t, z)\bar{N}(\mathrm{d}t, \mathrm{d}z). \qquad (1.1.12)$$

We call such processes *Itô–Lévy processes*.

Recall that a semimartingale $M(t)$ is called a *local* martingale up to time T (with respect to P) if there exists an increasing sequence of $\mathcal{F}_t$-stopping times τ_n such that $\lim_{n\to\infty}\tau_n = T$ a.s. and

$$M(t \wedge \tau_n) \quad \text{is a martingale with respect to } P \text{ for all } n.$$

Note that
  1. If

$$E\left[\int_0^T \int_{\mathbb{R}} \gamma^2(t,z)\nu(\mathrm{d}z)\mathrm{d}t\right] < \infty, \tag{1.1.13}$$

then the process

$$M(t) := \int_0^t \int_{\mathbb{R}} \gamma(t,z)\tilde{N}(\mathrm{d}t,\mathrm{d}z), \quad 0 \le t \le T$$

is a *martingale*.
  2. If

$$\int_0^T \int_{\mathbb{R}} \gamma^2(t,z)\nu(\mathrm{d}z)\mathrm{d}t < \infty \text{ a.s.,} \tag{1.1.14}$$

then $M(t)$ is a *local* martingale, $0 \le t \le T$.

1.2 The Itô Formula and Related Results

We now come to the important Itô formula for Itô–Lévy processes:

If $X(t)$ is given by (1.1.12) and $f : \mathbb{R}^2 \to \mathbb{R}$ is a C^2 function, is the process $Y(t) := f(t, X(t))$ again an Itô–Lévy process and if so, how do we represent it in the form (1.1.12)?

If we argue heuristically and use our knowledge of the classical Itô formula it is easy to guess what the answer is:

Let $X^{(c)}(t)$ be the continuous part of $X(t)$, i.e., $X^{(c)}(t)$ is obtained by removing the jumps from $X(t)$. Then an increment in $Y(t)$ stems from an increment in $X^{(c)}(t)$ plus the jumps (coming from $N(\cdot,\cdot)$). Hence in view of the classical Itô formula we would guess that

$$\mathrm{d}Y(t) = \frac{\partial f}{\partial t}(t, X(t))\mathrm{d}t + \frac{\partial f}{\partial x}(t, X(t))\mathrm{d}X^{(c)}(t) + \frac{1}{2}\frac{\partial^2 f}{\partial x^2}(t, X(t)) \cdot \beta^2(t)\mathrm{d}t$$

$$+ \int_{\mathbb{R}} \{f(t, X(t^-) + \gamma(t,z)) - f(t, X(t^-))\}N(\mathrm{d}t,\mathrm{d}z).$$

It can be proved that our guess is correct. Since

$$\mathrm{d}X^{(c)}(t) = \left(\alpha(t) - \int_{|z|<R} \gamma(t,z)\nu(\mathrm{d}z)\right)\mathrm{d}t + \beta(t)\mathrm{d}B(t),$$

this gives the following result.

Theorem 1.14 (The One-Dimensional Itô Formula [BL, A, P]). *Suppose $X(t) \in \mathbb{R}$ is an Itô–Lévy process of the form*

$$\mathrm{d}X(t) = \alpha(t, \omega)\mathrm{d}t + \beta(t, \omega)\mathrm{d}B(t) + \int_{\mathbb{R}} \gamma(t, z, \omega)\bar{N}(\mathrm{d}t, \mathrm{d}z), \qquad (1.2.1)$$

where

$$\bar{N}(\mathrm{d}t, \mathrm{d}z) = \begin{cases} N(\mathrm{d}t, \mathrm{d}z) - \nu(\mathrm{d}z)\mathrm{d}t & \text{if } |z| < R \\ N(\mathrm{d}t, \mathrm{d}z) & \text{if } |z| \geq R \end{cases} \qquad (1.2.2)$$

for some $R \in [0, \infty]$.

   Let $f \in C^2(\mathbb{R}^2)$ and define $Y(t) = f(t, X(t))$. Then $Y(t)$ is again an Itô–Lévy process and

$$\mathrm{d}Y(t) = \frac{\partial f}{\partial t}(t, X(t))\mathrm{d}t + \frac{\partial f}{\partial x}(t, X(t))[\alpha(t, \omega)\mathrm{d}t + \beta(t, \omega)\mathrm{d}B(t)]$$

$$+ \frac{1}{2}\beta^2(t, \omega)\frac{\partial^2 f}{\partial x^2}(t, X(t))\mathrm{d}t$$

$$+ \int_{|z| < R} \left\{ f(t, X(t^-) + \gamma(t, z)) - f(t, X(t^-)) \right.$$

$$\left. - \frac{\partial f}{\partial x}(t, X(t^-))\gamma(t, z) \right\} \nu(\mathrm{d}z)\mathrm{d}t$$

$$+ \int_{\mathbb{R}} \{f(t, X(t^-) + \gamma(t, z)) - f(t, X(t^-))\}\bar{N}(\mathrm{d}t, \mathrm{d}z). \qquad (1.2.3)$$

Note. If $R = 0$ then $\bar{N} = N$ everywhere.
       If $R = \infty$ then $\bar{N} = \tilde{N}$ everywhere.

Example 1.15 (The Geometric Lévy Process). Consider the stochastic differential equation (SDE)

$$\mathrm{d}X(t) = X(t^-)\left[\alpha\, \mathrm{d}t + \beta\, \mathrm{d}B(t) + \int_{\mathbb{R}} \gamma(t, z)\bar{N}(\mathrm{d}t, \mathrm{d}z)\right], \qquad (1.2.4)$$

where α, β are constants and $\gamma(t, z) \geq -1$. To find the solution $X(t)$ of this equation we rewrite it as follows:

$$\frac{\mathrm{d}X(t)}{X(t^-)} = \alpha\, \mathrm{d}t + \beta\, \mathrm{d}B(t) + \int_{\mathbb{R}} \gamma(t, z)\bar{N}(\mathrm{d}t, \mathrm{d}z).$$

Now define

$$Y(t) = \ln X(t).$$

Then by Itô's formula,

$$
\begin{aligned}
dY(t) = {} & \frac{X(t)}{X(t)} [\alpha \, dt + \beta \, dB(t)] - \frac{1}{2} \beta^2 X^{-2}(t) X^2(t) dt \\
& + \int_{|z|<R} \{\ln(X(t^-) + \gamma(t,z) X(t^-)) - \ln(X(t^-)) \\
& - X^{-1}(t^-) \gamma(t,z) X(t^-)\} \nu(dz) dt \\
& + \int_{\mathbb{R}} \{\ln(X(t^-) + \gamma(t,z) X(t^-)) - \ln(X(t^-))\} \bar{N}(dt, dz) \\
= {} & \left(\alpha - \frac{1}{2}\beta^2\right) dt + \beta \, dB(t) + \int_{|z|<R} \{\ln(1+\gamma(t,z)) - \gamma(t,z)\} \nu(dz) dt \\
& + \int_{\mathbb{R}} \ln(1+\gamma(t,z)) \bar{N}(dt, dz).
\end{aligned}
$$

Hence

$$
\begin{aligned}
Y(t) = {} & Y(0) + \left(\alpha - \frac{1}{2}\beta^2\right) t + \beta B(t) + \int_0^t \int_{|z|<R} \{\ln(1 + \gamma(s,z)) \\
& - \gamma(s,z)\} \nu(dz) ds + \int_0^t \int_{\mathbb{R}} \ln(1+\gamma(s,z)) \bar{N}(ds, dz)
\end{aligned}
$$

and this gives the solution

$$
\begin{aligned}
X(t) = {} & X(0) \exp \Bigg\{ \left(\alpha - \frac{1}{2}\beta^2\right) t + \beta B(t) \\
& + \int_0^t \int_{|z|<R} \{\ln(1 + \gamma(s,z)) - \gamma(s,z)\} \nu(dz) ds \\
& + \int_0^t \int_{\mathbb{R}} \ln(1 + \gamma(s,z)) \bar{N}(ds, dz) \Bigg\}. \tag{1.2.5}
\end{aligned}
$$

In analogy with the diffusion case ($N = 0$) we call this process $X(t)$ a *geometric Lévy process*. It is often used as a model for stock prices. See, e.g., [B].

Next we formulate the corresponding multidimensional version of Theorem 1.14.

Theorem 1.16 (The Multidimensional Itô Formula). *Let $X(t) \in \mathbb{R}^n$ be an Itô–Lévy process of the form*

$$
dX(t) = \alpha(t, \omega) dt + \sigma(t, \omega) dB(t) + \int_{\mathbb{R}^\ell} \gamma(t, z, \omega) \bar{N}(dt, dz), \tag{1.2.6}
$$

where $\alpha : [0, T] \times \Omega \to \mathbb{R}^n$, $\sigma : [0, T] \times \Omega \to \mathbb{R}^{n \times m}$, and $\gamma : [0, T] \times \mathbb{R}^\ell \times \Omega \to \mathbb{R}^{n \times \ell}$ are adapted processes such that the integrals exist. Here $B(t)$ is an m-dimensional Brownian motion and

$$\bar{N}(dt, dz)^T = (\bar{N}_1(dt, dz_1), \ldots, \bar{N}_\ell(dt, dz_\ell))$$

$$= (N_1(dt, dz_1) - \mathcal{X}_{|z_1|<R_1}\nu_1(dz_1)dt, \ldots, N_\ell(dt, dz_\ell)$$

$$- \mathcal{X}_{|z_\ell|<R_\ell}\nu_\ell(dz_\ell)dt),$$

where $\{N_j\}$ are independent Poisson random measures with Lévy measures ν_j coming from ℓ independent (one-dimensional) Lévy processes $\eta_1, \ldots, \eta_\ell$.

Note that each column $\gamma^{(k)}$ of the $n \times \ell$ matrix $\gamma = [\gamma_{ij}]$ depends on z only through the kth coordinate z_k, i.e.,

$$\gamma^{(k)}(t, z, \omega) = \gamma^{(k)}(t, z_k, \omega), \quad z = (z_1, \ldots, z_\ell) \in \mathbb{R}^\ell.$$

Thus the integral on the right of (1.2.6) is just a shorthand matrix notation. When written out in detail component number i of $X(t)$ in (1.2.6), $X_i(t)$ gets the form

$$dX_i(t) = \alpha_i(t, \omega)dt + \sum_{j=1}^m \sigma_{ij}(t, \omega)dB_j(t)$$

$$+ \sum_{j=1}^\ell \int_{\mathbb{R}} \gamma_{ij}(t, z_j, \omega)\bar{N}_j(dt, dz_j), \quad 1 \le i \le n. \quad (1.2.7)$$

Let $f \in C^{1,2}([0, T]) \times \mathbb{R}^n; \mathbb{R})$. Put $Y(t) = f(t, X(t))$. Then

$$dY(t) = \frac{\partial f}{\partial t}dt + \sum_{i=1}^n \frac{\partial f}{\partial x_i}(\alpha_i dt + \sigma_i dB(t)) + \frac{1}{2}\sum_{i,j=1}^n (\sigma\sigma^T)_{ij}\frac{\partial^2 f}{\partial x_i \partial x_j}dt$$

$$+ \sum_{k=1}^\ell \int_{|z_k|<R_k} \left\{ f(t, X(t^-) + \gamma^{(k)}(t, z_k)) - f(t, X(t^-)) \right.$$

$$\left. - \sum_{i=1}^n \gamma_i^{(k)}(t, z_k)\frac{\partial f}{\partial x_i}(X(t^-)) \right\} \nu_k(dz_k)dt$$

$$+ \sum_{k=1}^\ell \int_{\mathbb{R}} \{f(t, X(t^-) + \gamma^{(k)}(t, z_k)) - f(t, X(t^-))\}\bar{N}_k(dt, dz_k),$$

$$(1.2.8)$$

where $\gamma^{(k)} \in \mathbb{R}^n$ is column number k of the $n \times \ell$ matrix $\gamma = [\gamma_{ik}]$ and $\gamma_i^{(k)} = \gamma_{ik}$ is the coordinate number i of $\gamma^{(k)}$.

Theorem 1.17 (The Itô–Lévy Isometry). Let $X(t) \in \mathbb{R}^n$ be as in (1.2.6) but with $X(0) = 0$ and $\alpha = 0$. Then

$$E[X^2(T)] = E\left[\int_0^T \left\{\sum_{i=1}^n \sum_{j=1}^m \sigma_{ij}^2(t) + \sum_{i=1}^n \sum_{j=1}^\ell \int_{\mathbb{R}} \gamma_{ij}^2(t, z_j)\nu_j(\mathrm{d}z_j)\right\} \mathrm{d}t\right]$$

$$= \sum_{i=1}^n E\left[\int_0^T \left\{\sum_{j=1}^m \sigma_{ij}^2(t) + \sum_{j=1}^\ell \int_{\mathbb{R}} \gamma_{ij}^2(t, z_j)\nu_j(\mathrm{d}z_j)\right\} \mathrm{d}t\right] \quad (1.2.9)$$

provided that the right-hand side is finite.

Proof. This follows from the Itô formula applied to $f(t, x) = x^2 = |x|^2$. We omit the details. □

Remark 1.18. As a special case of Theorem 1.17 assume that

$$\mathrm{d}X(t) = \mathrm{d}\eta(t) = \int_{\mathbb{R}} z\tilde{N}(\mathrm{d}t, \mathrm{d}z) \in \mathbb{R}$$

with $E[X^2(T)] = T\int_{\mathbb{R}} z^2\nu(\mathrm{d}z) < \infty$. Then we get the isometry

$$E\left[\left(\int_0^T H(t)\mathrm{d}\eta(t)\right)^2\right] = E\left[\int_0^T H^2(t)\mathrm{d}t\right]\int_{\mathbb{R}} z^2\nu(\mathrm{d}z)$$

for all $H \in \mathbf{L}_{\mathrm{ucp}}$ (see Definition 1.12) such that $H \in L^2([0, T] \times \Omega)$, i.e., such that

$$\|H\|_{L^2([0,T]\times\Omega)}^2 := E\left[\int_0^T H^2(t)\mathrm{d}t\right] < \infty.$$

Using this we can in the usual way extend the definition of the integral

$$\int_0^T Y(t)\mathrm{d}\eta(t) \in L^2(\Omega)$$

to all processes $Y(t)$ which are limits in $L^2([0, T] \times \Omega)$ of processes $H_n(t) \in \mathbf{L}_{\mathrm{ucp}} \cap L^2([0, T] \times \Omega)$. We will call such processes $Y(t)$ *predictable processes.*

1.3 Lévy Stochastic Differential Equations

The geometric Lévy process is an example of a *Lévy diffusion*, i.e., the solution of a SDE driven by Lévy processes.

Theorem 1.19 (Existence and Uniqueness of Solutions of Lévy SDEs). *Consider the following Lévy SDE in $\mathbb{R}^n$: $X(0) = x_0 \in \mathbb{R}^n$ and*

$$\mathrm{d}X(t) = \alpha(t, X(t))\mathrm{d}t + \sigma(t, X(t))\mathrm{d}B(t) + \int_{\mathbb{R}^n} \gamma(t, X(t^-), z)\tilde{N}(\mathrm{d}t, \mathrm{d}z), \quad (1.3.1)$$

where $\alpha : [0,T] \times \mathbb{R}^n \to \mathbb{R}^n$, $\sigma : [0,T] \times \mathbb{R}^n \to \mathbb{R}^{n \times m}$ and $\gamma : [0,T] \times \mathbb{R}^n \times \mathbb{R}^n \to \mathbb{R}^{n \times \ell}$ satisfy the following conditions

(At most linear growth) There exists a constant $C_1 < \infty$ such that

$$\|\sigma(t,x)\|^2 + |\alpha(t,x)|^2 + \int_{\mathbb{R}} \sum_{k=1}^{\ell} |\gamma_k(t,x,z)|^2 \nu_k(\mathrm{d}z_k) \le C_1(1 + |x|^2)$$

for all $x \in \mathbb{R}^n$

(Lipschitz continuity) There exists a constant $C_2 < \infty$ such that

$$\|\sigma(t,x) - \sigma(t,y)\|^2 + |\alpha(t,x) - \alpha(t,y)|^2$$

$$+ \sum_{k=1}^{\ell} \int_{\mathbb{R}} |\gamma^{(k)}(t,x,z_k) - \gamma^{(k)}(t,y,z_k)|^2 \nu_k(\mathrm{d}z_k) \le C_2 |x-y|^2,$$

for all $x, y \in \mathbb{R}^n$.

Then there exists a unique càdlàg adapted solution $X(t)$ such that

$$E[|X(t)|^2] < \infty \quad \text{for all } t.$$

Solutions of Lévy SDEs in the *time homogeneous* case, i.e., when $\alpha(t,x) = \alpha(x)$, $\sigma(t,x) = \sigma(x)$, and $\gamma(t,x,z) = \gamma(x,z)$ are called *jump diffusions* (or *Lévy diffusions*).

Theorem 1.20. *A jump diffusion is a strong Markov process.*

Proof. See [P, Theorem V.32]. □

Definition 1.21. *Let $X(t) \in \mathbb{R}^n$ be a jump diffusion. Then the generator A of X is defined on functions $f : \mathbb{R}^n \to \mathbb{R}$ by*

$$Af(x) = \lim_{t \to 0^+} \frac{1}{t} \{E^x[f(X(t))] - f(x)\} \quad \text{(if the limit exists)},$$

where $E^x[f(X(t))] = E[f(X^{(x)}(t))]$, $X^{(x)}(0) = x$.

Theorem 1.22. *Suppose $f \in C_0^2(\mathbb{R}^n)$. Then $Af(x)$ exists and is given by*

$$Af(x) = \sum_{i=1}^{n} \alpha_i(x) \frac{\partial f}{\partial x_i}(x) + \frac{1}{2} \sum_{i,j=1}^{n} (\sigma \sigma^T)_{ij}(x) \frac{\partial^2 f}{\partial x_i \partial x_j}(x)$$

$$+ \int_{\mathbb{R}} \sum_{k=1}^{\ell} \{f(x + \gamma^{(k)}(x,z)) - f(x) - \nabla f(x) \cdot \gamma^{(k)}(x,z)\} \nu_k(\mathrm{d}z_k).$$

$$(1.3.2)$$

From now on we *define* $Af(x)$ by the expression (1.3.2) for all f such that the partial derivatives of f and the integrals in (1.3.2) exist at x.

Theorem 1.23 (The Dynkin Formula I). *Let $X(t) \in \mathbb{R}^n$ be a jump diffusion and let $f \in C_0^2(\mathbb{R}^n)$. Let τ be a stopping time such that*

$$E^x[\tau] < \infty.$$

Then

$$E^x[f(X(\tau))] = f(x) + E^x\left[\int_0^\tau Af(X(s))\mathrm{d}s\right].$$

Proof. This follows by combining the Itô formula (1.2.8) with the formula (1.3.2) for A and taking expectation.   □

This version is usually strong enough for applications in the case when there are no jumps ($N = 0$). However, for jump diffusions we need the following stronger, localized version.

Theorem 1.24 (The Dynkin Formula II). *Let $X(t) \in \mathbb{R}^n$ be a jump diffusion, $G \subset \mathbb{R}^n$ be an open set and let $f \in C^2(G) \cap C(\bar{G})$. Let $\tau < \infty$ be a stopping time. Suppose that*

$$\tau \leq \tau_G := \inf\{t > 0; X(t) \notin G\} \tag{1.3.3}$$
$$X(\tau) \in \bar{G} \quad a.s. \tag{1.3.4}$$
$$E^x\left[|f(X(\tau))| + \int_0^\tau |Af(X(t))|\mathrm{d}t\right] < \infty. \tag{1.3.5}$$

Then we have

$$E^x[f(X(\tau))] = f(x) + E^x\left[\int_0^\tau (Af)(X(t))\mathrm{d}t\right].$$

Definition 1.25. *In general, if $\{\psi_m\}_{m=1}^\infty$ and g are functions defined on a set $G \subset \mathbb{R}^n$, we say that $\psi_m \to g$ pointwise dominatedly in G if $\psi_m(x) \to g(x)$ for all $x \in G$ and there exists a constant $C < \infty$ such that*

$$|\psi_m(x)| \leq C|g(x)| \quad \text{for all } x \in G, \ m = 1, 2, \ldots$$

Proof of Theorem 1.24. Choose $f_m \in C_0^2(\mathbb{R}^n)$ such that $f_m \to f$ pointwise dominatedly in $\bar{G}$ and $\frac{\partial f_m}{\partial x_i} \to \frac{\partial f}{\partial x_i}, \frac{\partial^2 f_m}{\partial x_i \partial x_j} \to \frac{\partial^2 f}{\partial x_i \partial x_j}$, and $Af_m \to Af$ pointwise dominatedly in G for all $i, j = 1, \ldots, n$. Then apply Theorem 1.23 to each f_m and $\tau \wedge k$, $k = 1, 2, \ldots$ Let $m, k \to \infty$ and apply the dominated convergence theorem.   □

1.4 The Girsanov Theorem and Applications

The Girsanov theorem and the related concept of an equivalent local martingale measure (ELMM) are important in the applications of stochastic analysis to finance. In this chapter we first give a general semimartingale discussion and then we apply it to Itô–Lévy processes. We refer to [Ka] for more details.

Let $(\Omega, \mathcal{F}, \{\mathcal{F}_t\}_{t\geq 0}, P)$ be a filtered probability space. Let Q be another probability measure on $\mathcal{F}_T$. We say that Q is *equivalent* to $P \mid \mathcal{F}_T$ if $P \mid \mathcal{F}_T \ll Q$ and $Q \ll P \mid \mathcal{F}_T$, or, equivalently, if P and Q have the same zero sets in $\mathcal{F}_T$. By the Radon–Nikodym theorem this is the case if and only if we have

$$dQ(\omega) = Z(T)dP(\omega) \quad \text{and} \quad dP(\omega) = Z^{-1}(T)dQ(\omega) \quad \text{on } \mathcal{F}_T$$

for some $\mathcal{F}_T$-measurable random variable $Z(T) > 0$ a.s. P. In that case we also write

$$\frac{dQ}{dP} = Z(T) \quad \text{and} \quad \frac{dP}{dQ} = Z^{-1}(T) \quad \text{on } \mathcal{F}_T. \tag{1.4.1}$$

We first make a simple, but useful observation.

Lemma 1.26. *Suppose $Q \ll P$ with $\frac{dQ}{dP} = Z(T)$ on $\mathcal{F}_T$. Then*

$$Q \mid \mathcal{F}_t \ll P \mid \mathcal{F}_t \quad \text{for all } t \in [0,T] \text{ and}$$

$$Z(t) := \frac{d(Q \mid \mathcal{F}_t)}{d(P \mid \mathcal{F}_t)} = E_P[Z(T) \mid \mathcal{F}_t], \quad 0 \leq t \leq T. \tag{1.4.2}$$

In particular, $Z(t)$ is a P-martingale.

Proof of Theorem 1.24. Since $P(G) = 0 \Rightarrow Q(G) = 0$ for all $G \in \mathcal{F}_T \supseteq \mathcal{F}_t$, it is clear that $Q \mid \mathcal{F}_t \ll P \mid \mathcal{F}_t$. Choose $F \in \mathcal{F}_t$. Then

$$E_P\left[F \cdot E[Z(T) \mid \mathcal{F}_t]\right] = E_P\left[E_P[FZ(T) \mid \mathcal{F}_t]\right]$$
$$= E_P[FZ(T)] = E_Q[F] = E_P[F \cdot Z(t)].$$

Since this holds for all $F \in \mathcal{F}_t$ we conclude that

$$E_P[Z(T) \mid \mathcal{F}_t] = Z(t), \quad \text{as claimed.} \qquad \square$$

Lemma 1.27. *Suppose $Q \ll P$ with $\frac{dQ}{dP} = Z(T) > 0$ on $\mathcal{F}_T$. Let $X(t)$ be an adapted process such that $Z(t)X(t)$ is a martingale with respect to P. Then $X(t)$ is a martingale with respect to Q. Similarly, if $Z(t)X(t)$ is a local martingale with respect to P, then $X(t)$ is a local martingale with respect to Q.*

Proof of Theorem 1.24. We prove the last statement. Let $\tau \geq t$ be a stopping time, $\tau \leq T$. Then

$$E_Q[X(\tau) \mid \mathcal{F}_t] = \frac{E[Z(T)X(\tau) \mid \mathcal{F}_t]}{E[Z(T) \mid \mathcal{F}_t]} = \frac{E[Z(\tau)X(\tau) \mid \mathcal{F}_t]}{Z(t)} = \frac{Z(t)X(t)}{Z(t)}$$
$$= X(t) \quad \text{for all } t \in [0,T]. \qquad \square$$

Definition 1.28. *Let $X(t), Y(t) \in \mathbb{R}^n$ be two càdlàg semimartingales. The quadratic covariation of $X(\cdot)$ and $Y(\cdot)$, denoted by $[X, Y](\cdot)$, is the unique semimartingale such that*

$$X(t) \cdot Y(t) = X(0) \cdot Y(0) + \int_0^t X(s^-) \cdot \mathrm{d}Y(s) + \int_0^t Y(s^-) \cdot \mathrm{d}X(s) + [X, Y](t).$$

Example 1.29. Let

$$\mathrm{d}X_i(t) = \alpha_i(t, \omega)\mathrm{d}t + \sigma_i(t, \omega)\mathrm{d}B(t) + \int_{\mathbb{R}} \gamma_i(t, z)\widetilde{N}(\mathrm{d}t, \mathrm{d}z), \quad i = 1, 2$$

be two Itô–Lévy processes. Then by the Itô formula (Theorem 1.16) we have (see Exercise 1.7)

$$\mathrm{d}(X_1(t)X_2(t)) = X_1(t^-)\mathrm{d}X_2(t) + X_2(t^-)\mathrm{d}X_1(t) + \sigma_1(t)\sigma_2(t)\mathrm{d}t$$

$$+ \int_{\mathbb{R}} \gamma_1(t, z)\gamma_2(t, z)N(\mathrm{d}t, \mathrm{d}z).$$

Hence in this case the quadratic covariation is

$$[X_1, X_2](t) = \int_0^t \sigma_1(s)\sigma_2(s)\mathrm{d}s + \int_0^t \int_{\mathbb{R}} \gamma_1(s, z)\gamma_2(s, z)N(\mathrm{d}s, \mathrm{d}z)$$

$$= \int_0^t \left[\sigma_1(s)\sigma_2(s) + \int_{\mathbb{R}} \gamma_1(s, z)\gamma_2(s, z)\nu(\mathrm{d}z) \right] \mathrm{d}s$$

$$+ \int_0^t \int_{\mathbb{R}} \gamma_1(s, z)\gamma_2(s, z)\widetilde{N}(\mathrm{d}s, \mathrm{d}z).$$

Theorem 1.30 (Girsanov Theorem for Semimartingales). *Let Q be a probability measure on $\mathcal{F}_T$ and assume that Q is equivalent to P on $\mathcal{F}_T$, with*

$$\mathrm{d}Q(\omega) = Z(t)\mathrm{d}P(\omega) \quad on \ \mathcal{F}_t, \ t \in [0, T].$$

(See Lemma 1.26.) Let $M(t)$ be a local P-martingale. Then the process $\widehat{M}(t)$ defined by

$$\widehat{M}(t) := M(t) - \int_0^t \frac{\mathrm{d}[M, Z](s)}{Z(s)}$$

is a local Q-martingale.

Sketch of Proof. Since we will give a complete proof of the Itô–Lévy process version of this below (Theorem 1.31) we settle with a proof of Theorem 1.30 in the case when $Z(t)$ is continuous. By integration by parts we have

$$\widehat{M}(t)Z(t) - \widehat{M}(0)Z(0) = \int_0^t \widehat{M}(s^-)\mathrm{d}Z(s) + \int_0^t Z(s)\mathrm{d}\widehat{M}(s) + [M, Z](t)$$

$$= \int_0^t \widehat{M}(s^-)\mathrm{d}Z(s) + \int_0^t Z(s)\mathrm{d}M(s)$$

$$- \int_0^t Z(s)\frac{\mathrm{d}[M, Z](s)}{Z(s)} + [M, Z](t)$$

$$= \int_0^t \widehat{M}(s^-)\mathrm{d}Z(s) + \int_0^t Z(s)\mathrm{d}M(s).$$

Therefore, since $Z(t)$ is a local P-martingale (Lemma 1.26) we conclude that $\widehat{M}(t)Z(t)$ is a local P-martingale. But then $\widehat{M}(t)$ is a local Q-martingale, by Lemma 1.27. We refer to [P, Theorem 35, p. 132] for a proof in the general case.   □

Theorem 1.31 (Girsanov Theorem for Itô–Lévy Processes). *Let $X(t)$ be an n-dimensional Itô–Lévy process of the form*

$$\mathrm{d}X(t) = \alpha(t,\omega)\mathrm{d}t + \sigma(t,\omega)\mathrm{d}B(t) + \int_{\mathbb{R}} \gamma(t,z,\omega)\widetilde{N}(\mathrm{d}t,\mathrm{d}z), \quad 0 \leq t \leq T. \quad (1.4.3)$$

Assume there exist predictable processes $u(t) = u(t,\omega) \in \mathbb{R}^m$ and $\theta(t,z) = \theta(t,z,\omega) \in \mathbb{R}^\ell$ such that

$$\sigma(t)u(t) + \int_{\mathbb{R}^\ell} \gamma(t,z)\theta(t,z)\nu(\mathrm{d}z) = \alpha(t) \quad \text{for a.a. } (t,\omega) \in [0,T] \times \Omega \quad (1.4.4)$$

and such that the process

$$Z(t) := \exp\left[-\int_0^t u(s)\mathrm{d}B(s) - \frac{1}{2}\int_0^t u^2(s)\mathrm{d}s \right.$$

$$+ \sum_{j=1}^\ell \int_0^t \int_{\mathbb{R}} \ln(1 - \theta_j(s,z))\widetilde{N}_j(\mathrm{d}s,\mathrm{d}z)$$

$$\left. + \sum_{j=1}^\ell \int_0^t \int_{\mathbb{R}} \{\ln(1 - \theta_j(s,z)) + \theta_j(s,z)\}\nu_j(\mathrm{d}z)\mathrm{d}s \right], \quad 0 \leq t \leq T$$

$$(1.4.5)$$

is well defined and satisfies

$$E[Z(T)] = 1. \quad (1.4.6)$$

Define the probability measure Q on $\mathcal{F}_T$ by $\mathrm{d}Q(\omega) = Z(T)\mathrm{d}P(\omega)$. Then $X(t)$ is a local martingale with respect to Q.

Proof. By Lemma 1.27 it suffices to prove that $Z(t)X(t)$ is a local martingale with respect to P. By Exercises 1.7 and 1.9 (extended to include the Brownian motion case) we have

$$d(Z(t)X(t)) = Z(t^-)dX(t) + X(t^-)dZ(t) + \sigma(t)(-u(t))dt$$
$$+ \int_{\mathbb{R}} \gamma(t,z)(-Z(t^-)\theta(t,z))N(dt,dz)$$
$$= Z(t^-)\left(\alpha(t)dt + \sigma(t)dB(t) + \int_{\mathbb{R}} \gamma(t,z)\tilde{N}(dt,dz)\right)$$
$$- X(t^-)Z(t^-)\int_{\mathbb{R}} \theta(t,z)\tilde{N}(dt,dz) - \sigma(t)u(t)dt$$
$$- Z(t^-)\int_{\mathbb{R}} \gamma(t,z)\theta(t,z)\tilde{N}(dt,dz) - Z(t^-)\int_{\mathbb{R}} \gamma(t,z)\theta(t,z)\nu(dz)dt$$
$$= Z(t^-)\left(\alpha(t) - \left(\sigma(t)u(t) + \int_{\mathbb{R}} \gamma(t,z)\theta(t,z)\nu(dz)\right)\right)dt$$
$$+ \text{local martingale terms.}$$

By (1.4.4) this is a local martingale.   □

Remark 1.32. Such a measure Q is called an *equivalent local martingale measure* (ELMM) for $X(t)$. If $X(t)$ is a martingale with respect to Q, then Q is called an *equivalent martingale measure* (EMM) for $X(t)$.

We now apply this to two important special cases.

Corollary 1.33 (Girsanov Theorem I for Itô Processes). *Let $X(t)$ be an n-dimensional Itô process of the form*

$$dX(t) = \alpha(t,\omega)dt + \sigma(t,\omega)dB(t), \quad 0 \leq t \leq T,$$

where $\alpha(t) = \alpha(t,\omega) \in \mathbb{R}^n$, $\sigma(t) = \sigma(t,\omega) \in \mathbb{R}^{n \times m}$, and $B(t) \in \mathbb{R}^m$. Assume that there exists a process $\theta(t) \in \mathbb{R}^m$ such that

$$\sigma(t)\theta(t) = \alpha(t) \quad \text{for a.a. } (t,\omega) \in [0,T] \times \Omega \qquad (1.4.7)$$

and such that the process $Z(t)$ defined for $0 \leq t \leq T$ by

$$Z(t) := \exp\left\{-\int_0^t \theta(s)dB(s) - \frac{1}{2}\int_0^t \theta^2(s)ds\right\} \qquad (1.4.8)$$

exists. Define a measure Q on $\mathcal{F}_T$ by

$$dQ(\omega) = Z(T)dP(\omega) \quad \text{on } \mathcal{F}_T. \qquad (1.4.9)$$

Assume that

$$E_P[Z(T)] = 1. \qquad (1.4.10)$$

Then Q is a probability measure on $\mathcal{F}_T$, Q is equivalent to P and $X(t)$ is a local martingale with respect to Q.

Similarly, in the jump case we get:

Corollary 1.34 (Girsanov Theorem II for Jump Processes). *Let $X(t)$ be an n-dimensional Itô–Lévy process of the form*

$$dX(t) = \alpha(t)dt + \int_{\mathbb{R}^n} \gamma(t,z)\tilde{N}(dt,dz),$$

where $\alpha(t) = \alpha(t,\omega) \in \mathbb{R}^n$, $\gamma(t,z) \in \mathbb{R}^{n\times\ell}$, and $\tilde{N}(dt,dz) = (\tilde{N}_1(dt,dz_1),\ldots,$ $\tilde{N}_\ell(dt,dz_\ell))$ is ℓ-dimensional. Assume that there exists a process $\theta(t,z) = (\theta_1(t,z),\ldots,\theta_\ell(t,z))^T \in \mathbb{R}^\ell$ such that $\theta_j(s,z) \le 1$ and

$$\sum_{j=1}^{\ell} \int_{\mathbb{R}} \gamma_{ij}(t,z_j)\theta_j(t,z_j)\nu_j(dz_j) = \alpha_i(t), \quad i = 1,\ldots,n, \quad t \in [0,T] \quad (1.4.11)$$

and such that the process

$$Z(t) := \exp\left\{ \sum_{j=1}^{\ell} \int_0^t \int_{\mathbb{R}} \left[\ln(1 - \theta_j(s,z_j))\tilde{N}_j(ds,dz_j) \right.\right.$$
$$\left.\left. + (\ln(1 - \theta_j(s,z_j)) + \theta_j(s,z_j))\nu_j(dz_j)ds \right] \right\}$$

exists for $0 \le t \le T$. Define a measure Q on $\mathcal{F}_T$ by

$$dQ(\omega) = Z(T)dP(\omega).$$

Assume that

$$E[Z(T)] = 1.$$

Then Q is an ELMM for $X(t)$.

By the above we deduce the following version, which is sometimes useful. For simplicity we only present the one-dimensional case.

Theorem 1.35 (Girsanov Theorem III). *Let $u(t)$ and $\theta(t,z) \le 1$ be predictable processes such that the process*

$$Z(t) := \exp\left(-\int_0^t u(s)dB(s) - \frac{1}{2}\int_0^t u^2(s)ds \right.$$
$$+ \int_0^t \int_{\mathbb{R}} \ln(1 - \theta(s,z))\tilde{N}(ds,dz)$$
$$\left. + \int_0^t \int_{\mathbb{R}} \{\ln(1 - \theta(s,z)) + \theta(s,z)\}\nu(dz)ds \right) \quad (1.4.12)$$

exists for $0 \le t \le T$ and satisfies

$$E[Z(T)] = 1.$$

Define the probability measure Q on $\mathcal{F}_T$ by

$$dQ(\omega) = Z(T)dP(\omega).$$

Define the process $B_Q(t)$ and the random measure $\tilde{N}_Q(dt, dz)$ by

$$dB_Q(t) = u(t)dt + dB(t), \qquad (1.4.13)$$

and

$$\tilde{N}_Q(dt, dz) = \theta(t, z)\nu(dz)dt + \tilde{N}(dt, dz). \qquad (1.4.14)$$

Then $B_Q(\cdot)$ is a Brownian motion with respect to Q and $\tilde{N}_Q(\cdot, \cdot)$ is the Q-compensated Poisson random measure of $N(\cdot, \cdot)$, in the sense that the process

$$M(t) := \int_0^t \int_{\mathbb{R}} \gamma(s, z)\tilde{N}_Q(ds, dz), \quad 0 \le t \le T \qquad (1.4.15)$$

is a local Q-martingale, for all predictable processes $\gamma(s, z)$ such that

$$\int_0^T \int_{\mathbb{R}} \gamma(s, z)^2 \theta(s, z)^2 \nu(dz)ds < \infty \ a.s. \qquad (1.4.16)$$

Proof. With $M(t)$ as in (1.4.15) and $\epsilon \in [0, 1]$ define

$$X_\epsilon(t) := \epsilon B_Q(t) + M(t). \qquad (1.4.17)$$

We have

$$dX_\epsilon(t) = \epsilon u(t)dt + \epsilon dB(t) + \int_{\mathbb{R}} \gamma(t, z)\theta(t, z)\nu(dz)dt + \int_{\mathbb{R}} \gamma(t, z)\tilde{N}(dt, dz)$$

$$= \alpha_\epsilon(t)dt + \epsilon dB(t) + \int_{\mathbb{R}} \gamma(t, z)\tilde{N}(dt, dz),$$

where

$$\alpha_\epsilon(t) = \epsilon u(t) + \int_{\mathbb{R}} \gamma(t, z)\theta(t, z)\nu(dz).$$

By Theorem 1.31 $X_\epsilon(t)$ is a local Q-martingale, for all $\epsilon \in [0, T]$. In particular, $X_0(t) = M(t)$ is a local Q-martingale, but then $B_Q(t) = X(t) - M(t)$ is a local Q-martingale also. Since $[B_Q, B_Q](t) = t$ we conclude that $B_Q(t)$ is a Q-Brownian motion. □

Example 1.36.

1. Suppose $X(t) = \pi(t) := \int_{\mathbb{R}} z\tilde{N}(t, dz) \in \mathbb{R}$ is a Poisson process with intensity $\lambda > 0$. Then

$$\nu(dz) = \lambda\delta_1(dz)$$

where $\delta_1(dz)$ is the unit point mass at 1 and condition (1.4.11) gets the form

$$\lambda\gamma(t, 1)\theta(t, 1) = \alpha(t), \quad t \in [0, T]. \qquad (1.4.18)$$

This corresponds to (1.4.7) in the Brownian motion case. Note that there is *at most one* solution $\theta(t, 1)$ of (1.4.18) (unless $\alpha(t) = \gamma(t, 1) = 0$).

2. Next, suppose ν is supported on n points $z_1, \ldots, z_n$. Then (1.4.11) gets the form

$$\sum_{j=1}^{n} \gamma(t, z_j)\theta(t, z_j)\nu(\{z_j\}) = \alpha(t), \quad t \in [0, T].$$

For each $t \in [0, T]$ this is one linear equation in the n unknowns $\theta(t, z_1), \ldots, \theta(t, z_n)$. So unless in degenerate cases this equation will have infinitely many solutions and – under some conditions on $\{\gamma(t, z_j)\}_{j=1}^{n}$ – also infinitely many solutions satisfying $\theta(t, z_j) \leq 1$. This corresponds to infinitely many ELMMs, which again is equivalent to incompleteness of the associated financial market, according to the Second Fundamental Theorem of Asset Pricing (see, e.g., [DS, LS]).

1.5 Application to Finance

It has been argued (see, e.g., [EK, B-N, Sc, Eb, CT]) that Lévy processes are relevant in mathematical finance, in particular in the modeling of stock prices.

Consider the following Lévy version of the Black–Scholes market:

(Bond price) $dS_0(t) = rS_0(t)dt;$ $S_0(0) = 1$

(Stock price) $dS_1(t) = S_1(t^-)[\mu\,dt + \gamma\,d\eta(t)];$ $S_1(0) = x > 0$

where r, μ, and $\gamma \neq 0$ are constants and

$$\eta(t) = \int_0^t \int_{\mathbb{R}} z\tilde{N}(dt, dz)$$

is a pure jump Lévy martingale. To ensure that $S_1(t) \geq 0$ for all $t \geq 0$ we assume as before that $\gamma z \geq -1$ a.s. ν. Assume in addition that

$$\int_{\mathbb{R}} |z|\nu(dz) > \left|\frac{\mu - r}{\gamma}\right|, \tag{1.5.1}$$

i.e., that the total mass of the jump measure (Lévy measure) ν exceeds $\left|\frac{\mu - r}{\gamma}\right|$.

The *normalized* stock price $\bar{S}_1(t)$ is given by

$$\bar{S}_1(t) = \frac{1}{S_0(t)}S_1(t) = e^{-rt}S_1(t).$$

Note that

$$d\bar{S}_1(t) = \bar{S}_1(t^-)[(\mu - r)dt + \gamma\,d\eta(t)], \quad \bar{S}_1(0) = x.$$

We seek an ELMM Q of the process $\bar{S}_1(t)$.

To this end we apply Corollary 1.34 and try to find a solution $\theta(z) \le 1$ of (1.4.11), which in this case gets the form

$$\int_{\mathbb{R}} z\theta(z)\nu(\mathrm{d}z) = \frac{\mu - r}{\gamma}. \tag{1.5.2}$$

By (1.5.1) we see that if $A \in \mathbf{B}_0$ with $\left|\frac{\mu-r}{\gamma}\right| < \int_A |z|\nu(\mathrm{d}z) := K < \infty$ then

$$\theta(z) = \frac{\mu - r}{\gamma K}\mathcal{X}_A(z)\,\mathrm{sign}\,z$$

is a possible solution. If ν is concentrated on one point z_0, i.e., if

$$\nu(\mathbb{R}) = \nu(\{z_0\})$$

(which means that $\eta(t)$ is a Poisson process multiplied by z_0) then this is the only solution. On the other hand, if there exist two sets $A, B \in \mathbf{B}_0$ such that $A \cap B = \emptyset$ and

$$\nu(A) > 0, \quad \nu(B) > 0 \tag{1.5.3}$$

then we see that there are infinitely many solutions $\theta(z)$ of (1.5.2) such that $\theta(z) < 1$.

Fix a solution $\theta(z) < 1$ of (1.5.2) and define

$$Z(t) = Z^\theta(t) = \exp\left\{ \int_0^t \int_{\mathbb{R}} \ln(1 - \theta(z))\tilde{N}(\mathrm{d}s, \mathrm{d}z) \right.$$
$$\left. + t\int_{\mathbb{R}} \{\ln(1 - \theta(z)) + \theta(z)\}\nu(\mathrm{d}z) \right\}, \quad 0 \le t \le T$$

and

$$\mathrm{d}Q = \mathrm{d}Q^\theta = Z^\theta(T)\mathrm{d}P \quad \text{on } \mathcal{F}_T.$$

Then by Corollary 1.34 $\bar{S}_1(t)$ is a local martingale with respect to Q.

We now discuss the concept of *arbitrage* in this market. For more information on the mathematics of finance see, e.g., [KS] or [Ø1, Chapter 12] and the references therein.

A *portfolio* in this market is a predictable process $\phi(t) = (\phi_0(t), \phi_1(t)) \in \mathbb{R}^2$ such that

$$\int_0^T \{\phi_0^2(s) + \phi_1^2(s)\}\mathrm{d}s < \infty \text{ a.s. } P. \tag{1.5.4}$$

The corresponding *wealth process* $V^\phi(t)$ is defined by

$$V^\phi(t) = \phi_0(t)S_0(t) + \phi_1(t)S_1(t), \quad 0 \le t \le T. \tag{1.5.5}$$

We say that (ϕ_0, ϕ_1) is *self-financing* if $V^\phi(t)$ is also given by

$$V^\phi(t) = V^\phi(0) + \int_0^t \phi_0(s)\mathrm{d}S_0(s) + \int_0^t \phi_1(s)\mathrm{d}S_1(s). \qquad (1.5.6)$$

If, in addition,

$$\{V^\phi(t)\}_{t \in [0,T]} \quad \text{is lower bounded} \qquad (1.5.7)$$

we say that ϕ is *admissible* and write $\phi \in \mathcal{A}_0$. A portfolio $\phi \in \mathcal{A}_0$ is called an *arbitrage* if

$$V^\phi(0) = 0, \quad V^\phi(T) \geq 0, \quad \text{and} \quad P[V^\phi(T) > 0] > 0. \qquad (1.5.8)$$

Does this market have an arbitrage? To answer this we combine (1.5.5) and (1.5.6) to get

$$\phi_0(t) = \mathrm{e}^{-rt}(V^\phi(t) - \phi_1(t)S_1(t))$$

and

$$\mathrm{d}V^\phi(t) = rV^\phi(t)\mathrm{d}t + \phi_1(t)S_1(t^-)\left[(\mu - r)\mathrm{d}t + \gamma \int_{\mathbb{R}} z\widetilde{N}(\mathrm{d}t, \mathrm{d}z)\right].$$

From this we obtain

$$\mathrm{d}(\mathrm{e}^{-rt}V^\phi(t)) = \mathrm{e}^{-rt}\phi_1(t)S_1(t^-)\left[(\mu - r)\mathrm{d}t + \gamma \int_{\mathbb{R}} z\widetilde{N}(\mathrm{d}t, \mathrm{d}z)\right]$$

or

$$\mathrm{e}^{-rt}V^\phi(t) = V^\phi(0) + \int_0^t \mathrm{e}^{-rs}\phi_1(s)S_1(s^-)\left[(\mu - r)\mathrm{d}s + \gamma \int_{\mathbb{R}} z\widetilde{N}(\mathrm{d}s, \mathrm{d}z)\right].$$

Therefore $\mathrm{e}^{-rt}V^\phi(t)$ is a lower bounded local martingale, and hence a supermartingale, with respect to Q. But then

$$0 = E_Q[V^\phi(0)] \geq E_Q[\mathrm{e}^{-rT}V^\phi(T)],$$

which shows that (1.5.8) cannot hold.

We conclude that *there is no arbitrage in this market* (if (1.5.1) holds).

This example illustrates the First Fundamental Theorem of Asset Pricing, which states the connection between (1) the existence of an ELMM and (2) the nonexistence of arbitrage, or No Free Lunch with Vanishing Risk (NFLVR) to be more precise. See, e.g., [DS, LS].

1.6 Exercises

Exercise* 1.1. Suppose

$$\mathrm{d}X(t) = \alpha\,\mathrm{d}t + \sigma\,\mathrm{d}B(t) + \int_{\mathbb{R}} \gamma(z)\bar{N}(\mathrm{d}t, \mathrm{d}z), \quad X(0) = x \in \mathbb{R},$$

where α, σ are constants, $\gamma : \mathbb{R} \to \mathbb{R}$ is a given function.

1. Use Itô's formula to find $dY(t)$ when

$$Y(t) = \exp(X(t)).$$

2. How do we choose α, σ, and $\gamma(z)$ if we want $Y(t)$ to solve the SDE

$$dY(t) = Y(t^-) \left[\beta\,dt + \theta\,dB(t) + \lambda \int_{\mathbb{R}} z\bar{N}(dt, dz) \right],$$

for given constants, β, θ, and λ?

Exercise* 1.2. Solve the following Lévy SDEs:

1. $dX(t) = (m - X(t))dt + \sigma\,dB(t) + \gamma \int_{\mathbb{R}} z\bar{N}(dt, dz),$ $X(0) = x \in \mathbb{R}$

   $(m, \sigma, \gamma \text{ constants})$ (the mean-reverting Lévy–Ornstein–Uhlenbeck process)

2. $dX(t) = \alpha\,dt + \gamma\,X(t^-) \int_{\mathbb{R}} z\bar{N}(dt, dz),$ $X(0) = x \in \mathbb{R}$

   $(\alpha, \gamma \text{ constants}, \gamma z > -1 \text{ a.s. } \nu).$

[Hint: Try to multiply the equation by

$$F(t) := \exp\left\{ -\int_0^t \int_{\mathbb{R}} \theta(z)\bar{N}(dt, dz) + \int_{|z|<R} (e^{\theta(z)} - 1 - \theta(z))\nu(dz) \cdot t \right\},$$

for suitable $\theta(z)$.]

Exercise 1.3 (Geometric Lévy Martingales). Let $h \in L^2(\mathbb{R})$ be deterministic and define

$$Y(t) = \exp\left\{ \int_0^t \int_{\mathbb{R}} h(s)z\tilde{N}(ds, dz) - \int_0^t \int_{\mathbb{R}} \left(e^{h(s)z} - 1 - h(s)z \right) \nu(dz)ds \right\}.$$

Show that

$$dY(t) = Y(t^-) \int_{\mathbb{R}} \left(e^{h(t)z} - 1 \right) \tilde{N}(dt, dz).$$

Exercise 1.4. Find the generator A of the following jump diffusions:

(a) (The geometric Lévy process)

$$dX(t) = X(t^-) \left[\mu\,dt + \sigma\,dB(t) + \gamma \int_{\mathbb{R}} z\tilde{N}(dt, dz) \right].$$

(b) (The mean-reverting Lévy–Ornstein–Uhlenbeck process)

$$dX(t) = (m - X(t))dt + \sigma\,dB(t) + \gamma \int_{\mathbb{R}} z\tilde{N}(dt, dz).$$

(c) (The graph of the geometric Lévy process)

$$dY(t) = \begin{bmatrix} dt \\ dX(t) \end{bmatrix}, \text{ where } X(t) \text{ is as in (a).}$$

(d) (The n-dimensional geometric Lévy process)

$$X(t) = \begin{bmatrix} X(t) \\ \vdots \\ X_n(t) \end{bmatrix},$$

where

$$dX_i(t) = X_i(t^-) \left[\mu_i dt + \sum_{j=1}^n \sigma_{ij} dB_j(t) + \sum_{j=1}^n \gamma_{ij} \int_{\mathbb{R}} z_j \widetilde{N}_j(dt, dz_j) \right],$$

$$1 \le i \le n.$$

Exercise 1.5 (The First Exit Time from a Ball).
Let $K = \{x \in \mathbb{R}^n, |x| \le R\}$ be the open ball of radius R in $\mathbb{R}^n$ and let

$$\eta(t) = (\eta_1(t), \dots, \eta_n(t)),$$

where

$$\eta_i(t) = a_i + \int_0^t \int_{\mathbb{R}} z_i \widetilde{N}_i(dt, dz_i), \quad 1 \le i \le n$$

are independent one-dimensional pure jump Lévy processes and $a = (a_1, \dots, a_n) \in K$ is constant. We assume that $0 < E\left[\sum_{i=1}^n (\eta_i(t) - a_i)^2\right] < \infty$ for all $t > 0$.

(a) Find the generator A of $\eta(\cdot)$ and show that if $f(x) = |x|^2$ then

$$Af(x) = \sum_{i=1}^n \int_{\mathbb{R}} |\zeta|^2 \nu_i(d\zeta) := \rho(n) \in (0, \infty) \quad \text{(constant).}$$

(b) Let

$$\tau = \inf\{t > 0, \eta(t) \notin K\} \le \infty$$

and put

$$\tau_k = \tau \wedge k, \quad k = 1, 2, \dots$$

Let $\{f_m\}_{m=1}^\infty$ be a sequence of functions in $C^2(\mathbb{R}^n)$ such that
$f_m(x) = |x|^2$ for $|x| \le R, 0 \le f_m(x) \le 2R^2$ for all $x \in \mathbb{R}$,
supp $f_m \subset \{x \in \mathbb{R}^n, |x| \le R + \frac{1}{m}\}$ for all m and
$f_m(x) \to |x|^2 \cdot \chi_K(x)$ as $m \to \infty$, for all $x \in \mathbb{R}^n$.
Use the Dynkin formula I to show that

$$E^a\left[f_m\left(\eta(\tau_k)\right)\right] = |a|^2 + \rho(n) \cdot E^a[\tau_k] \quad \text{for all } m, k.$$

(c) Show that

$$E^a[\tau] = \frac{1}{\rho(n)}\left(R^2 P^a[\eta(\tau) \in K] - |a|^2\right) \leq \frac{1}{\rho(n)}\left(R^2 - |a|^2\right).$$

In particular, $\tau < \infty$, a.s.

Remark. If we replace $\eta(\cdot)$ by an n-dimensional Brownian motion $B(\cdot)$, then the corresponding exit time $\tau^{(B)}$ satisfies

$$E^a\left[\tau^{(B)}\right] = \frac{1}{n}\left(R^2 - |a|^2\right)$$

(see, e.g., [Ø1, Example 7.4.2]).

Exercise* 1.6. Show that, under some conditions on $\gamma(s, z)$ (deterministic),

$$E\left[\exp\left(\int_0^t \int_{\mathbb{R}} \gamma(s, z)\widetilde{N}(ds, dz)\right)\right] = \exp\left(\int_0^t \int_{\mathbb{R}} \{e^{\gamma(s,z)} - 1 - \gamma(s, z)\}\nu(dz)ds\right).$$

Exercise* 1.7. Let

$$dX_i(t) = \int_{\mathbb{R}} \gamma_i(t, z)\widetilde{N}(dt, dz), \quad i = 1, 2$$

be two one-dimensional Itô–Lévy processes. Use the two-dimensional Itô formula (Theorem 1.16) to prove the following *integration by parts formula*:

$$X_1(t)X_2(t) = X_1(0)X_2(0) + \int_0^t X_1(s^-)dX_2(s) + \int_0^t X_2(s^-)dX_1(s)$$

$$+ \int_0^t \int_{\mathbb{R}} \gamma_1(s, z)\gamma_2(s, z)N(ds, dz). \tag{1.6.1}$$

Remark. The process

$$[X_1, X_2](t) := \int_0^t \int_{\mathbb{R}} \gamma_1(s, z)\gamma_2(s, z)N(ds, dz)$$

$$= \int_0^t \int_{\mathbb{R}} \gamma_1(s, z)\gamma_2(s, z)\nu(dz)ds + \int_0^t \int_{\mathbb{R}} \gamma_1(s, z)\gamma_2(s, z)\widetilde{N}(ds, dz)$$

$$\tag{1.6.2}$$

is called the *quadratic covariation* of X_1 and X_2. See Definition 1.28.

Exercise* 1.8. Consider the following market

(Bond price) $dS_0(t) = 0;$ $S_0(0) = 0$
(Stock price 1) $dS_1(t) = S_1(t^-)[\mu_1 dt + \gamma_{11}d\eta_1(t) + \gamma_{12}d\eta_2(t)];$ $S_1(0) = x_1 > 0$
(Stock price 2) $dS_2(t) = S_2(t^-)[\mu_2 dt + \gamma_{21}d\eta_1(t) + \gamma_{22}d\eta_2(t)];$ $S_2(0) = x_2 > 0$

where μ_i and γ_{ij} are constants and $\eta_1(t), \eta_2(t)$ are independent Lévy martingales of the form

$$d\eta_i(t) = \int_{\mathbb{R}} z \tilde{N}_i(dt, dz), \quad i = 1, 2.$$

Assume that the matrix $\gamma := [\gamma_{ij}]_{1 \leq i,j \leq 2} \in \mathbb{R}^2$ is invertible, with inverse

$$\gamma^{-1} = \lambda = [\lambda_{ij}]_{1 \leq i,j \leq 2}$$

and assume that

$$\int_{\mathbb{R}} |z| \nu_i(dz) > |\lambda_{i1}\mu_1 + \lambda_{i2}\mu_2| \quad \text{for} \quad i = 1, 2. \tag{1.6.3}$$

Find an ELMM Q for $(S_1(t), S_2(t))$ and use this to deduce that there is no arbitrage in this market.

Exercise* 1.9. Define, with suitable conditions on $\theta(s, z)$,

$$Z(t) = \exp\left(\int_0^t \int_{\mathbb{R}} \ln(1 - \theta(s, z))\tilde{N}(ds, dz) \right.$$
$$\left. + \int_0^t \int_{\mathbb{R}} \{\ln(1 - \theta(s, z)) + \theta(s, z)\}\nu(dz)ds\right).$$

Show that

$$dZ(t) = -Z(t^-)\int_{\mathbb{R}} \theta(s, z)\tilde{N}(ds, dz).$$

In particular, $Z(t)$ is a local martingale.

Optimal Stopping of Jump Diffusions

2.1 A General Formulation and a Verification Theorem

Fix an open set $\mathcal{S} \subset \mathbb{R}^k$ (the *solvency region*) and let $Y(t)$ be a jump diffusion in $\mathbb{R}^k$ given by

$$dY(t) = b(Y(t))dt + \sigma(Y(t))dB(t) + \int_{\mathbb{R}^k} \gamma(Y(t^-), z)\bar{N}(dt, dz), \quad Y(0) = y \in \mathbb{R}^k,$$

where $b : \mathbb{R}^k \to \mathbb{R}^k$, $\sigma : \mathbb{R}^k \to \mathbb{R}^{k \times m}$, and $\gamma : \mathbb{R}^k \times \mathbb{R}^k \to \mathbb{R}^{k \times \ell}$ are given functions such that a unique solution $Y(t)$ exists (see Theorem 1.19). Let

$$\tau_{\mathcal{S}} = \tau_{\mathcal{S}}(y, \omega) = \inf\{t > 0; Y(t) \notin \mathcal{S}\} \qquad (2.1.1)$$

be the *bankruptcy time* and let $\mathcal{T}$ denote the set of all stopping times $\tau \leq \tau_{\mathcal{S}}$.

The results below remain valid, with the natural modifications, if we allow $\mathcal{S}$ to be any Borel set such that $\mathcal{S} \subset \overline{\mathcal{S}^0}$ where $\mathcal{S}^0$ denotes the interior of $\mathcal{S}$, $\overline{\mathcal{S}^0}$ its closure.

Let $f : \mathbb{R}^k \to \mathbb{R}$ and $g : \mathbb{R}^k \to \mathbb{R}$ be continuous functions satisfying the conditions

$$E^y\left[\int_0^{\tau_{\mathcal{S}}} f^-(Y(t))dt\right] < \infty \quad \text{for all } y \in \mathbb{R}^k. \qquad (2.1.2)$$

The family $\{g^-(Y(\tau)) \cdot \mathcal{X}_{\{\tau < \infty\}}, \tau \in \mathcal{T}\}$ is uniformly integrable, for all $y \in \mathbb{R}^k$.
$$(2.1.3)$$

(If x is a real number, then $x^- := \max(-x, 0)$ denotes the *negative part* of x.)

The general *optimal stopping problem* is the following:
Find $\Phi(y)$ and $\tau^* \in \mathcal{T}$ such that

$$\Phi(y) = \sup_{\tau \in \mathcal{T}} J^\tau(y) = J^{\tau^*}(y), \quad y \in \mathbb{R}^k,$$

where

$$J^{\tau}(y) = E^y \left[\int_0^{\tau} f(Y(t))dt + g(Y(\tau)) \cdot \mathcal{X}_{\{\tau < \infty\}} \right], \quad \tau \in \mathcal{T}$$

is the *performance criterion*.

The function Φ is called the *value function* and the stopping time τ^* (if it exists) is called an *optimal stopping time*.

In the following we let A be the integrodifferential operator which coincides with the generator of $Y(t)$ on $C_0^2(\mathbb{R}^k)$, i.e.,

$$A\phi(y) = \sum_{i=1}^{k} b_i(y) \frac{\partial \phi}{\partial y_i}(y) + \frac{1}{2} \sum_{i,j=1}^{k} (\sigma\sigma^T)_{ij}(y) \frac{\partial^2 \phi}{\partial y_i \partial y_j}(y)$$

$$+ \sum_{j=1}^{\ell} \int_{\mathbb{R}} \{ \phi(y + \gamma^{(j)}(y, z_j))$$

$$- \phi(y) - \nabla\phi(y) \cdot \gamma^{(j)}(y, z_j) \} \nu_j(dz_j) \qquad (2.1.4)$$

for all $\phi : \mathbb{R}^k \to \mathbb{R}$ and $y \in \mathbb{R}^k$ such that (2.1.4) exists. (See Theorems 1.22 and 1.24.)

We will need the following result. A proof of a related result (in the no jump case) can be found in [Ø1].

Theorem 2.1 (Approximation Theorem).
   Let D be an open set, $D \subset \mathcal{S}$. Assume that $Y(\tau_S) \in \partial\mathcal{S}$ a.s. on $\{\tau_S < \infty\}$ and

$$\partial D \text{ is a Lipschitz surface} \qquad (2.1.5)$$

(i.e., ∂D is locally the graph of a Lipschitz continuous function) and let $\varphi : \bar{\mathcal{S}} \to \mathbb{R}$ be a function with the following properties:

$$\varphi \in C^1(\mathcal{S}) \cap C(\bar{\mathcal{S}}) \qquad (2.1.6)$$

and

$$\varphi \in C^2(\mathcal{S} \backslash \partial D) \qquad (2.1.7)$$

and the second-order derivatives of φ are locally bounded near ∂D.
   Then there exists a sequence $\{\varphi_m\}_{m=1}^{\infty} \subset C^2(\mathcal{S}) \cap C(\bar{\mathcal{S}})$ such that, with A as in (2.1.4),

$$\varphi_m \to \varphi \text{ pointwise dominatedly in } \bar{\mathcal{S}} \text{ as } m \to \infty \qquad (2.1.8)$$

$$\frac{\partial \varphi_m}{\partial x_i} \to \frac{\partial \varphi}{\partial x_i} \text{ pointwise dominatedly in } \mathcal{S} \text{ as } m \to \infty \qquad (2.1.9)$$

$$\frac{\partial^2 \varphi_m}{\partial x_i \partial x_j} \to \frac{\partial^2 \varphi}{\partial x_i \partial x_j} \text{ and } A\varphi_m \to A\varphi$$

pointwise dominatedly in $\mathcal{S} \backslash \partial D$ as $m \to \infty$. $\qquad (2.1.10)$

We now formulate a set of sufficient conditions that a given function ϕ actually coincides with the value function Φ and that a corresponding stopping time, τ_D, actually is optimal. The result is analogous to the variational inequality verification theorem for optimal stopping of continuous diffusions. See, e.g., [Ø1, Theorem 10.4.1].

Theorem 2.2 (Integrovariational Inequalities for Optimal Stopping).

(a) *Suppose we can find a function $\phi : \bar{\mathcal{S}} \to \mathbb{R}$ such that*
   (i) $\phi \in C^1(\mathcal{S}) \cap C(\bar{\mathcal{S}})$.
   (ii) $\phi \geq g$ *on* $\mathcal{S}$.
   Define

$$D = \{y \in \mathcal{S}; \phi(y) > g(y)\} \quad \text{(the continuation region).}$$

   Suppose
   (iii) $E^y \left[\int_0^{\tau_S} \mathcal{X}_{\partial D}(Y(t)) \mathrm{d}t \right] = 0.$
   (iv) ∂D *is a Lipschitz surface.*
   (v) $\phi \in C^2(\mathcal{S} \setminus \partial D)$ *with locally bounded derivatives near* ∂D.
   (vi) $A\phi + f \leq 0$ *on* $\mathcal{S} \setminus \partial D$.
   (vii) $Y(\tau_S) \in \partial \mathcal{S}$ *a.s. on* $\{\tau_S < \infty\}$ *and* $\lim_{t \to \tau_S^-} \phi(Y(t)) = g(Y(\tau_S)) \cdot \mathcal{X}_{\{\tau_S < \infty\}}.$
   (viii) $E^y \left[|\phi(Y(\tau))| + \int_0^{\tau_S} |A\phi(Y(t))| \mathrm{d}t \right] < \infty$ *for all* $\tau \in \mathcal{T}.$
   Then $\phi(y) \geq \Phi(y)$ *for all* $y \in \bar{\mathcal{S}}.$
(b) *Moreover, assume*
   (ix) $A\phi + f = 0$ *on* D.
   (x) $\tau_D := \inf\{t > 0; Y(t) \notin D\} < \infty$ *a.s. for all* y.
   (xi) $\{\phi(Y(\tau)); \tau \in \mathcal{T}, \tau \leq \tau_D\}$ *is uniformly integrable, for all* y.
   Then

$$\phi(y) = \Phi(y)$$

   and

$$\tau^* = \tau_D \text{ is an optimal stopping time.}$$

Proof of Theorem 2.2 (Sketch). (a) Let $\tau \leq \tau_S$ be a stopping time. By Theorem 2.1 we can assume that $\phi \in C^2(\mathcal{S})$. Then by (vii) and (viii) and the Dynkin formula (Theorem 1.24) applied to $\tau_m := \min(\tau, m)$, $m = 1, 2, \ldots$ we have, by (vi),

$$E^y[\phi(Y(\tau_m))] = \phi(y) + E^y \left[\int_0^{\tau_m} A\phi(Y(t)) \mathrm{d}t \right] \leq \phi(y) - E^y \left[\int_0^{\tau_m} f(Y(t)) \mathrm{d}t \right].$$

Hence by (ii) and the Fatou lemma

$$\phi(y) \geq \liminf_{m \to \infty} E^y \left[\int_0^{\tau_m} f(Y(t))dt + \phi(Y(\tau_m)) \right]$$

$$\geq E^y \left[\int_0^\tau f(Y(t))dt + g(Y(\tau))\chi_{\{\tau < \infty\}} \right] = J^\tau(y).$$

Hence

$$\phi(y) \geq \Phi(y). \tag{2.1.11}$$

(b) Moreover, if we apply the above argument to $\tau = \tau_D$ then by (ix)–(xi) and the definition of D we get *equality* in (2.1.11), so that

$$\phi(y) = J^{\tau_D}(y) \leq \Phi(y). \tag{2.1.12}$$

From (2.1.11) and (2.1.12) we conclude that $\phi(y) = \Phi(y)$ and τ_D is optimal.
□

The following result is sometimes helpful.

Proposition 2.3. *Suppose the conditions of Theorem 2.2 hold. Suppose $g \in C^2(\mathbb{R}^k)$ and that $\phi = g$ satisfies (viii). Define*

$$U = \{y \in \mathcal{S}; Ag(y) + f(y) > 0\}.$$

Suppose that for all $y \in U$ there exists a neighborhood W of y such that $\tau_W := \inf\{t > 0; Y(t) \notin W\} < \infty$ a.s. Then

$$U \subset \{y \in \mathcal{S}; \Phi(y) > g(y)\} = D.$$

Hence it is never optimal to stop while $Y(t) \in U$.

Proof. Choose $y \in U$ and let $W \subset U$ be a neighborhood of y with $\tau_W < \infty$ a.s. Then by the Dynkin formula (Theorem 1.24)

$$E^y[g(Y(\tau_W))] = g(y) + E^y \left[\int_0^{\tau_W} Ag(Y(t))dt \right] > g(y) - E^y \left[\int_0^{\tau_W} f(Y(t))dt \right].$$

Hence

$$g(y) < E^y \left[\int_0^{\tau_W} f(Y(t))dt + g(Y_{\tau_W}) \right] \leq \Phi(y),$$

as claimed.
□

Another useful observation is:

Proposition 2.4. *Let U be as in Proposition 2.3. Suppose $U = \emptyset$. Then*

$$\Phi(y) = g(y) \quad \text{and} \quad \tau^* = 0 \quad \text{is optimal.}$$

Proof. If $U = \emptyset$ then $Ag(y) + f(y) \leq 0$ for all $y \in \mathcal{S}$. Hence the function $\phi = g$ satisfies all the conditions of Theorem 2.2. Therefore $D = \emptyset$, $g(y) = \Phi(y)$, and $\tau^* = 0$ is optimal. $\qquad\square$

2.2 Applications and Examples

Example 2.5 (The Optimal Time to Sell). Suppose the price $X(t)$ at time t of an asset (a property, a stock...) is a geometric Lévy process given by

$$dX(t) = X(t^-)\left[\alpha\,dt + \beta\,dB(t) + \gamma\int_{\mathbb{R}} z\widetilde{N}(dt,dz)\right], \quad X(0) = x > 0,$$

where α, β, and γ are constants, and we assume that

$$-1 < \gamma z \leq 0 \quad \text{a.s. } \nu. \tag{2.2.1}$$

If we sell the asset at time $s + \tau$ we get the expected discounted net payoff

$$J^\tau(s,x) := E^{s,x}\left[e^{-\rho(s+\tau)}(X(\tau) - a)\mathcal{X}_{\{\tau < \infty\}}\right],$$

where $\rho > 0$ (the discounting exponent) and $a > 0$ (the transaction cost) are constants.

We seek the value function $\Phi(s,x)$ and an optimal stopping time $\tau^* \leq \infty$ such that

$$\Phi(s,x) = \sup_{\tau \leq \infty} J^\tau(s,x) = J^{\tau^*}(s,x). \tag{2.2.2}$$

We apply Theorem 2.2 to solve this problem as follows: Put $\mathcal{S} = \mathbb{R} \times (0,\infty)$ and

$$Y(t) = \begin{bmatrix} s+t \\ X(t) \end{bmatrix}, \quad t \geq 0.$$

Then

$$dY(t) = \begin{bmatrix} 1 \\ \alpha X(t) \end{bmatrix}dt + \begin{bmatrix} 0 \\ \beta X(t) \end{bmatrix}dB(t) + \begin{bmatrix} 0 \\ \gamma X(t^-)\int_{\mathbb{R}} z\widetilde{N}(dt,dz) \end{bmatrix}, \quad Y(0) = \begin{bmatrix} s \\ x \end{bmatrix}$$

and the generator A of $Y(t)$ is

$$A\phi(s,x) = \frac{\partial\phi}{\partial s} + \alpha x\frac{\partial\phi}{\partial x} + \frac{1}{2}\beta^2 x^2\frac{\partial^2\phi}{\partial x^2}$$
$$+ \int_{\mathbb{R}}\left\{\phi(s, x + \gamma xz) - \phi(s,x) - \gamma xz\frac{\partial\phi}{\partial x}\right\}\nu(dz). \tag{2.2.3}$$

If we try a function ϕ of the form

$$\phi(s,x) = e^{-\rho s}x^\lambda \quad \text{for some constant } \lambda \in \mathbb{R}$$

we get

$$A\phi(s,x) = e^{-\rho s}\left[-\rho x^\lambda + \alpha x \lambda x^{\lambda-1} + \frac{1}{2}\beta^2 x^2 \lambda(\lambda-1)x^{\lambda-2}\right.$$

$$\left. + \int_{\mathbb{R}}\{(x+\gamma xz)^\lambda - x^\lambda - \gamma xz\lambda x^{\lambda-1}\}\nu(dz)\right]$$

$$= e^{-\rho s}x^\lambda h(\lambda),$$

where

$$h(\lambda) = -\rho + \alpha\lambda + \frac{1}{2}\beta^2\lambda(\lambda-1) + \int_{\mathbb{R}}\{(1+\gamma z)^\lambda - 1 - \lambda\gamma z\}\nu(dz).$$

Note that

$$h(1) = \alpha - \rho \quad\text{and}\quad \lim_{\lambda\to\infty} h(\lambda) = \infty.$$

Therefore, if we assume that

$$\alpha < \rho, \tag{2.2.4}$$

then we get that there exists $\lambda_1 \in (1, \rho/\alpha)$ such that

$$h(\lambda_1) = 0. \tag{2.2.5}$$

With this value of λ_1 we put

$$\phi(s,x) = \begin{cases} e^{-\rho s}Cx^{\lambda_1} & \text{for} \quad (s,x) \in D \\ e^{-\rho s}(x-a) & \text{for} \quad (s,x) \notin D \end{cases} \tag{2.2.6}$$

for some constant C, to be determined.

To find a reasonable guess for the continuation region D we use Proposition 2.3. In this case we have $f = 0$ and $g(s,x) = e^{-\rho s}(x-a)$ and hence by (2.2.3)

$$Ag + f = e^{-\rho s}(-\rho(x-a) + \alpha x) = e^{-\rho s}((\alpha-\rho)x + \rho a).$$

Therefore

$$U = \{(s,x); (\alpha-\rho)x + \rho a > 0\}.$$

Case 1: $\alpha \geq \rho$. In this case $U = \mathbb{R}^2$ and it is easily seen that $\Phi = \infty$. We can get as high expected payoff as we wish by waiting long enough before stopping.

Case 2: $\alpha < \rho$. In this case

$$U = \left\{(s,x); x < \frac{\rho a}{\rho-\alpha}\right\}. \tag{2.2.7}$$

Therefore, in view of Proposition 2.3 we now guess that the continuation region D has the form

$$D = \{(s,x); 0 < x < x^*\} \qquad (2.2.8)$$

for some x^* such that $U \subseteq D$, i.e.,

$$x^* \geq \frac{\rho a}{\rho - \alpha}. \qquad (2.2.9)$$

Hence, by (2.2.6) we now put

$$\phi(s,x) = \begin{cases} e^{-\rho s} C x^{\lambda_1} & \text{for} \quad 0 < x < x^* \\ e^{-\rho s}(x - a) & \text{for} \quad x^* \leq x, \end{cases} \qquad (2.2.10)$$

for some constant $C > 0$ (to be determined). We guess that the value function is C^1 at $x = x^*$ and this gives the following "high contact" conditions:

$$C(x^*)^{\lambda_1} = x^* - a \quad \text{(continuity at } x = x^*\text{)}$$

and

$$C\lambda_1 (x^*)^{\lambda_1 - 1} = 1 \quad \text{(differentiability at } x = x^*\text{)}.$$

It is easy to see that the solution of these equations is

$$x^* = \frac{\lambda_1 a}{\lambda_1 - 1}, \quad C = \frac{1}{\lambda_1}(x^*)^{1-\lambda_1}. \qquad (2.2.11)$$

It remains to verify that with these values of x^* and C the function ϕ given by (2.2.10) satisfies all the conditions (i)–(xi) of Theorem 2.2.

To this end, first note that (i) and (ix) hold by construction of ϕ and by (2.2.1). Moreover, $\phi = g$ outside D. Therefore, to verify (ii) we only need to prove that $\phi \geq g$ on D, i.e., that

$$C x^{\lambda_1} \geq x - a \quad \text{for} \quad 0 < x < x^*. \qquad (2.2.12)$$

Define $k(x) = C x^{\lambda_1} - x + a$. By our chosen values of C and x^* we have $k(x^*) = k'(x^*) = 0$. Moreover, $k''(x) = C\lambda_1(\lambda_1 - 1)x^{\lambda_1 - 2} > 0$ for $x < x^*$. Therefore $k(x) > 0$ for $0 < x < x^*$ and (2.2.12) holds and hence (ii) is proved.

(iii): In this case $\partial D = \{(s,x); x = x^*\}$ and hence

$$E^y\left[\int_0^\infty \chi_{\partial D}(Y(t))dt\right] = \int_0^\infty P^x[X(t) = x^*]dt = 0.$$

(iv) and (v) are trivial.

(vi): Outside D we have $\phi(s,x) = e^{-\rho s}(x-a)$ and therefore

$$A\phi + f(s,x) = e^{-\rho s}(-\rho(x-a) + \alpha x)$$

$$+ \int_{\mathbb{R}} \left\{ \phi(s, x+\gamma xz) - \phi(s,x) - \frac{\partial \phi}{\partial x}(s,x)\gamma xz \right\} \nu(dz)$$

$$= e^{-\rho s}\Big((\alpha - \rho)x + \rho a$$

$$+ \int_{x+\gamma xz < x^*} \{ C(x+\gamma xz)^{\lambda_1} - (x-a) - \gamma xz \} \nu(dz) \Big)$$

$$\le e^{-\rho s}\left((\alpha - \rho)x + (\rho + \|\nu\|)a \right),$$

where $\|\nu\| = \nu(\mathbb{R})$. By (2.2.4) we get that

$$(\alpha - \rho)x + (\rho + \|\nu\|)a \le 0 \qquad \text{for all} \quad x \ge x^*$$

$$\Updownarrow$$

$$(\alpha - \rho)x^* + (\rho + \|\nu\|)a \le 0$$

$$\Updownarrow$$

$$x^* \ge \frac{(\rho + \|\nu\|)a}{\rho - \alpha}$$

$$\Updownarrow$$

$$\lambda_1 \le \frac{\rho + \|\nu\|}{\alpha + \|\nu\|}. \tag{2.2.13}$$

(vii): holds since we have assumed that $\gamma z > -1$ a.s. ν.

For (viii) to hold it suffices that

$$E^x\left[\int_0^\infty e^{-2\rho t}\left\{ X^2(t) + \gamma^2 \int_{\mathbb{R}} z^2 \nu(dz)t \right\} dt\right] < \infty.$$

By the above this holds if

$$2\alpha - 2\rho + \beta^2 + \gamma^2 \int_{\mathbb{R}} z^2 \nu(dz) < 0. \tag{2.2.14}$$

(x): To check if $\tau_D < \infty$ a.s. we consider the solution $X(t)$ of (2.2.1), which by (1.2.5) is given by

$$X(t) = x \exp\left\{ \left(\alpha - \frac{1}{2}\beta^2 - \gamma \int_{\mathbb{R}} z\nu(dz)\right)t \right.$$

$$\left. + \int_0^t \int_{\mathbb{R}} \ln(1+\gamma z)N(dt, dz) + \beta B(t) \right\}.$$

By the law of iterated logarithm for Brownian motion (see the argument in [Ø1, Chapter 5] we see that if

$$\alpha > \frac{1}{2}\beta^2 + \gamma \int_{\mathbb{R}} z\nu(dz) \qquad (2.2.15)$$

and

$$z \geq 0 \quad \text{a.s. } \nu \qquad (2.2.16)$$

then

$$\lim_{t\to\infty} X(t) = \infty \quad \text{a.s.}$$

and in particular $\tau_D < \infty$ a.s.

(xi): Since ϕ is bounded on $[0, x^*]$ it suffices to check that

$$\{e^{-\rho\tau}X(\tau)\}_{\tau\in\mathcal{T}} \quad \text{is uniformly integrable.}$$

For this to hold it suffices that there exists a constant K such that

$$E[e^{-2\rho\tau}X^2(\tau)] \leq K \quad \text{for all } \tau \in \mathcal{T}. \qquad (2.2.17)$$

By (2.2.13) and Exercise 1.6 we have

$$E[e^{-2\rho T}X^2(\tau)] = x^2 E\left[\exp\left\{\left(2\alpha - 2\rho - \beta^2 - 2\gamma\int_{\mathbb{R}} z\nu(dz)\right)\tau\right.\right.$$
$$\left.\left. + 2\int_0^\tau \int_{\mathbb{R}} \ln(1+\gamma z)N(dt, dz) + 2\beta B(\tau)\right\}\right]$$
$$= x^2 E\left[\exp\left\{\left(2\alpha - 2\rho + \beta^2 + 2\int_{\mathbb{R}}(\ln(1+\gamma z) - \gamma z)\nu(dz)\right)\tau\right.\right.$$
$$\left.\left. + 2\int_0^\tau \int_{\mathbb{R}} \ln(1+\gamma z)\tilde{N}(dt, dz)\right\}\right]$$
$$= x^2 E\left[\exp\left\{\left(2\alpha - 2\rho + \beta^2 + \int_{\mathbb{R}}[2\ln(1+\gamma z) - 2\gamma z + (1+\gamma z)^2\right.\right.\right.$$
$$\left.\left.\left. - 1 - 2\ln(1+\gamma z)]\nu(dz)\right)\tau\right\}\right]$$
$$= x^2 E\left[\exp\left\{\left(2\alpha - 2\rho + \beta^2 + \int_{\mathbb{R}}[(1+\gamma z)^2\right.\right.\right.$$
$$\left.\left.\left. - 1 - 2\gamma z]\nu(dz)\right)\tau\right\}\right].$$

We conclude that if

$$2\alpha - 2\rho + \beta^2 + \gamma^2 \int_{\mathbb{R}} z^2\nu(dz) \leq 0$$

then (2.2.17) holds and hence (xi) holds also.

We summarize what we have proved.

Theorem 2.6. *Suppose that* (2.2.1), (2.2.4), (2.2.13), (2.2.15), *and* (2.2.14) *hold. Then, with* λ_1, C, *and* x^* *given by* (2.2.5) *and* (2.2.11), *the function* ϕ *given by* (2.2.10) *coincides with the value function* Φ *of problem* (2.2.2) *and* $\tau^* = \tau_D$ *is an optimal stopping time, where* D *is given by* (2.2.8).

Remark 2.7.
(1) If condition (2.2.1) is relaxed to

$$-1 < \gamma z \quad \text{a.s. } \nu$$

  then the situation becomes more complicated. See [Mo] for a solution of a related problem in this case. See also [Ky].
(2) The C^1-property of the value function assumed in Theorem 2.2 (often call the "high contact" or "smooth pasting" assumption) need not hold in general, although we found that it holds in Example 2.5 (under some conditions). See [AKy] for a discussion of this. See also Sect. 9.1.
(3) For other applications of optimal stopping to jump diffusions we refer to [Ma].

2.3 Optimal Stopping with Delayed Information

This presentation is based on [Ø4]. Let $Y(t)$ be a jump diffusions in $\mathbb{R}^k$. Let $\delta \geq 0$ be a fixed constant. In this section we consider optimal stopping problems of the form

$$\Phi_\delta(y) := \sup_{\alpha \in \mathcal{T}_\delta} E^y \left[\int_0^\alpha f(Y(t))\mathrm{d}t + g(Y(\alpha)) \right], \qquad (2.3.1)$$

where we interpret $g(Y(\alpha))$ as 0 if $\alpha = \infty$. Here $\mathcal{T}_\delta$ is the set of δ-*delayed stopping times*, defined as follows.

Definition 2.8. *A function* $\alpha : \Omega \rightarrow [\delta, \infty]$ *is called a* δ-delayed stopping time *if*

$$\{\omega; \alpha(\omega) \leq t\} \in \mathcal{F}_{t-\delta} \quad \text{for all } t \geq \delta \qquad (2.3.2)$$

or, equivalently,

$$\{\omega; \alpha(\omega) \leq s + \delta\} \in \mathcal{F}_s \quad \text{for all } s \geq 0. \qquad (2.3.3)$$

The set of all δ-*delayed stopping times is denoted by* $\mathcal{T}_\delta$.

In other words, if we interpret $\alpha(\omega)$ as the time to stop, then $\alpha \in \mathcal{T}_\delta$ if the decision whether or not to stop at or before time t is based on the information represented by $\mathcal{F}_{t-\delta}$. In particular, if $\delta = 0$ then $\mathcal{T}_\delta = \mathcal{T}_0$ is the family of classical stopping times and (2.3.1) becomes the classical optimal stopping problem, discussed in previous sections.

In the delayed case problem (2.3.1) models the situation when there is a delay $\delta > 0$ in the flow of information available to the agent searching for the optimal time to stop. An alternative way of stating this is that there is a delay $\delta > 0$ from the decided stopping time $\tau \in \mathcal{T}_0$ (based on the complete current information available from the system) to the time $\alpha = \tau + \delta \in \mathcal{T}_\delta$ when the system actually stops. This new formulation is based on the following simple observation.

Lemma 2.9. (i) $\tau \in \mathcal{T}_0 \Longleftrightarrow \alpha := \tau + \delta \in \mathcal{T}_\delta$,

(ii) $\alpha \in \mathcal{T}_\delta \Longleftrightarrow \tau := \alpha - \delta \in \mathcal{T}_0$.

Proof. It suffices to prove (i).

First, assume $\tau \in \mathcal{T}_0$. Then, for $t \geq \delta$,

$$\{\omega; \tau(\omega) + \delta \leq t\} = \{\omega; \tau(\omega) \leq t - \delta\} \in \mathcal{F}_{t-\delta},$$

and hence $\alpha := \tau + \delta \in \mathcal{T}_\delta$.
Conversely, if $\alpha := \tau + \delta \in \mathcal{T}_\delta$ then

$$\{\omega; \tau(\omega) \leq t\} = \{\omega; \tau(\omega) + \delta \leq t + \delta\} = \{\omega; \alpha(\omega) \leq t + \delta\} \in \mathcal{F}_{(t+\delta)-\delta} = \mathcal{F}_t,$$

and hence $\tau \in \mathcal{T}_0$. $\qquad\square$

Remark 2.10. In view of this result we see that it is possible to give another interpretation of problem (2.3.1), namely

$$\Phi_\delta(y) = \sup_{\tau \in \mathcal{T}_0} E^y \left[\int_0^{\tau+\delta} f(Y(t)) \mathrm{d}t + g(Y(\tau + \delta)) \right]. \qquad (2.3.4)$$

In this formulation the problem appears as an optimal stopping problem over classical stopping times $\tau \in \mathcal{T}_0$, but with *delayed effect* of the stopping. If the stopping time $\tau \in \mathcal{T}_0$ is chosen, then the system itself is stopped at time $\tau + \delta$, i.e., after a delay $\delta > 0$.
Note that $\mathcal{T}_\delta \subset \mathcal{T}_0$ for $\delta > 0$ and hence

$$\Phi_\delta(y) \leq \Phi_0(y)$$

and we can interpret $\Phi_0(y) - \Phi_\delta(y)$ as the loss of value due to the delay of information.

In this section we show that the delayed optimal stopping problem (2.3.1) can be reduced to a classical optimal stopping problem by a simple transformation (Theorem 2.11).
We call $\alpha^* \in \mathcal{T}_\delta$ an *optimal stopping time* for the problem (2.3.1) if

$$\Phi_\delta(y) = E^y \left[\int_0^{\alpha^*} f(Y(t)) \mathrm{d}t + g(Y(\alpha^*)) \right]. \qquad (2.3.5)$$

The result of this section may be regarded as a partial extension of [AK], where the geometric Brownian motion case is studied and solved (see Exercise 2.12), with a more general (Markovian) delay $\delta(X) \leq 0$. See also [AK]. For a related type of problem involving impulse control with delivery lags, see [BS].

We are now ready to state and prove the main result of this section.

Theorem 2.11. (a) *Consider the two optimal stopping problems:*

$$\Phi_\delta(y) := \sup_{\alpha \in \mathcal{T}_\delta} E^y \left[\int_0^\alpha f(Y(t))dt + g(Y(\alpha)) \right], \tag{2.3.6}$$

$$\tilde{\Phi}(y) := \sup_{\tau \in \mathcal{T}_0} E^y \left[\int_0^\tau f(Y(t))dt + \tilde{g}_\delta(Y(\tau)) \right], \tag{2.3.7}$$

where

$$\tilde{g}_\delta(y) = E^y \left[\int_0^\delta f(Y(t))dt + g(Y(\delta)) \right]. \tag{2.3.8}$$

Then we have

$$\Phi_\delta(y) = \tilde{\Phi}(y) \quad \text{for all } y \in \mathbb{R}^k, \, \delta \geq 0.$$

(b) *Moreover, $\alpha^* \in \mathcal{T}_\delta$ is an optimal stopping time for the delayed problem (2.3.6) if and only if*

$$\alpha^* := \tau^* + \delta \tag{2.3.9}$$

where $\tau^* \in \mathcal{T}_0$ is an optimal stopping time for the nondelayed problem (2.3.7).

Proof. (a) Define

$$J^{(\alpha)}(y) = E^y \left[\int_0^\alpha f(Y(t))dt + g(Y(\alpha)) \right], \quad \alpha \in \mathcal{T}_\delta, \tag{2.3.10}$$

and

$$\tilde{J}^{(\tau)}(y) = E^y \left[\int_0^\tau f(Y(t))dt + \tilde{g}_\delta(Y(\tau)) \right], \quad \tau \in \mathcal{T}_0. \tag{2.3.11}$$

Choose $\alpha \in \mathcal{T}_\delta$ and put

$$\tau = \alpha - \delta \in \mathcal{T}_0.$$

Then $\alpha = \tau + \delta$ and hence

$$J^{(\alpha)}(y) = E^y \left[\int_0^\alpha f(Y(t))dt + g(Y(\alpha)) \right]$$

$$= E^y \left[\int_0^{\tau+\delta} f(Y(t))dt + g(Y(\tau + \delta)) \right]$$

$$= E^y \left[\int_0^\tau f(Y(t))dt + \int_\tau^{\tau+\delta} f(Y(t))dt + g(Y(\tau+\delta)) \right]$$

$$= E^y \left[\int_0^\tau f(Y(t))dt + E^y \left[\theta_\tau \left\{ \int_0^\delta f(Y(t))dt + g(Y(\delta)) \right\} \right] \right], \quad (2.3.12)$$

where θ_τ is the *shift operator*, defined by

$$\theta_\tau \{ h(Y(s)) \} = h(Y(\tau+s)) \text{ for } s \geq 0, \text{ for all measurable } h : \mathbb{R}^k \longrightarrow \mathbb{R},$$

and we have used that

$$\theta_\tau \left(\int_0^\delta f(Y(t))dt \right) = \int_\tau^{\tau+\delta} f(Y(t))dt.$$

We refer to [BG] for more information about Markov processes. By the strong Markov property we now get from (2.3.12) that

$$J^{(\alpha)}(y) = E^y \left[\int_0^\tau f(Y(t))dt + E^y \left[\theta_\tau \left\{ \int_0^\tau f(Y(t))dt + g(Y(\delta)) \right\} \Big| \mathcal{F}_\tau \right] \right]$$

$$= E^y \left[\int_0^\tau f(Y(t))dt + E^{Y(\tau)} \left[\int_0^\delta f(Y(t))dt + g(Y(\delta)) \right] \right]$$

$$= E^y \left[\int_0^\tau f(Y(t))dt + \tilde{g}_\delta(Y(\tau)) \right] = \tilde{J}^{(\tau)}(y). \quad (2.3.13)$$

Hence, by Lemma 2.9 (ii),

$$\Phi_\delta(y) = \sup_{\alpha \in \mathcal{T}_\delta} J^{(\alpha)}(y) = \sup_{\alpha - \delta \in \mathcal{T}_0} J^{(\alpha)}(y)$$

$$= \sup_{\alpha - \delta \in \mathcal{T}_0} \tilde{J}^{(\alpha - \delta)}(y) = \sup_{\tau \in \mathcal{T}_0} \tilde{J}^{(\tau)}(y) = \tilde{\Phi}(y),$$

as claimed.

(b) Suppose $\tau^* \in \mathcal{T}_0$ is optimal for (2.3.7). Define

$$\alpha^* := \tau^* + \delta.$$

Then $\alpha^* \in \mathcal{T}_\delta$ by Lemma 2.9 and by (2.3.13) combined with (a) we have

$$J^{(\alpha^*)}(y) = \tilde{J}^{(\tau^*)}(y) = \tilde{\Phi}(y) = \Phi_\delta(y).$$

Hence α^* is optimal for (2.3.6).

Conversely, if $\alpha^* \in \mathcal{T}_\delta$ is optimal for (2.3.6) a similar argument gives that $\tau^* := \alpha^* - \delta$ is optimal for (2.3.7). $\qquad \square$

We illustrate Theorem 2.11 by solving the following problem.

In the no delay case the following example is discussed in [Ø1] (continuous case) and Exercise 2.2 (jump diffusion case). Our example models the situation when there is a time lag $\delta > 0$ between the decided stopping time $\tau \in \mathcal{T}_0$ and the time $\alpha = \tau + \delta \in \mathcal{T}_\delta$ when the result of the stopping decision comes into effect.

Example 2.12 (Optimal Time to Stop Resource Extraction). Suppose the price $P(t)$ at time t per unit of a resource (oil, gas, . . .) is given by

$$dP(t) = P(t^-)\left[\mu\,dt + \sigma\,dB(t) + \int_{\mathbb{R}} z\tilde{N}(dt, dz)\right], \quad P(0) = p > 0, \quad (2.3.14)$$

where μ and σ are given constants and we assume that $z \geq 0$ a.s. with respect to ν.

Let $Q(t)$ denote the amount of remaining resources at time t. As long as the extraction field is open, we assume that the extraction rate is proportional to the remaining amount, i.e.,

$$dQ(t) = -\lambda Q(t)dt, \quad Q(0) = q > 0, \quad (2.3.15)$$

where $\lambda > 0$ is a known constant.

If we decide to stop the extraction and close the field at a (delayed) stopping time $\alpha \in \mathcal{T}_\delta$, then the expected total discounted net profit $J^\alpha(s, p, q)$ is assumed to have the form

$$J^\alpha(s, p, q) = E^{(s,p,q)}\left[\int_0^\alpha e^{-\rho(s+t)}(\lambda P(t)Q(t) - K)dt + \theta e^{-\rho(s+\alpha)} P(\alpha)Q(\alpha)\right],$$
$$(2.3.16)$$

where $K > 0$ is the (constant) running cost rate and $\rho > 0$, $\theta > 0$ are other constants. The expectation $E^{(s,p,q)}$ is taken with respect to the probability law $P^{(s,p,q)}$ of the strong Markov process

$$Y(t) := \begin{bmatrix} s + t \\ P(t) \\ Q(t) \end{bmatrix}, \text{ which starts at } y = \begin{bmatrix} s \\ p \\ q \end{bmatrix} \text{ at time } t = 0. \quad (2.3.17)$$

The explanation of the quantity $J^\alpha(s, p, q)$ in (2.3.16) is the following.

As long as the field is open (i.e., as long as $t < \alpha$) the gross income rate from the production is price times production rate, i.e., $P(t)\lambda Q(t)$. Subtracting the running cost rate K we get the net profit rate

$$\lambda P(t)Q(t) - K \quad \text{for } 0 \leq t < \alpha.$$

If the field is closed at time α the net value of the remaining resources is estimated to be $\theta P(\alpha)Q(\alpha)$. Discounting and integrating/adding these quantities and taking expectation we get (2.3.16).

We want to find the value function $\Phi_\delta(s, p, q)$ and the corresponding optimal delayed stopping time $\alpha^* \in \mathcal{T}_\delta$ such that

$$\Phi_\delta(y) = \Phi_\delta(s, p, q) = \sup_{\alpha \in \mathcal{T}_\delta} J^\alpha(s, p, q) = J^{\alpha^*}(s, p, q). \tag{2.3.18}$$

In the case of *no delay* ($\delta = 0$) it is shown in Exercise 2.2 that if the following relations between the parameters hold

$$0 < \theta(\lambda + \rho - \mu) < \lambda \tag{2.3.19}$$

then the optimal stopping time $\tau_0^* \in \mathcal{T}_0$ is

$$\tau_0^* = \inf\{t > 0; P(t)Q(t) \le w_0^*\}, \tag{2.3.20}$$

where

$$w_0^* = \frac{(-r_2)K(\lambda + \rho - \mu)}{(1 - r_2)\rho(\lambda - \theta(\lambda + \rho - \mu))}, \tag{2.3.21}$$

$r_2 < 0$ being the negative solution of the equation

$$h(r) := -\rho + (\mu - \lambda)r + \frac{1}{2}\sigma^2 r(r-1) + \int_{\mathbb{R}} \{(1+z)^r - 1 - rz\}\nu(dz) = 0. \tag{2.3.22}$$

In this case we have

$$f(y) = f(s, p, q) = e^{-\rho s}(\lambda pq - K)$$

and

$$g(y) = g(s, p, q) = \theta e^{-\rho s} pq.$$

Thus

$$\tilde{g}_\delta(y) = E^y\left[\int_0^\delta e^{-\rho(s+t)}(\lambda P(t)Q(t) - K)dt\right] + E^y[\theta e^{-\rho(s+\delta)}P(\delta)Q(\delta)]$$

$$= \int_0^\delta e^{-\rho(s+t)}(\lambda E[P(t)Q(t)] - K)dt + \theta e^{-\rho(s+\delta)}E^y[P(\delta)Q(\delta)]$$

$$= \int_0^\delta e^{-\rho(s+t)}(\lambda pq e^{(\mu-\lambda)t} - K)dt + \theta e^{-\rho(s+\delta)}pq e^{(\mu-\lambda)\delta}$$

$$= e^{-\rho s}\left[\{(\lambda + \rho - \mu)^{-1}\lambda(1 - e^{-(\lambda+\rho-\mu)\delta}) + \theta e^{-(\lambda+\rho-\mu)\delta}\}pq\right.$$

$$\left. - \frac{K}{\rho}(1 - e^{-\rho\delta})\right]$$

$$= e^{-\rho s}[F_1 pq - F_2], \tag{2.3.23}$$

where

$$F_1 = (\lambda + \rho - \mu)^{-1}\lambda(1 - e^{-(\lambda+\rho-\mu)\delta}) + \theta e^{-(\lambda+\rho-\mu)\delta} \qquad (2.3.24)$$

and

$$F_2 = \frac{K}{\rho}(1 - e^{-\rho\delta}). \qquad (2.3.25)$$

Therefore, according to Theorem 2.11 we have

$$\Phi_\delta(y) = \sup_{\tau \in \mathcal{T}_0} E^y \left[\int_0^\tau e^{-\rho(s+t)}(\lambda P(t)Q(t) - K)dt \right]$$
$$+ E^y[e^{-\rho(s+\tau)}(F_1 P(\tau)Q(\tau) + F_2)]. \qquad (2.3.26)$$

The method used in Exercise 2.2 to provide the solutions (2.3.20)–(2.3.22) in the no delay case can easily be modified to find the optimal stopping time τ^* for the problem (2.3.26). The result is

$$w_\delta^* = \frac{(-r_2)K(\lambda + \rho - \mu)e^{(\lambda-\mu)\delta}}{(1 - r)\rho[\lambda - \theta(\lambda + \rho - \mu)]} = w_0^* e^{(\lambda-\mu)\delta}. \qquad (2.3.27)$$

We have proved.

Theorem 2.13. *The optimal stopping time $\alpha^* \in \mathcal{T}_\delta$ for the delayed optimal stopping problem (2.3.18) is*

$$\alpha^* = \tau_\delta^* + \delta, \qquad (2.3.28)$$

where

$$\tau_\delta^* = \inf\{t > 0; P(t)Q(t) \leq w_\delta^*\}, \qquad (2.3.29)$$

with w_δ^* given by (2.3.27).

Remark 2.14. Note that the threshold w_δ^* for the decision to close down in the case of a time lag in the action only differs from the corresponding threshold w_0^* in the no delay case by the factor $e^{(\lambda-\mu)\delta}$.

Assume, for example, that $\lambda > \mu$. Then we should decide to stop *sooner* in the delay case than in the no delay case, because of the anticipation that $P(t)Q(t)$ will probably decrease during the extra time δ it takes before the closing down actually takes place.

2.4 Exercises

Exercise* 2.1. Solve the optimal stopping problem

$$\Phi(s, x) = \sup_{\tau \geq 0} E^{(s,x)}\left[e^{-\rho(s+\tau)}(X(\tau) - a) \right],$$

where

$$dX(t) = dB(t) + \gamma \int_{\mathbb{R}} z\bar{N}(dt, dz), \quad X(0) = x \in \mathbb{R}$$

and $\rho > 0$, $a > 0$, and γ are constants, $\gamma z \leq 0$ a.s. ν.

Exercise* 2.2 (An Optimal Resource Extraction Stopping Problem). Suppose the price $P(t)$ per unit of a resource (oil, gas...) at time t is given by

(1) $\mathrm{d}P(t) = \alpha P(t)\mathrm{d}t + \beta P(t)\mathrm{d}B(t) + \gamma P(t^-) \int_{\mathbb{R}} z\tilde{N}(\mathrm{d}t, \mathrm{d}z), \quad P(0) = p > 0$

and the remaining amount of resources $Q(t)$ at time t is

(2) $\mathrm{d}Q(t) = -\lambda Q(t)\mathrm{d}t, \quad Q(0) = q > 0,$

where $\lambda > 0$ is the (constant) relative extraction rate and α, β, γ are constants. We assume that $\gamma z \geq 0$ a.s. ν.

If we decide to stop extraction and close the field at a stopping time $\tau \geq 0$, the expected discounted total net profit $J^\tau(s, p, q)$ is given by

$$J^\tau(s, p, q) = E^{(s,p,q)}\left[\int_0^\tau e^{-\rho(s+t)}(\lambda P(t)Q(t) - K)\mathrm{d}t + \theta e^{-\rho(s+\tau)}P(\tau)Q(\tau)\right],$$

where $K > 0$ is the (constant) running cost rate, $\rho > 0$ is the (constant) discounting exponent, and $\theta > 0$ another constant. Find Φ and τ^* such that

$$\Phi(s, p, q) = \sup_{\tau \geq 0} J^\tau(s, p, q) = J^{\tau^*}(s, p, q).$$

[Hint: Try $\phi(s, p, q) = e^{-\rho s}\psi(p \cdot q)$ for some function $\psi : \mathbb{R} \to \mathbb{R}$.]

Exercise* 2.3. Solve the optimal stopping problem

$$\Phi(s, x) = \sup_{\tau \geq 0} E^x[e^{-\rho(s+\tau)}|X(\tau)|]$$

where

$$\mathrm{d}X(t) = \mathrm{d}B(t) + \int_{\mathbb{R}} \theta(X(t), z)\tilde{N}(\mathrm{d}t, \mathrm{d}z)$$

and $\rho > 0$ is a constant. Assume that there exists $\xi > 0$ such that

$$\theta(x, z) = 0 \quad \text{for a.a. } z \text{ if } |x| < \xi \tag{2.4.1}$$
$$\theta(x, z)(x - \xi) \geq 0 \quad \text{for a.a. } z \text{ if } |x| \geq \xi \tag{2.4.2}$$

and

$$\mathrm{tgh}(\sqrt{2\rho}\,\xi) \geq \frac{1}{\sqrt{2\rho}\,\xi}. \tag{2.4.3}$$

Exercise* 2.4 (The Optimal Time with Delay to Sell an Asset). This case (without the jump part) was first solved by [AK], with a more general (Markovian) delay $\delta(X) \geq 0$.

Suppose the value $X(t)$ of an asset at time t is modeled by a geometric Lévy process of the form

$$dX(t) = X(t^-)\left[\mu\,dt + \sigma\,dB(t) + \int_{\mathbb{R}} z\tilde{N}(dt, dz)\right], \quad X(0) = x > 0, \quad (2.4.4)$$

where μ, σ, and x are constants.

We assume that

$$-1 < z < 0 \quad \text{a.s. } \nu. \quad (2.4.5)$$

This guarantees that $X(t)$ never jumps down to a negative value. For convenience, we also assume that

$$E[\eta^2(t)] < \infty \quad \text{for all } t \geq 0. \quad (2.4.6)$$

Then by the Itô formula for Lévy processes the solution of (2.4.4) is

$$X(t) = x\exp\left[\left(\mu - \frac{1}{2}\sigma^2\right)t + \sigma B(t)\right.$$

$$\left. + \int_0^t \int_{\mathbb{R}} \{\ln(1 + z) - z\}\nu(dz)ds + \int_0^t \int_{\mathbb{R}} \ln(1 + z)\tilde{N}(ds, dz)\right],$$

$$t \geq 0. \quad (2.4.7)$$

Solve the following delayed optimal stopping problem

$$\Phi_\delta(s, x) = \sup_{\alpha \in \mathcal{T}_\delta} E^{s,x}[e^{-\rho(s+\alpha)}(X(\alpha) - q)], \quad (2.4.8)$$

where $E^{s,x}$ denotes expectation with respect to the probability law $P^{s,x}$ of the time–space process

$$dY(t) = \begin{bmatrix} dt \\ dX(t) \end{bmatrix}, \quad Y(0) = \begin{bmatrix} s \\ x \end{bmatrix}$$

and $\rho > 0$, $q > 0$ are constants. We assume that

$$\rho > \mu. \quad (2.4.9)$$

One possible interpretation of this problem is that $\Phi_\delta(s, x)$ represents the maximal expected discounted net payment obtained by selling the asset at a δ-delayed stopping time (ρ is the discounting exponent and q is the transaction cost).

3

Stochastic Control of Jump Diffusions

3.1 Dynamic Programming

Fix a domain $\mathcal{S} \subset \mathbb{R}^k$ (our solvency region) and let $Y(t) = Y^{(u)}(t)$ be a stochastic process of the form

$$dY(t) = b(Y(t), u(t))dt + \sigma(Y(t), u(t))dB(t)$$
$$+ \int_{\mathbb{R}^k} \gamma(Y(t^-), u(t^-), z)\bar{N}(dt, dz), \quad Y(0) = y \in \mathbb{R}^k, \qquad (3.1.1)$$

where

$$b : \mathbb{R}^k \times U \to \mathbb{R}^k, \quad \sigma : \mathbb{R}^k \times U \to \mathbb{R}^{k \times m}, \quad \text{and} \quad \gamma : \mathbb{R}^k \times U \times \mathbb{R}^k \to \mathbb{R}^{k \times \ell}$$

are given functions, $U \subset \mathbb{R}^p$ is a given set. The process $u(t) = u(t, \omega) :$ $[0, \infty) \times \Omega \to U$ is our *control process*, assumed to be càdlàg and adapted. We call $Y(t) = Y^{(u)}(t)$ a *controlled jump diffusion*.

We consider a performance criterion $J = J^{(u)}(y)$ of the form

$$J^{(u)}(y) = E^y \left[\int_0^{\tau_\mathcal{S}} f(Y(t), u(t))dt + g(Y(\tau_\mathcal{S})) \cdot \mathcal{X}_{\{\tau_\mathcal{S} < \infty\}} \right],$$

where
$$\tau_\mathcal{S} = \inf\{t > 0; Y^{(u)}(t) \notin \mathcal{S}\} \quad \text{(the bankruptcy time)}$$

and $f : \mathcal{S} \to \mathbb{R}$ and $g : \mathbb{R}^k \to \mathbb{R}$ are given continuous functions.

We say that the control process u is *admissible* and write $u \in \mathcal{A}$ if (3.1.1) has a unique, strong solution $Y(t)$ for all $y \in \mathcal{S}$ and

$$E^y \left[\int_0^{\tau_\mathcal{S}} f^-(Y(t), u(t))dt + g^-(Y(\tau_\mathcal{S})) \cdot \mathcal{X}_{\{\tau_\mathcal{S} < \infty\}} \right] < \infty.$$

The *stochastic control problem* is to find the *value function* $\Phi(y)$ and an *optimal control* $u^* \in \mathcal{A}$ defined by

$$\Phi(y) = \sup_{u \in \mathcal{A}} J^{(u)}(y) = J^{(u^*)}(y). \tag{3.1.2}$$

It turns out that – under mild conditions (see, e.g., [Ø1, Theorem 11.2.3]) – it suffices to consider *Markov* controls, i.e., controls $u(t)$ of the form

$$u(t) = u_0(Y(t^-))$$

for some function $u_0 : \mathbb{R}^k \to U$. Therefore, from now on we will only consider Markov controls and we will, with a slight abuse of notation, write $u(t) = u(Y(t^-))$.

Note that if $u = u(y)$ is a Markov control then $Y(t) = Y^{(u)}(t)$ is a Lévy diffusion with generator

$$A\phi(y) = A^u\phi(y) = \sum_{i=1}^{k} b_i(y, u(y)) \frac{\partial \phi}{\partial y_i}(y) + \frac{1}{2} \sum_{i,j=1}^{k} (\sigma\sigma^T)_{ij}(y, u(y)) \cdot \frac{\partial^2 \phi}{\partial y_i \partial y_j}(y)$$

$$+ \sum_{j=1}^{\ell} \int_{\mathbb{R}} \{\phi(y + \gamma^{(j)}(y, u(y), z_j)) - \phi(y)$$

$$- \nabla\phi(y) \cdot \gamma^{(j)}(y, u(y), z_j)\} \nu_j(dz_j).$$

We now formulate a verification theorem for the optimal control problem (3.1.2), analogous to the classical Hamilton–Jacobi–Bellman (HJB) for (continuous) Itô diffusions.

Theorem 3.1 (HJB for Optimal Control of Jump Diffusions).

(a) *Suppose $\phi \in C^2(\mathcal{S}) \cap C(\bar{\mathcal{S}})$ satisfies the following:*
  (i) $A^v\phi(y) + f(y, v) \leq 0$ *for all $y \in \mathcal{S}$, $v \in U$.*
  (ii) $Y(\tau_{\mathcal{S}}) \in \partial\mathcal{S}$ *a.s. on $\{\tau_{\mathcal{S}} < \infty\}$ and*
    $$\lim_{t \to \tau_{\mathcal{S}}^-} \phi(Y(t)) = g(Y(\tau_{\mathcal{S}})) \cdot \mathcal{X}_{\{\tau_{\mathcal{S}} < \infty\}} \text{ a.s., for all } u \in \mathcal{A}.$$
  (iii) $E^y \left[|\phi(Y(\tau))| + \int_0^{\tau_{\mathcal{S}}} |A\phi(Y(t))| dt \right] < \infty$, *for all $u \in \mathcal{A}$ and all $\tau \in \mathcal{T}$.*
  (iv) $\{\phi^-(Y(\tau))\}_{\tau \leq \tau_{\mathcal{S}}}$ *is uniformly integrable for all $u \in \mathcal{A}$ and $y \in \mathcal{S}$.*
  Then

$$\phi(y) \geq \Phi(y) \quad \text{for all } y \in \mathcal{S}. \tag{3.1.3}$$

(b) *Moreover, suppose that for each $y \in \mathcal{S}$ there exists $v = \hat{u}(y) \in U$ such that*
  (v) $A^{\hat{u}(y)}\phi(y) + f(y, \hat{u}(y)) = 0$
  and
  (vi) $\{\phi(Y^{(\hat{u})}(\tau))\}_{\tau \leq \tau_{\mathcal{S}}}$ *is uniformly integrable.*
  Suppose $u^*(t) := \hat{u}(Y(t^-)) \in \mathcal{A}$. Then u^* is an optimal control and

$$\phi(y) = \Phi(y) = J^{(u^*)}(y) \quad \text{for all } y \in \mathcal{S}. \tag{3.1.4}$$

Proof. (a) Let $u \in \mathcal{A}$. For $n = 1, 2, \ldots$ put $\tau_n = \min(n, \tau_S)$. Then by the Dynkin formula (Theorem 1.24) we have

$$E^y[\phi(Y(\tau_n))] = \phi(y) + E^y\left[\int_0^{\tau_n} A^u \phi(Y(t)) \mathrm{d}t\right] \leq \phi(y) - E\left[\int_0^{\tau_n} f(Y(t), u(t)) \mathrm{d}t\right].$$

Hence

$$\phi(y) \geq \liminf_{n \to \infty} E^y\left[\int_0^{\tau_n} f(Y(t), u(t)) \mathrm{d}t + \phi(Y(\tau_n))\right]$$

$$\geq E^y\left[\int_0^{\tau_S} f(Y(t), u(t)) \mathrm{d}t + g(Y(\tau_S)) \cdot \mathcal{X}_{\{\tau_S < \infty\}}\right] = J^{(u)}(y). \quad (3.1.5)$$

Since $u \in \mathcal{A}$ was arbitrary we conclude that

$$\phi(y) \geq \Phi(y) \quad \text{for all } y \in \mathcal{S}. \tag{3.1.6}$$

(b) Now apply the above argument to $u(t) = \hat{u}(Y(t))$, where $\hat{u}$ is as in (v). Then we get *equality* in (3.1.5) and hence

$$\phi(y) = J^{(\hat{u})}(y) \leq \Phi(y) \quad \text{for all } y \in \mathcal{S}. \tag{3.1.7}$$

Combining (3.1.6) and (3.1.7) we get (3.1.4). $\qquad\square$

Example 3.2 (Optimal Consumption and Portfolio in a Lévy Type Black–Scholes Market [Aa, FØS1]).

Suppose we have a market with two possible investments:

(i) A *safe* investment (bond, bank account) with price dynamics

$$\mathrm{d}P_1(t) = rP_1(t)\mathrm{d}t, \quad P_1(0) = p_1 > 0.$$

(ii) A *risky* investment (stock) with price dynamics

$$\mathrm{d}P_2(t) = P_2(t^-)\left[\mu\,\mathrm{d}t + \sigma\,\mathrm{d}B(t) + \int_{-1}^{\infty} z\tilde{N}(\mathrm{d}t, \mathrm{d}z)\right], \quad P_2(0) = p_2 > 0,$$

where $r > 0$, $\mu > 0$, and $\sigma \in \mathbb{R}$ are constants. We assume that

$$\int_{-1}^{\infty} |z|\mathrm{d}\nu(z) < \infty \quad \text{and} \quad \mu > r.$$

Assume that at any time t the investor can choose a consumption rate $c(t) \geq 0$ (adapted and càdlàg) and is also free to transfer money from one investment to the other without transaction cost. Let $X_1(t)$ and $X_2(t)$ be the amounts of money invested in the bonds and the stocks, respectively. Let

$$\theta(t) = \frac{X_2(t)}{X_1(t) + X_2(t)}$$

be the fraction of the total wealth invested in stocks at time t. Define the *performance criterion* by

$$J^{(c,\theta)}(s, x_1, x_2) = E^{x_1,x_2}\left[\int_0^\infty e^{-\delta(s+t)} \frac{c^\gamma(t)}{\gamma} dt\right],$$

where $\delta > 0$, $\gamma \in (0,1)$ are constants and E^{x_1,x_2} is the expectation w.r.t. the probability law P^{x_1,x_2} of $(X_1(t), X_2(t))$ when $X_1(0) = x_1$, $X_2(0) = x_2$. Call the control $u(t) = (c(t), \theta(t)) \in [0,\infty) \times [0,1]$ *admissible* and write $u \in \mathcal{A}$ if the corresponding *total wealth*

$$W(t) = W^{(u)}(t) = X_1^{(u)}(t) + X_2^{(u)}(t)$$

is *nonnegative* for all $t \geq 0$.

The problem is to find $\Phi(s, x_1, x_2)$ and $u^*(c^*, \theta^*) \in \mathcal{A}$ such that

$$\Phi(s, x_1, x_2) = \sup_{u \in \mathcal{A}} J^{(u)}(s, x_1, x_2) = J^{(u^*)}(s, x_1, x_2).$$

Case 1: $\nu = 0$.

In this case the problem was solved by Merton [M]. He proved that if

$$\delta > \gamma\left[r + \frac{(\mu - r)^2}{2\sigma^2(1-\gamma)}\right], \tag{3.1.8}$$

then the value function is

$$\Phi_0(s, x_1, x_2) = K_0 e^{-\delta s}(x_1 + x_2)^\gamma, \tag{3.1.9}$$

where

$$K_0 = \frac{1}{\gamma}\left[\frac{1}{1-\gamma}\left(\delta - \gamma r - \frac{\gamma(\mu - r)^2}{2\sigma^2(1-\gamma)}\right)\right]^{\gamma-1}. \tag{3.1.10}$$

Moreover, the optimal consumption rate $c_0^*(t)$ is given by

$$c_0^*(t) = (K_0 \gamma)^{1/(\gamma-1)}(X_1(t) + X_2(t)) \tag{3.1.11}$$

and the optimal portfolio $\theta_0^*(t)$ is (the constant)

$$\theta_0^*(t) = \frac{\mu - r}{\sigma^2(1-\gamma)} \quad \text{for all } t \in [0, \infty). \tag{3.1.12}$$

In other words, it is optimal to keep the state $(X_1(t), X_2(t))$ on the line

$$x_2 = \frac{\theta_0^*}{1 - \theta_0^*} x_1 \tag{3.1.13}$$

in the (x_1, x_2)-plane at all times (the Merton line). See Fig. 3.1.

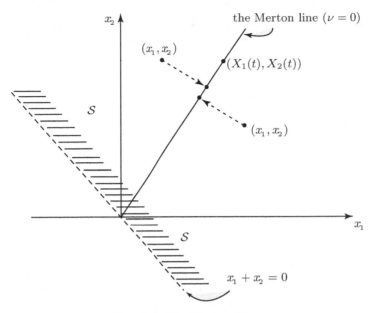

Fig. 3.1. The Merton line

Case 2: $\nu \neq 0$

We now ask: How does the presence of jumps influence the optimal strategy?

As in [M] we reduce the dimension by introducing

$$W(t) = X_1(t) + X_2(t).$$

Then we see that

$$dW(t) = ([r(1 - \theta(t)) + \mu\theta(t)]W(t) - c(t))\, dt + \sigma\theta(t)W(t)dB(t)$$
$$+ \theta(t)W(t^-) \int_{-1}^{\infty} z\tilde{N}(dt, dz), \quad W(0) = x_1 + x_2 = w \geq 0.$$

The generator $A^{(u)}$ of the controlled process

$$Y(t) = \begin{bmatrix} s+t \\ W(t) \end{bmatrix}; \quad t \geq 0, \quad Y(0) = y = \begin{bmatrix} s \\ w \end{bmatrix}$$

is

$$A^{(u)}\phi(y) = \frac{\partial\phi}{\partial s} + ([r(1-\theta) + \mu\theta]w - c)\frac{\partial\phi}{\partial w} + \frac{1}{2}\sigma^2\theta^2 w^2 \frac{\partial^2\phi}{\partial w^2}$$
$$+ \int_{-1}^{\infty} \left\{ \phi(s, w + \theta wz) - \phi(s, w) - \frac{\partial\phi}{\partial w}(s, w)\theta wz \right\} \nu(dz).$$

If we try

$$\phi(y) = \phi(s, w) = e^{-\delta s} \psi(w)$$

we get

$$A^{(u)} \phi(y) = e^{-\delta s} A_0^{(u)} \psi(w), \quad \text{where}$$

$$A_0^{(u)} \psi(w) = -\delta \psi(w) + ([r(1 - \theta) + \mu \theta] w - c) \psi'(w) + \frac{1}{2} \sigma^2 \theta^2 w^2 \psi''(w)$$

$$+ \int_{-1}^{\infty} \{\psi((1 + \theta z) w) - \psi(w) - \psi'(w) \theta w z\} \nu(dz).$$

In particular, if we try

$$\psi(w) = K w^{\gamma}$$

we get

$$A_0^{(u)} \psi(w) + f(w, u) = -\delta K w^{\gamma} + ([r(1 - \theta) + \mu \theta] w - c) K \gamma w^{\gamma - 1}$$

$$+ K \cdot \frac{1}{2} \sigma^2 \theta^2 w^2 \gamma (\gamma - 1) w^{\gamma - 2}$$

$$+ K w^{\gamma} \int_{-1}^{\infty} \{(1 + \theta z)^{\gamma} - 1 - \gamma \theta z\} \nu(dz) + \frac{c^{\gamma}}{\gamma}.$$

Let $h(c, \theta)$ be the expression on the right-hand side. Then h is concave in (c, θ) and the maximum of h is attained at the critical points, i.e., when

$$\frac{\partial h}{\partial c} = -K \gamma w^{\gamma - 1} + c^{\gamma - 1} = 0 \tag{3.1.14}$$

and

$$\frac{\partial h}{\partial \theta} = (\mu - r) K \gamma w^{\gamma} + K \sigma^2 \theta \gamma (\gamma - 1) w^{\gamma} + K w^{\gamma} \int_{-1}^{\infty} \{\gamma (1 + \theta z)^{\gamma - 1} z - \gamma z\} \nu(dz) = 0. \tag{3.1.15}$$

From (3.1.14) we get

$$c = \hat{c} = (K \gamma)^{1/(\gamma - 1)} w \tag{3.1.16}$$

and from (3.1.15) we get that $\theta = \hat{\theta}$ should solve the equation

$$\Lambda(\theta) := \mu - r - \sigma^2 \theta (1 - \gamma) - \int_{-1}^{\infty} \{1 - (1 + \theta z)^{\gamma - 1}\} z \nu(dz) = 0. \tag{3.1.17}$$

Since $\Lambda(0) = \mu - r > 0$ we see that if

$$\sigma^2 (1 - \gamma) + \int_{-1}^{\infty} \{1 - (1 + z)^{\gamma - 1}\} z \nu(dz) \geq \mu - r \tag{3.1.18}$$

then there exists an optimal $\theta = \hat{\theta} \in (0, 1]$.

With this choice of $c = \hat{c} = (K\gamma)^{1/(\gamma-1)} w$ and $\theta = \hat{\theta}$ (constant) we require that

$$A_0^{(\hat{u})}\psi(w) + f(w, \hat{u}) = 0, \quad \text{i.e.,}$$

$$-\delta K + \left([r(1-\hat{\theta}) + \mu\hat{\theta}] - (K\gamma)^{1/(\gamma-1)}\right)K\gamma$$

$$+ K\frac{1}{2}\sigma^2\hat{\theta}^2\gamma(\gamma-1) + K\int_{-1}^{\infty}\{(1+\hat{\theta}z)^\gamma - 1 - \gamma\hat{\theta}z\}\nu(dz)$$

$$+ (K\gamma)^{\gamma/(\gamma-1)}\frac{1}{\gamma} = 0$$

or

$$-\delta + \gamma[r(1-\hat{\theta}) + \mu\hat{\theta}] - (K\gamma)^{1/(\gamma-1)}\gamma$$

$$-\frac{1}{2}\sigma^2\hat{\theta}^2(1-\gamma)\gamma + \int_{-1}^{\infty}\{(1+\hat{\theta}z)^\gamma - 1 - \gamma\hat{\theta}z\}\nu(dz) + K^{1/(\gamma-1)} \cdot \gamma^{\gamma/(\gamma-1)} \cdot \frac{1}{\gamma}$$

or

$$(K\gamma)^{1/(\gamma-1)}[1-\gamma] = \delta - \gamma[r(1-\hat{\theta}) + \mu\hat{\theta}] + \frac{1}{2}\sigma^2\hat{\theta}^2(1-\gamma)\gamma$$

$$-\int_{-1}^{\infty}\{(1+\hat{\theta}z)^\gamma - 1 - \gamma\hat{\theta}z\}\nu(dz)$$

or

$$K = \frac{1}{\gamma}\left[\frac{1}{1-\gamma}\left(\delta - \gamma\{r(1-\hat{\theta}) + \mu\hat{\theta}\} + \frac{1}{2}\sigma^2\hat{\theta}^2(1-\gamma)\gamma\right.\right.$$

$$\left.\left. -\int_{\mathbb{R}}\{(1+\hat{\theta}z)^\gamma - 1 - \gamma\hat{\theta}z\}\nu(dz)\right)\right]^{\gamma-1}. \qquad (3.1.19)$$

We now study condition (iii):
Here $\sigma^T\nabla\phi(y) = e^{-\delta s}\sigma\theta w K\gamma w^{\gamma-1} = e^{-\delta s}\sigma\theta K w^\gamma$ and

$$\phi(Y(t) + \gamma(Y(t), u(t))) - \phi(Y(t)) = KW(t)^\gamma e^{-\delta s}[(1+\theta z)^\gamma - 1].$$

So (iii) holds if

$$E\left[\int_0^T e^{-2\delta t}W^{2\gamma}(t)dt\right] + \int_{\mathbb{R}}[(1+\theta z)^\gamma - 1]\nu(dz) < \infty. \qquad (3.1.20)$$

We refer to [FØS1] for sufficient conditions on the parameters for (3.1.20) to hold.

We conclude that the value function is

$$\Phi(s, w) = \Phi(s, x_1, x_2) = e^{-\delta s}K(x_1 + x_2)^\gamma \qquad (3.1.21)$$

with optimal control $u^*(t) = (c^*(t), \theta^*(t))$ where $c^* = \hat{c} = (K\gamma)^{1/\gamma-1}(x_1 + x_2)$ is given by (3.1.16) and $\theta^* = \hat{\theta}$ is given by (3.1.17), with K given by (3.1.19).

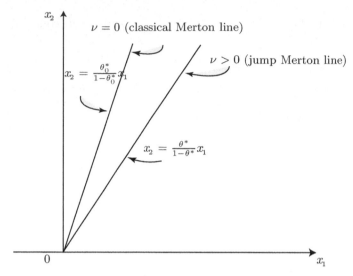

Fig. 3.2. The Merton line for $\nu = 0$ and $\nu > 0$

Finally we compare the solution in the jump case ($\nu \neq 0$) with Merton's solution in the no jump case ($\nu = 0$).

As before let Φ_0, c_0^*, and θ_0^* be the solution when there are no jumps ($\nu = 0$). Then it can be seen that

$$K < K_0 \quad \text{and hence} \quad \Phi(s, w) = e^{-\delta s} K w^\gamma < e^{-\delta s} K_0 w^\gamma = \Phi_0(s, w)$$

$$c^*(s, w) \geq c_0^*(s, w)$$

$$\theta^* \leq \theta_0^*.$$

So with jumps it is optimal to place a *smaller* wealth fraction in the risky investment, consume *more* relative to the current wealth and the resulting value is *smaller* than in the no jump case. See Fig.3.2.

For more details we refer to [FØS1].

Remark 3.3. For more information and other applications of stochastic control of jump diffusions, see [GS, BKR1, BKR2, BKR3, BKR4, BKR5, BKR6, Ma] and the references therein.

3.2 The Maximum Principle

Suppose the state $X(t) = X^{(u)}(t)$ of a controlled jump diffusion in $\mathbb{R}^n$ is given by

$$dX(t) = b(t, X(t), u(t))dt + \sigma(t, X(t), u(t))dB(t)$$

$$+ \int_{\mathbb{R}^\ell} \gamma(t, X(t^-), u(t^-), z)\tilde{N}(dt, dz). \tag{3.2.1}$$

As before $\tilde{N}(dt, dz) = (\tilde{N}_1(dt, dz_1), \ldots, \tilde{N}_\ell(dt, dz_\ell))^T$, where

$$\tilde{N}_j(dt, dz_j) = N_j(dt, dz_j) - \nu_j(dz_j)dt, \quad 1 \le j \le \ell$$

(see the notation of Theorem 1.16).

The process $u(t) = u(t, \omega) \in U \subset \mathbb{R}^k$ is our *control*. We assume that u is adapted and càdlàg, and that the corresponding (3.2.1) has a unique strong solution $X^{(u)}(t)$, $t \in [0, T]$. Such controls are called *admissible*. The set of admissible controls is denoted by $\mathcal{A}$.

Suppose the performance criterion has the form

$$J(u) = E\left[\int_0^T f(t, X(t), u(t))dt + g(X(T))\right], \quad u \in \mathcal{A},$$

where $f : [0, T] \times \mathbb{R}^n \times U \to \mathbb{R}$ is continuous, $g : \mathbb{R}^n \to \mathbb{R}$ is C^1, $T < \infty$ is a fixed deterministic time and

$$E\left[\int_0^T f^-(t, X(t), u(t))dt + g^-(X(T))\right] < \infty \quad \text{for all } u \in \mathcal{A}.$$

Consider the problem to find $u^* \in \mathcal{A}$ such that

$$J(u^*) = \sup_{u \in \mathcal{A}} J(u). \tag{3.2.2}$$

In Chap. 2 we saw how to solve such a problem using dynamic programming and the associated HJB equation. Here we present an alternative approach, based on what is called *the maximum principle*. In the deterministic case this principle was first introduced by Pontryagin and his group [PBGM]. A corresponding maximum principle for Itô diffusions was formulated by Kushner [Ku], Bismut [Bi], and subsequently further developed by Bensoussan [Ben1, Ben2, Ben3], Haussmann [H1], and others. For *jump diffusions* a sufficient maximum principle has recently been formulated in [FØS3] and it is this approach that is presented here, in a somewhat simplified version.

Define the *Hamiltonian* $H : [0, T] \times \mathbb{R}^n \times U \times \mathbb{R}^n \times \mathbb{R}^{n \times m} \times \mathcal{R} \to \mathbb{R}$ by

$$H(t, x, u, p, q, r) = f(t, x, u) + b^T(t, x, u)p + \text{tr}(\sigma^T(t, x, u)q)$$

$$+ \sum_{j=1}^\ell \sum_{i=1}^n \int_{\mathbb{R}} \gamma_{ij}(t, x, u, z_j)r_{ij}(t, z)\nu_j(dz_j), \tag{3.2.3}$$

where $\mathcal{R}$ is the set of functions $r : \mathbb{R}^{\ell+1} \to \mathbb{R}^{n \times \ell}$ such that the integrals in (3.2.3) converge. From now on we assume that H is differentiable with respect to x.

The *adjoint* equation (corresponding to u and $X^{(u)}$) in the unknown processes $p(t) \in \mathbb{R}^n$, $q(t) \in \mathbb{R}^{n \times m}$, and $r(t,z) \in \mathbb{R}^{n \times \ell}$ is the backward stochastic differential equation

$$\begin{cases} \mathrm{d}p(t) = -\nabla_x H(t, X(t), u(t), p(t), q(t), r(t, \cdot))\mathrm{d}t \\ \qquad\qquad + q(t)\mathrm{d}B(t) + \displaystyle\int_{\mathbb{R}^\ell} r(t^-, z)\tilde{N}(\mathrm{d}t, \mathrm{d}z), \qquad t < T \qquad (3.2.4) \\ p(T) = \nabla g(X(T)). \end{cases}$$

Theorem 3.4 (A Sufficient Maximum Principle [FØS3]). *Let $\hat{u} \in \mathcal{A}$ with corresponding solution $\hat{X} = X^{(\hat{u})}$ and suppose there exists a solution $(\hat{p}(t), \hat{q}(t), \hat{r}(t, z))$ of the corresponding adjoint equation (3.2.4) satisfying*

$$E\left[\int_0^T (\hat{X}(t) - X^{(u)}(t))^T \left\{ \hat{q}\hat{q}^T(t) + \int_{\mathbb{R}^\ell} r\,r^T(t, z)\nu(\mathrm{d}z) \right\} (\hat{X}(t) - X^{(u)}(t))\mathrm{d}t \right] < \infty$$

$$(3.2.5)$$

and

$$E\left[\int_0^T \hat{p}(t)^T \left\{ \sigma\sigma^T(t, X^{(u)}(t), u(t)) + \int_{\mathbb{R}^\ell} \gamma\,\gamma^T(t, X^{(u)}(t), u(t), z)\nu(\mathrm{d}z) \right\} \hat{p}(t)\mathrm{d}t \right]$$
$$< \infty \quad \text{for all } u \in \mathcal{A}.$$

Moreover, suppose that

$$H(t, \hat{X}(t), \hat{u}(t), \hat{p}(t), \hat{q}(t), \hat{r}(t, \cdot)) = \sup_{v \in U} H(t, \hat{X}(t), v, \hat{p}(t), \hat{q}(t), \hat{r}(t, \cdot))$$

for all t, that $g(x)$ is a concave function of x and that

$$\hat{H}(x) := \max_{v \in U} H(t, x, v, \hat{p}(t), \hat{q}(t), \hat{r}(t, \cdot)) \text{ exists and is}$$

$$(3.2.6)$$

a concave function of x, for all $t \in [0, T]$ (the Arrow condition).

Then $\hat{u}$ is an optimal control.

Remark 3.5. For (3.2.6) to hold it suffices that the function

$$(x, v) \to H(t, x, v, \hat{p}(t), \hat{q}(t), \hat{r}(t, \cdot)) \quad \text{is concave, for all } t \in [0, T]. \quad (3.2.7)$$

To prove Theorem 3.4 we first establish the following.

Lemma 3.6 (Integration by Parts). *Suppose $E[(Y^{(j)}(T)^2] < \infty$ for $j = 1, 2$, where*

$$\mathrm{d}Y^{(j)}(t) = b^{(j)}(t,\omega)\mathrm{d}t + \sigma^{(j)}(t,\omega)\mathrm{d}B(t) + \int_{\mathbb{R}^\ell} \gamma^{(j)}(t,z,\omega)\widetilde{N}(\mathrm{d}t,\mathrm{d}z)$$

$$Y^{(j)}(0) = y^{(j)} \in \mathbb{R}^n, \quad j = 1, 2$$

where $b^{(j)} \in \mathbb{R}^n$, $\sigma^{(j)} \in \mathbb{R}^{n\times m}$, and $\gamma^{(j)} \in \mathbb{R}^{n\times\ell}$. Then

$$E[Y^{(1)}(T) \cdot Y^{(2)}(T)]$$

$$= y_1 \cdot y_2 + E\left[\int_0^T Y^{(1)}(t^-)\mathrm{d}Y^{(2)}(t)\right.$$

$$+ \int_0^T Y^{(2)}(t^-)\mathrm{d}Y^{(1)}(t) + \int_0^T \mathrm{tr}[\sigma^{(1)^T}\sigma^{(2)}](t)\mathrm{d}t$$

$$+ \left.\int_0^T \left[\sum_{j=1}^\ell \left(\sum_{i=1}^n \int_{\mathbb{R}} \gamma_{ij}^{(1)}(t,z_j)\gamma_{ij}^{(2)}(t,x)\right)\nu_j(\mathrm{d}z_j)\right]\mathrm{d}t\right].$$

Proof. This follows from the Itô formula (Theorem 1.16). (See also Exercise 1.7.) $\qquad\square$

Proof of Theorem 3.4. Let $u \in \mathcal{A}$ be an admissible control with corresponding state process $X(t) = X^{(u)}(t)$. Then

$$J(\hat{u}) - J(u) = E\left[\int_0^T \{f(t,\hat{X}(t),\hat{u}(t)) - f(t,X(t),u(t))\}\mathrm{d}t + g(\hat{X}(T)) - g(X(T))\right].$$

Since g is concave we get by Lemma 3.6

$$E[g(\hat{X}(T)) - g(X(T))] \geq E[(\hat{X}(T) - X(T))^T \nabla g(\hat{X}(T))]$$

$$= E[(\hat{X}(T) - X(T))^T \hat{p}(T)]$$

$$= E\left[\int_0^T (\hat{X}(t^-) - X(t^-))^T \mathrm{d}\hat{p}(t)\right.$$

$$+ \int_0^T \hat{p}(t^-)^T (\mathrm{d}\hat{X}(t) - \mathrm{d}X(t))$$

$$+ \int_0^T \mathrm{tr}\left[\{\sigma(t,\hat{X}(t),\hat{u}(t)) - \sigma(t,X(t),u(t))\}^T \hat{q}(t)\right]\mathrm{d}t$$

$$+ \int_0^T \left(\sum_{j=1}^\ell \left(\sum_{i=1}^n \int_{\mathbb{R}} \{\gamma_{ij}(t,\hat{X}(t),\hat{u}(t),z_j)\right.\right.$$

$$\left.\left.- \gamma_{ij}(t,X(t),u(t),z_j)\}\hat{r}_{ij}(t,z_j)\right)\nu_j(\mathrm{d}z_j)\right)\mathrm{d}t\right]$$

$$= E\left[\int_0^T (\hat{X}(t) - X(t))^T \times \left(-\nabla_x H(t, \hat{X}(t), \hat{u}(t), \hat{p}(t), \hat{q}(t), \hat{r}(t, \cdot))\right) \mathrm{d}t\right.$$

$$+ \int_0^T \hat{p}^T(t^-)\{b(t, \hat{X}(t), \hat{u}(t)) - b(t, X(t), u(t))\} \mathrm{d}t$$

$$+ \int_0^T \mathrm{tr}[\{\sigma(t, \hat{X}(t), \hat{u}(t)) - \sigma(t, X(t), u(t))\}^T \hat{q}(t)] \mathrm{d}t$$

$$+ \int_0^T \left(\sum_{j=1}^\ell \left(\sum_{i=1}^n \int_{\mathbb{R}} \{\gamma_{ij}(t, \hat{X}(t), \hat{u}(t), z_j)\right.\right.$$

$$\left.\left.\left. - \gamma_{ij}(t, X(t), u(t), z_j)\}\hat{r}_{ij}(t, z_j)\right)\nu_j(\mathrm{d}z_j)\right) \mathrm{d}t\right]. \tag{3.2.8}$$

By the definition of H we find

$$E\left[\int_0^T \{f(t, \hat{X}(t), \hat{u}(t)) - f(t, X(t), u(t))\} \mathrm{d}t\right]$$

$$= E\left[\int_0^T \{H(t, \hat{X}(t), \hat{u}(t), \hat{p}(t), \hat{q}(t), \hat{r}(t, \cdot))\right.$$

$$- H(t, X(t), u(t), \hat{p}(t), \hat{q}(t), \hat{r}(t, \cdot))\} \mathrm{d}t$$

$$- \int_0^T \{b(t, \hat{X}(t), \hat{u}(t)) - b(t, X(t), u(t))\}^T \hat{p}(t) \mathrm{d}t$$

$$- \int_0^T \mathrm{tr}[\{\sigma(t, \hat{X}(t), \hat{u}(t)) - \sigma(t, X(t), u(t))\}^T \hat{q}(t)] \mathrm{d}t$$

$$- \int_0^T \left(\sum_{j=1}^\ell \left(\sum_{i=1}^n \int_{\mathbb{R}} \{\gamma_{ij}(t, \hat{X}(t), \hat{u}(t), z_j)\right.\right.$$

$$\left.\left.\left. - \gamma_{ij}(t, X(t), u(t), z_j)\}\hat{r}_{ij}(t, z_j)\right)\nu_j(\mathrm{d}z_j)\right) \mathrm{d}t\right]. \tag{3.2.9}$$

Adding (3.2.8) and (3.2.9) we get

$$J(\hat{u}) - J(u) \geq E\left[\int_0^T \{H(t, \hat{X}(t), \hat{u}(t), \hat{p}(t), \hat{q}(t), \hat{r}(t, \cdot))\right.$$

$$- H(t, X(t), u(t), \hat{p}(t), \hat{q}(t), \hat{r}(t, \cdot))$$

$$\left. - (\hat{X}(t) - X(t))^T \nabla_x H(t, \hat{X}(t), \hat{u}(t), \hat{p}(t), \hat{q}(t), \hat{r}(t, \cdot))\} \mathrm{d}t\right].$$

If (3.2.6) (or (3.2.7)) holds then $J(\hat{u}) - J(u) \geq 0$. This follows from the proof in [SeSy, p. 108]. For details we refer to [FØS3]. $\qquad\square$

We mention briefly the relation to dynamic programming. Define

$$J^{(u)}(s,x) = E\left[\int_0^{T-s} f(s+t, X^x(t), u(t))dt + g(X^x(T-s))\right], \quad u \in \mathcal{A},$$

where $X^x(t)$ is the solution of (3.2.1) for $t \geq 0$ with initial value $X(0) = x$.
Then put

$$V(s,x) = \sup_{u \in \mathcal{A}} J^{(u)}(s,x). \tag{3.2.10}$$

Theorem 3.7 ([FØS3]). *Assume that $V(s,x) \in C^{1,3}(\mathbb{R} \times \mathbb{R}^n)$ and that there exists an optimal Markov control $u^*(t,x)$ for problem (3.2.2), with corresponding solution $X^*(t)$ of (3.2.1). Define*

$$p_i(t) = \frac{\partial V}{\partial x_i}(t, X^*(t)), \quad 1 \leq i \leq n,$$

$$q_{jk}(t) = \sum_{i=1}^n \sigma_{ik}(t, X^*(t), u^*(t)) \frac{\partial^2 V}{\partial x_i \partial x_j}(t, X^*(t)), \quad 1 \leq j \leq n, 1 \leq k \leq m,$$

$$r_{ik}(t,z) = \frac{\partial V}{\partial x_i}(t, X^*(t) + \gamma^{(k)}(t, X^*(t), u^*(t), z_k)) - \frac{\partial V}{\partial x_i}(t, X^*(t)),$$

$$1 \leq i \leq n, 1 \leq k \leq \ell.$$

Then $p(t), q(t), r(t, \cdot)$ solve the adjoint equation (3.2.4).

For a proof see [FØS3].

Remark 3.8. A general discussion of impulse control for jump diffusions can be found in [F]. A study with vanishing impulse costs is given in [ØUZ].

3.3 Application to Finance

The following example is from [FØS3].

Consider a financial market with two investment possibilities, a risk free (e.g., a bond or bank account) and risky (e.g., a stock), whose prices $S_0(t), S_1(t)$ at time $t \in [0, T]$ are given by

$$(\text{bond}) \quad dS_0(t) = \rho_t S_0(t)dt, \quad S_0(0) = 1, \tag{3.3.1}$$

$$(\text{stock}) \quad dS_1(t) = S_1(t^-)\left[\mu_t dt + \sigma_t dB(t) + \int_{\mathbb{R}} \gamma(t,z)\tilde{N}(dt,dz)\right], \quad S_1(0) > 0, \tag{3.3.2}$$

where $\rho_t > 0$, μ_t, σ_t, and $\gamma(t,z) \geq -1$ are given bounded deterministic functions. We assume that the function

$$t \to \int_{\mathbb{R}} \gamma^2(t, z)\nu(\mathrm{d}z) \text{ is locally bounded.} \tag{3.3.3}$$

We may regard this market as a jump diffusion extension of the classical Black–Scholes market (see Sect. 1.5).

A *portfolio* in this market is a two-dimensional càdlàg and adapted process $\theta(t) = (\theta_0(t), \theta_1(t))$ giving the number of units of bonds and stocks, respectively, held at time t by an agent.

The corresponding *wealth* process $X(t) = X^{(\theta)}(t)$ is defined by

$$X(t) = \theta_0(t)S_0(t) + \theta_1(t)S_1(t), \quad t \in [0, T]. \tag{3.3.4}$$

The portfolio θ is called *self-financing* if

$$X(t) = X(0) + \int_0^t \theta_0(s)\mathrm{d}S_0(s) + \int_0^t \theta_1(s)\mathrm{d}S_1(s) \tag{3.3.5}$$

or, in shorthand notation,

$$\mathrm{d}X(t) = \theta_0(t)\mathrm{d}S_0(t) + \theta_1(t)\mathrm{d}S_1(t). \tag{3.3.6}$$

Alternatively, the portfolio can also be expressed in terms of the *amounts* $w_0(t)$ and $w_1(t)$ invested in the bond and stock, respectively. They are given by

$$w_i(t) = \theta_i(t)S_i(t), \quad i = 0, 1. \tag{3.3.7}$$

Now put

$$u(t) = w_1(t). \tag{3.3.8}$$

Then $w_0(t) = X(t) - u(t)$ and (3.3.6) gets the form

$$\mathrm{d}X(t) = [\rho_t X(t) + (\mu_t - \rho_t)u(t)]\mathrm{d}t + \sigma_t u(t)\mathrm{d}B(t) + u(t^-) \int_{\mathbb{R}} \gamma(t, z)\tilde{N}(\mathrm{d}t, \mathrm{d}z). \tag{3.3.9}$$

We call $u(t)$ *admissible* and write $u(t) \in \mathcal{A}$ if (3.3.9) has a unique solution $X(t) = X^{(u)}(t)$ such that $E[(X^{(u)}(T))^2] < \infty$.

The *mean–variance portfolio selection problem* is to find $u(t)$ which minimizes

$$\mathrm{Var}[X(T)] := E\big[(X(T) - E[X(T)])^2\big] \tag{3.3.10}$$

under the condition that

$$E[X(T)] = A, \quad \text{a given constant.} \tag{3.3.11}$$

By the Lagrange multiplier method the problem can be reduced to minimizing, for a given constant $a \in \mathbb{R}$,

$$E[(X(T) - a)^2]$$

without constraints. To see this, consider

$$E[(X(T) - A)^2 - \lambda(E[X(T)] - A)]$$

$$= E\left[X^2(T) - 2\left(A + \frac{\lambda}{2}\right)X(T) + A^2 + \lambda A\right]$$

$$= E\left[\left(X(T) - \left(A + \frac{\lambda}{2}\right)\right)^2\right] - \frac{\lambda^2}{4}, \quad \text{where } \lambda \in \mathbb{R} \text{ is constant.}$$

We will consider the equivalent problem

$$\sup_{u \in \mathcal{A}} E\left[-\frac{1}{2}(X^{(u)}(T) - a)^2\right]. \tag{3.3.12}$$

In this case the Hamiltonian (3.2.3) gets the form

$$H(t, x, u, p, q, r) = \{\rho_t x + (\mu_t - \rho_t)u\}p + \sigma_t u q + u \int_{\mathbb{R}} \gamma(t, z) r(t, z) \nu(dz).$$

Hence the adjoint equations (3.2.4) are

$$\begin{cases} dp(t) = -\rho_t p(t)dt + q(t)dB(t) + \int_{\mathbb{R}} r(t^-, z)\tilde{N}(dt, dz), \quad t < T \\ p(T) = -(X(T) - a). \end{cases} \tag{3.3.13}$$

We try a solution of the form

$$p(t) = \phi_t X(t) + \psi_t, \tag{3.3.14}$$

where ϕ_t, ψ_t are deterministic C^1 functions. Substituting in (3.3.13) and using (3.3.9) we get

$$dp(t) = \phi_t\left[\{\rho_t X(t) + (\mu_t - \rho_t)u(t)\}dt + \sigma_t u(t)dB(t)\right.$$

$$\left. + u(t^-)\int_{\mathbb{R}} \gamma(t, z)\tilde{N}(dt, dz)\right] + X(t)\phi_t' \, dt + \psi_t' \, dt$$

$$= [\phi_t \rho_t X(t) + \phi_t(\mu_t - \rho_t)u(t) + X(t)\phi_t' + \psi_t']dt$$

$$+ \phi_t \sigma_t u(t)dB(t) + \phi_t u(t^-)\int_{\mathbb{R}} \gamma(t, z)\tilde{N}(dt, dz). \tag{3.3.15}$$

Comparing with (3.3.13) we get

$$\phi_t \rho_t X(t) + \phi_t(\mu_t - \rho_t)u(t) + X(t)\phi_t' + \psi_t' = -\rho_t(\phi_t X(t) + \psi_t), \tag{3.3.16}$$

$$q(t) = \phi_t \sigma_t u(t), \tag{3.3.17}$$

$$r(t, z) = \phi_t u(t)\gamma(t, z). \tag{3.3.18}$$

Let $\hat{u} \in \mathcal{A}$ be a candidate for the optimal control with corresponding $\hat{X}$ and $\hat{p}, \hat{q}, \hat{r}$. Then

$$H(t, \hat{X}(t), u, \hat{p}(t), \hat{q}(t), \hat{r}(t, \cdot))$$

$$= \rho_t \hat{X}(t)\hat{p}(t) + u\left[(\mu_t - \rho_t)\hat{p}(t) + \sigma_t \hat{q}(t) + \int_{\mathbb{R}} \gamma(t, z)\hat{r}(t, z)\nu(dz)\right].$$

Since this is a linear expression in u, it is natural to guess that the coefficient of u vanishes, i.e.,

$$(\mu_t - \rho_t)\hat{p}(t) + \sigma_t \hat{q}(t) + \int_{\mathbb{R}} \gamma(t, z)\hat{r}(t, z)\nu(dz) = 0. \qquad (3.3.19)$$

Using that by (3.3.17) and (3.3.18) we have

$$\hat{q}(t) = \phi_t \sigma_t \hat{u}(t), \quad \hat{r}(t, z) = \phi_t \hat{u}(t)\gamma(t, z)$$

we get from (3.3.19) that

$$\hat{u}(t) = \frac{(\rho_t - \mu_t)\hat{p}(t)}{\phi_t \Lambda_t} = \frac{(\rho_t - \mu_t)(\phi_t \hat{X}(t) + \psi_t)}{\phi_t \Lambda_t}, \qquad (3.3.20)$$

where

$$\Lambda_t = \sigma_t^2 + \int_{\mathbb{R}} \gamma^2(t, z)\nu(dz). \qquad (3.3.21)$$

On the other hand, from (3.3.16) we have

$$\hat{u}(t) = \frac{(\phi_t \rho_t + \phi_t')\hat{X}(t) + \rho_t(\phi_t \hat{X}(t) + \psi_t) + \psi_t'}{\phi_t(\rho_t - \mu_t)}. \qquad (3.3.22)$$

Combining (3.3.20) and (3.3.22) we get the equations

$$(\rho_t - \mu_t)^2 \phi_t - [2\rho_t \phi_t + \phi_t']\Lambda_t = 0, \quad \phi_T = -1,$$

$$(\rho_t - \mu_t)^2 \psi_t - [\rho_t \psi_t + \psi_t']\Lambda_t = 0, \quad \psi_T = a,$$

which have the solutions

$$\phi_t = -\exp\left(\int_t^T \left\{\frac{(\rho_s - \mu_s)^2}{\Lambda_s} - 2\rho_s\right\}ds\right), \quad 0 \le t \le T, \qquad (3.3.23)$$

$$\psi_t = a\exp\left(\int_t^T \left\{\frac{(\rho_s - \mu_s)^2}{\Lambda_s} - \rho_s\right\}ds\right), \quad 0 \le t \le T. \qquad (3.3.24)$$

With this choice of ϕ_t and ψ_t the processes

$$\hat{p}(t) := \phi_t \hat{X}(t) + \psi_t, \quad \hat{q}(t) := \phi_t \sigma_t \hat{u}(t), \quad \text{and} \quad \hat{r}(t, z) := \phi_t \hat{u}(t)\gamma(t, z)$$

solve the adjoint equation, and by (3.3.19) we see that all the conditions of the sufficient maximum principle (Theorem 3.4) are satisfied. We conclude that $\hat{u}(t)$ given by (3.3.20) is an optimal control. In feedback form the control can be written

$$\hat{u}(t,x) = \frac{(\rho_t - \mu_t)(\phi_t x + \psi_t)}{\phi_t \Lambda_t}. \tag{3.3.25}$$

3.4 Exercises

Exercise* 3.1. Suppose the wealth $X(t) = X^{(u)}(t)$ of a person with consumption rate $u(t) \geq 0$ satisfies the following Lévy type mean reverting Ornstein–Uhlenbeck SDE

$$dX(t) = (\mu - \rho X(t) - u(t))dt + \sigma\, dB(t) + \theta \int_{\mathbb{R}} z\tilde{N}(dt, dz), \quad t > 0,$$

$$X(0) = x > 0.$$

Fix $T > 0$ and define

$$J^{(u)}(s,x) = E^{s,x}\left[\int_0^{T-s} e^{-\delta(s+t)} \frac{u^\gamma(t)}{\gamma} dt + \lambda X(T-s) \right].$$

Use dynamic programming to find the value function $\Phi(s,x)$ and the optimal consumption rate (control) $u^*(t)$ such that

$$\Phi(s,x) = \sup_{u(\cdot)} J^{(u)}(s,x) = J^{(u^*)}(s,x).$$

In the above $\mu, \rho, \sigma, \theta, T, \delta > 0$, $\gamma \in (0,1)$, and $\lambda > 0$ are constants.

Exercise* 3.2. Solve the problem of Exercise 3.1 by using the stochastic maximum principle.

Exercise* 3.3. Define

$$dX^{(u)}(t) = dX(t) = \begin{bmatrix} dX_1(t) \\ dX_2(t) \end{bmatrix} = \begin{bmatrix} u(t,\omega) \int_{\mathbb{R}} z\tilde{N}(dt, dz) \\ \int_{\mathbb{R}} z^2 \tilde{N}(dt, dz) \end{bmatrix} \in \mathbb{R}^2$$

and, for fixed $T > 0$ (deterministic)

$$J(u) = E\left[-(X_1(T) - X_2(T))^2 \right].$$

Use the stochastic maximum principle to find u^* such that

$$J(u^*) = \sup_u J(u).$$

Interpretation. Put $F(\omega) = \int_{\mathbb{R}} z^2 \tilde{N}(T, dz)$. We may regard F as a given T-claim in the normalized market with the two investment possibilities bond and stock, whose prices are

(bond) $dS_0(t) = 0, \quad S_0(0) = 1,$

(stock) $dS_1(t) = \int_{\mathbb{R}} z \tilde{N}(dt, dz),$ a Lévy martingale.

Then $-J(u)$ is the variance of the difference between $F = X_2(T)$ and the wealth $X_1(T)$ generated by a self-financing portfolio $u(t, \omega)$. See [BDLØP] for more information on minimal variance hedging in markets driven by Lévy martingales.

Exercise* 3.4. Solve the stochastic control problem

$$\Phi_1(s, x) = \inf_{u \geq 0} E^{s,x} \left[\int_0^\infty e^{-\rho(s+t)} (X^2(t) + \theta u^2(t)) dt \right],$$

where

$$dX(t) = u(t)dt + \sigma dB(t) + \int_{\mathbb{R}} z \tilde{N}(dt, dz), X(0) = x,$$

where $\rho > 0$, $\theta > 0$, and $\sigma > 0$ are constants.

The interpretation of this problem is that we want to push the process $X(t)$ as close as possible to 0 by using a minimum of energy, its rate being measured by $\theta u^2(t)$.

[Hint: Try $\varphi(s, x) = e^{-\rho s}(ax^2 + b)$ for some constants a, b.]

Exercise* 3.5 (The Stochastic Linear Regulator Problem).
   Solve the stochastic control problem

$$\Phi_0(x) = \inf_u E^x \left[\int_0^T (X^2(t) + \theta u(t)^2) dt + \lambda X^2(T) \right],$$

where

$$dX(t) = u(t)dt + \sigma dB(t) + \int_{\mathbb{R}} z \tilde{N}(dt, dz), X(0) = x$$

and

$$T > 0 \text{ is a constant.}$$

(a) By using dynamic programming (Theorem 3.1).
(b) By using the stochastic maximum principle (Theorem 3.4).

Exercise* 3.6. Solve the stochastic control problem

$$\Phi(s, x) = \sup_{c(t) \geq 0} E^{s,x} \left[\int_0^{T_0} e^{-\delta(s+t)} \ln c(t) dt \right],$$

where the supremum is taken over all $\mathcal{F}_t$-adapted processes $c(t) \geq 0$ and

$$\tau_0 = \inf\{t > 0; X(t) \leq 0\},$$

where

$$dX(t) = X(t^-)\left[\mu\,dt + \sigma\,dB(t) + \theta\int_{\mathbb{R}} z\tilde{N}(dt, dz)\right] - c(t)dt, \quad X(0) = x > 0,$$

where $\delta > 0$, μ, σ, and θ are constants, and

$$\theta z > -1 \text{ for a.a. } z \text{ w.r.t. } \nu.$$

We may interpret $c(t)$ as the consumption rate, $X(t)$ as the corresponding wealth, and τ_0 as the bankruptcy time. Thus Φ represents the maximal expected total discounted logarithmic utility of the consumption up to bankruptcy time.

[Hint: Try $\varphi(s, x) = e^{-\delta s}(a \ln x + b)$ as a candidate for $\Phi(s, x)$, where a and b are suitable constants.]

Exercise 3.7. Use the stochastic maximum principle (Theorem 3.4) to solve the problem

$$\sup_{c(t)\geq 0} E\left[\int_0^T e^{-\delta t} \ln c(t)dt + \lambda e^{-\delta T} \ln X(T)\right],$$

where

$$dX(t) = X(t^-)\left[\mu\,dt + \sigma\,dB(t) + \theta\int_{\mathbb{R}} z\tilde{N}(dt, dz)\right] - c(t)dt, \quad X(0) > 0.$$

Here $\delta > 0$, $\lambda > 0$, μ, σ, and θ are constants, and

$$\theta z > -1 \text{ for a.a. } z(\nu).$$

(see Exercise 3.6 for an interpretation of this problem).

[Hint: Try $p(t) = ae^{-\delta t}X^{-1}(t)$ and $c(t) = X(t)/a$, for some constant $a > 0$.]

4

Combined Optimal Stopping
and Stochastic Control of Jump Diffusions

4.1 Introduction

In this chapter we discuss combined optimal stopping and stochastic control problems and their associated Hamilton–Jacobi–Bellman (HJB) variational inequalities. This is a subject which deserves to be better known because of its many applications. A thorough treatment of such problems (but without the associated HJB variational inequalities) can be found in Krylov [K].

This chapter may also serve as a brief review of the theory of optimal stopping and their variational inequalities on one hand, and the theory of stochastic control and their HJB equations on the other. An introduction to these topics separately can be found in [Ø1].

As an illustration of how combined optimal stopping and stochastic control problems may appear in economics, let us consider the following example which is an extension of Exercise 2.2.

Example 4.1 (An Optimal Resource Extraction Control and Stopping Problem). Suppose the price $P_t = P(t)$ of one unit of a resource (e.g., gas or oil) at time t is varying like a geometric Lévy process, i.e.,

$$\mathrm{d}P(t) = P(t^-)\left(\alpha\,\mathrm{d}t + \beta\,\mathrm{d}B(t) + \gamma\int_{\mathbb{R}} z\bar{N}(\mathrm{d}t,\mathrm{d}z)\right), \quad P_0 = p \geq 0, \quad (4.1.1)$$

where $\alpha, \beta \neq 0, \gamma$ are constants and $\gamma z \geq 0$ a.s. ν.

Let Q_t denote the amount of remaining resources at time t. If we extract the resources at the "intensity" $u_t = u_t(\omega) \in [0, m]$ at time t, then the dynamics of Q_t is

$$\mathrm{d}Q_t = -u_t Q_t \mathrm{d}t, \quad Q_0 = q \geq 0. \tag{4.1.2}$$

(m is a constant giving the maximal intensity.)

We assume as before that our *control* $u_t(\omega)$ is adapted to the filtration $\{\mathcal{F}\}_{t\geq 0}$. If the running cost is given by $K_0 + K_1 u_t$ (with $K_0, K_1 \geq 0$ constants) as long as the field is open and if we decide to stop the extraction for good at time $\tau(\omega) \geq 0$ let us assume that the expected total discounted profit is

$$J^{(u,\tau)}(s,p,q) = E^{(p,q)}\left[\int_0^\tau e^{-\rho(s+t)}(u_t(P_tQ_t - K_1) - K_0)dt\right.$$

$$\left. + e^{-\rho(s+\tau)}(\theta P_\tau Q_\tau - a)\right], \tag{4.1.3}$$

where $\rho > 0$ is the discounting exponent and $\theta > 0$, $a \geq 0$ are constants. Thus $e^{-\rho(s+t)}(u_t(P_tQ_t - K_1) - K_0)$ gives the discounted net profit rate when the field is in operation, while $e^{-\rho(s+\tau)}(\theta P_\tau Q_\tau - a)$ gives the discounted net value of the remaining resources at time τ. (We may interpret $a \geq 0$ as a transaction cost.) We assume that the closing time τ is a *stopping time* with respect to the filtration $\{\mathcal{F}_t\}_{t \geq 0}$, i.e., that

$$\{\omega; \tau(\omega) \leq t\} \in \mathcal{F}_t \quad \text{for all } t.$$

Thus both the extraction intensity u_t and the decision whether to close before or at time t must be based on the information $\mathcal{F}_t$ only, not on any future information.

The problem is to find the *value function* $\Phi(s,p,q)$ and the *optimal control* $u_t^* \in [0,m]$ and the *optimal stopping time* τ^* such that

$$\Phi(s,p,q) = \sup_{u_t,\tau} J^{(u,\tau)}(s,p,q) = J^{(u^*,\tau^*)}(s,p,q). \tag{4.1.4}$$

This problem is an example of a combined optimal stopping and stochastic control problem. It is a modification of a problem discussed in [BØ1, DZ].

We will return to this and other examples after presenting a general theory for problems of this type.

4.2 A General Mathematical Formulation

Consider a controlled stochastic system of the same type as in Chap. 3, where the state $Y^{(u)}(t) = Y(t) \in \mathbb{R}^k$ at time t is given by

$$dY(t) = b(Y(t), u(t))dt + \sigma(Y(t), u(t))dB(t) + \int_{\mathbb{R}^k} \gamma(Y(t^-), u(t^-), z)\bar{N}(dt, dz),$$

$$Y(0) = y \in \mathbb{R}^k. \tag{4.2.1}$$

Here $b : \mathbb{R}^k \times U \to \mathbb{R}^k$, $\sigma : \mathbb{R}^k \times U \to \mathbb{R}^{k \times m}$, and $\gamma : \mathbb{R}^k \times U \times \mathbb{R}^k \to \mathbb{R}^{k \times \ell}$ are given continuous functions and $u(t) = u(t, \omega)$ is our *control*, assumed to be $\mathcal{F}_t$-adapted and with values in a given closed, convex set $U \subset \mathbb{R}^p$.

Associated to a control $u = u(t, \omega)$ and an $\mathcal{F}_t$-stopping time $\tau = \tau(\omega)$ belonging to a given set $\mathcal{T}$ of *admissible* stopping times, we assume there is a *performance criterion* of the form

$$J^{(u,\tau)}(y) = E^y \left[\int_0^\tau f(Y(t), u(t)) dt + g(Y(\tau)) \chi_{\{\tau < \infty\}} \right], \qquad (4.2.2)$$

where $f : \mathbb{R}^k \times U \to \mathbb{R}$ (the profit rate) and $g : \mathbb{R}^k \to \mathbb{R}$ (the bequest function) are given functions.

We assume that we are given a set $\mathcal{U} = \mathcal{U}(y)$ of admissible controls u which is contained in the set of controls u such that a unique strong solution $Y(t) = Y^{(u)}(t)$ of (4.2.1) exists and the following, (4.2.3) and (4.2.4), hold:

$$- E^y \left[\int_0^{\tau_S} |f(Y(t), u(t))| dt \right] < \infty, \quad \text{for all } y \in \mathcal{S} \qquad (4.2.3)$$

where $\tau_S = \tau_S(y, u) = \inf\{t > 0; Y^{(u)}(t) \notin \mathcal{S}\}$.

$-$ The family $\{g^-(Y^{(u)}(\tau)); \tau \in \mathcal{T}\}$ is uniformly P^y-integrable for all $y \in \mathcal{S}$, where $g^-(y) = \max(0, -g(y))$. $\qquad (4.2.4)$

We interpret $g(Y(\tau(\omega)))$ as 0 if $\tau(\omega) = \infty$. Here, and in the following, E^y denotes expectation with respect to P when $Y(0) = y$ and $\mathcal{S} \subset \mathbb{R}^k$ is a fixed Borel set such that

$$\mathcal{S} \subset \overline{\mathcal{S}^0}.$$

We can think of $\mathcal{S}$ as the "universe" or "solvency set" of our system, in the sense that we are only interested in the system up to time T, which may be interpreted as the time of bankruptcy.

We now consider the following *combined optimal stopping and control problem.*

Let $\mathcal{T}$ be the set of $\mathcal{F}_t$-stopping times $\tau \leq \tau_S$. Find $\Phi(y)$ and $u^* \in \mathcal{U}$, $\tau^* \in \mathcal{T}$ such that

$$\Phi(y) = \sup\{J^{(u,\tau)}(y); u \in \mathcal{U}, \tau \in \mathcal{T}\} = J^{(u^*,\tau^*)}(y). \qquad (4.2.5)$$

We will prove a verification theorem for this problem. The theorem can be regarded as a combination of the variational inequalities for optimal stopping (Theorem 2.2) and the HJB equation for stochastic control (Theorem 3.1).

We say that the control u is *Markov* or *Markovian* if it has the form

$$u(t) = u_0(Y(t))$$

for some function $u_0 : \bar{\mathcal{S}} \to U$. If this is the case we usually do not distinguish notationally between u and u_0 and write (with abuse of notation)

$$u(t) = u(Y(t)).$$

If $u \in \mathcal{U}$ is Markovian then $Y^{(u)}(t)$ is a Markov process whose genera-tor coincides on $C_0^2(\mathbb{R}^k)$ with the differential operator $L = L^u$ defined for $y \in \mathbb{R}^k$ by

$$L^u \psi(y) = \sum_{i=1}^{k} b_i(y, u(y)) \frac{\partial \psi}{\partial y_i} + \frac{1}{2} \sum_{i,j=1}^{k} (\sigma \sigma^T)_{ij}(y, u(y)) \frac{\partial^2 \psi}{\partial y_i \partial y_j}$$

$$+ \sum_{j=1}^{\ell} \int_{\mathbb{R}} \{ \psi(y + \gamma^{(j)}(y, u(y), z_j)) - \psi(y)$$

$$- \nabla \psi(y) . \gamma^{(j)}(y, u(y), z_j) \} \nu_j(dz_j) \qquad (4.2.6)$$

for all functions $\psi : \mathbb{R}^k \to \mathbb{R}$ which are twice differentiable at y.

Typically the value function Φ will be C^2 outside the boundary ∂D of the continuation region D (see (ii) below) and it will satisfy an HJB equation in D and an HJB inequality outside $\bar{D}$. Across ∂D the function Φ will not be C^2, but it will usually be C^1, and this feature is often referred to as the "high contact" – or "smooth fit" – principle. This is the background for the verification theorem given below (Theorem 4.2). Note, however, that there are cases when Φ is not even C^1 at ∂D. To handle such cases one can use a verification theorem based on the viscosity solution concept. See Chap. 9 and in particular Sect. 9.2.

Theorem 4.2 (HJB-Variational Inequalities for Optimal Stopping and Control).

(a) *Suppose we can find a function $\varphi : \bar{\mathcal{S}} \to \mathbb{R}$ such that*
    (i) *$\varphi \in C^1(\mathcal{S}^0) \cap C(\bar{\mathcal{S}})$*
    (ii) *$\varphi \geq g$ on $\mathcal{S}^0$*
       Define

$$D = \{ y \in \mathcal{S}; \varphi(y) > g(y) \} \quad \text{(the continuation region)}.$$

       Suppose $Y^{(u)}(t)$ spends 0 time on ∂D a.s., i.e.,
    (iii) *$E^y \left[\int_0^{\tau_{\mathcal{S}}} \chi_{\partial D}(Y^{(u)}(t)) dt \right] = 0 \quad$ for all $y \in \mathcal{S}$, $u \in \mathcal{U}$*
       and suppose that
    (iv) *∂D is a Lipschitz surface*
    (v) *$\varphi \in C^2(\mathcal{S}^0 \setminus \partial D)$ and the second-order derivatives of φ are locally bounded near ∂D*
    (vi) *$L^v \varphi(y) + f(y, v) \leq 0$ on $\mathcal{S}^0 \setminus \partial D$ for all $v \in U$*
    (vii) *$Y^{(u)}(\tau_{\mathcal{S}}) \in \partial \mathcal{S}$ a.s. on $\{ \tau_{\mathcal{S}} < \infty \}$ and*

$$\lim_{t \to \tau_{\mathcal{S}}^-} \varphi(Y^{(u)}(t)) = g(Y^{(u)}(\tau_{\mathcal{S}})) \chi_{\{\tau_{\mathcal{S}} < \infty\}} \quad a.s.$$

(viii) $E^y \left[|\varphi(Y^{(u)}(\tau))| + \int_0^{\tau_S} |L^u \varphi(Y^{(u)}(t))| dt \right] < \infty$ *for all* $u \in \mathcal{U}$, $\tau \in \mathcal{T}$.

Then
$$\varphi(y) \geq \Phi(y) \quad \text{for all } y \in \mathcal{S}.$$

(b) *Suppose, in addition to (i)–(viii) above, that*
   (ix) *for each* $y \in D$ *there exists* $\hat{u}(y) \in U$ *such that*
$$L^{\hat{u}(y)} \varphi(y) + f(y, \hat{u}(y)) = 0,$$

   (x) $\tau_D := \inf\{t > 0; Y^{(\hat{u})}(t) \notin D\} < \infty$ *a.s. for all* $y \in \mathcal{S}$,
      and
   (xi) *the family* $\{\varphi(Y^{(\hat{u})}(\tau)); \tau \in \mathcal{T}\}$ *is uniformly integrable with respect to* P^y *for all* $y \in D$.
      Suppose $\hat{u} \in \mathcal{U}$. *Then*
$$\varphi(y) = \Phi(y) \quad \text{for all } y \in \mathcal{S}.$$

   Moreover, $u^* := \hat{u}$ *and* $\tau^* := \tau_D$ *are optimal control and stopping times, respectively.*

Proof. The proof is a synthesis of the proofs of Theorems 2.2 and 3.1. For completeness we give some details:

(a) By Theorem 2.1 we may assume that $\varphi \in C^2(\mathcal{S}^0) \cap C(\bar{\mathcal{S}})$. Choose $u \in \mathcal{U}$ and put $Y(t) = Y^{(u)}(t)$. Let $\tau \leq \tau_S$ be a stopping time. Then by Dynkin's formula (Theorem 1.24) we have, for $m = 1, 2, \ldots$,

$$E^y \left[\varphi(Y(\tau \wedge m)) \right] = \varphi(y) + E^y \left[\int_0^{\tau \wedge m} L^u \varphi(Y(t)) dt \right]. \tag{4.2.7}$$

Hence by (vii) and the Fatou lemma

$$\begin{aligned} \varphi(y) &= \lim_{m \to \infty} E^y \left[\int_0^{\tau \wedge m} -L^u \varphi(Y(t)) dt + \varphi(Y(\tau \wedge m)) \right] \\ &\geq E^y \left[\int_0^{\tau} -L^u \varphi(Y(t)) dt + g(Y(\tau)) \chi_{\{\tau < \infty\}} \right]. \end{aligned} \tag{4.2.8}$$

If we now use (vi) we can conclude that

$$\varphi(y) \geq E^y \left[\int_0^{\tau} f(Y(t), u(t)) dt + g(Y(\tau)) \chi_{\{\tau < \infty\}} \right] = J^{u,\tau}(y). \tag{4.2.9}$$

Since $u \in \mathcal{U}$ and $\tau \leq \tau_S$ were arbitrary we conclude that

$$\varphi(y) \geq \sup_{u,\tau} J^{(u,\tau)}(y) = \Phi(y), \tag{4.2.10}$$

which proves (a).

To prove the opposite inequality, assume that (ix)–(xi) hold. Choose a point $y \in D$. Then apply the argument above to the Markovian control $\hat{u}(t) := \hat{u}(Y(t))$ and the stopping time $\hat{\tau} = \tau_D = \inf\{t > 0; \hat{Y}(t) \notin D\}$, where $\hat{Y}(t) = Y^{(\hat{u})}(t)$. We get

$$E^y\left[\varphi(\hat{Y}(\hat{\tau}))\right] = \varphi(y) + E^y\left[\int_0^{\hat{\tau}} L^{\hat{u}}\varphi(\hat{Y}(t))\mathrm{d}t\right], \tag{4.2.11}$$

which implies that

$$\varphi(y) = E^y\left[\int_0^{\hat{\tau}} f(\hat{Y}(t), \hat{u}(t))\mathrm{d}t + g(\hat{Y}(\hat{\tau}))\right] = J^{(\hat{u},\hat{\tau})}(y).$$

Combined with (a) this shows that $\varphi(y) = \Phi(y)$ and that $(\hat{u}, \hat{\tau})$ is optimal if $y \in D$.

Finally, if $y \notin D$ then $\varphi(y) = g(y) \leq \Phi(y)$ and hence by (a) we have $\varphi(y) = \Phi(y)$ also if $y \notin D$. In this case $\tau^* = 0$ is an optimal stopping time. □

Remark 4.3. If we neglect all the technical conditions of Theorem 4.2 and concentrate on conditions (ii), (vi), and (ix), then we can write the conditions of Theorem 4.2 in the following condensed form:

$$\max\left(\sup_{v \in U}\{L^v\phi(y) + f(y,v)\}, g(y) - \phi(y)\right) = 0, \quad y \in \mathcal{S}^0. \tag{4.2.12}$$

Since this is a combination of the HJB equation of stochastic control and the variational inequality (VI) of optimal stopping, we call (4.2.12) an HJBVI.

One can prove that, under some conditions, the value function ϕ is indeed a solution of (4.2.12) in the weak sense of *viscosity*. See Chap. 9 for a discussion of this concept.

Remark 4.4. Note that the problem (4.2.5) contains the general optimal stopping problem as a special case.

More precisely, if the functions b, σ, and f do not depend on u, then the problem reduces to the optimal stopping problem

$$\Phi(y) = \sup_{\tau \in \mathcal{T}} E^y\left[\int_0^{\tau} f(Y(t))\mathrm{d}t + g(Y(\tau))\chi_{\{\tau < \infty\}}\right]$$

discussed in Chap. 2 and the HJBVI (4.2.12) becomes the VI

$$\max(L\Phi(y) + f(y), g(y) - \Phi(y)) = 0, \quad y \in \mathcal{S}^0. \tag{4.2.13}$$

The problem (4.2.5) is also closely related to the general *stochastic control problem* discussed in Chap. 3. In such a problem the stopping time τ is fixed to $\tau = \tau_S$ and hence the problem is to find $\Phi(y)$ and $u^* \in \mathcal{U}$ such that

$$\Phi(y) = \sup_{u \in \mathcal{U}} J^{(u)}(y) = J^{(u^*)}(y), \tag{4.2.14}$$

where

$$J^{(u)}(y) = E^y\left[\int_0^{\tau_{\mathcal{S}}} f(Y(t), u(t))\mathrm{d}t + g(Y(\tau_{\mathcal{S}}))\chi_{\{\tau_{\mathcal{S}} < \infty\}}\right].$$

4.3 Applications

To illustrate Theorem 4.2 let us apply it to the problem of Example 4.1.
In this case the generator L^u of the time–space state process

$$\mathrm{d}Y(t) = (\mathrm{d}t, \mathrm{d}P_t, \mathrm{d}Q_t), \quad Y(0) = (s, p, q) \in [0, \infty)^3$$

is given by (see Theorem 1.22)

$$L^u\psi(s, p, q) = \frac{\partial\psi}{\partial s} + \alpha p\frac{\partial\psi}{\partial p} - uq\frac{\partial\psi}{\partial q} + \frac{1}{2}\beta^2 p^2\frac{\partial^2\psi}{\partial p^2}$$

$$+ \int_{\mathbb{R}}\left\{\psi(s, p + \gamma pz, q) - \psi(s, p, q) - \frac{\partial\psi}{\partial p}(s, p, q).\gamma zp\right\}\nu(\mathrm{d}z).$$

In view of Theorem 4.2 we are looking for a subset D of $\mathcal{S} = [0, \infty)^3$ and a function $\varphi(s, p, q) : \mathcal{S} \to \mathbb{R}$ such that

$$\varphi(s, p, q) = \mathrm{e}^{-\rho s}(\theta pq - a) \quad \text{for all } (s, p, q) \notin D, \tag{4.3.1}$$

$$\varphi(s, p, q) \geq \mathrm{e}^{-\rho s}(\theta pq - a) \quad \text{for all } (s, p, q) \in \mathcal{S}, \tag{4.3.2}$$

$$L^v\varphi(s, p, q) + \mathrm{e}^{-\rho s}(v(pq - K_1) - K_0) \leq 0 \quad \text{for all } (s, p, q) \in \mathcal{S}^0 \setminus \bar{D}$$

$$\text{and all } v \in [0, m], \tag{4.3.3}$$

$$\sup_{v \in [0, m]}\{L^v\varphi(s, p, q) + \mathrm{e}^{-\rho s}(v(pq - K_1) - K_0)\} = 0 \quad \text{for all } (s, p, q) \in D. \tag{4.3.4}$$

Let us try a function φ of the form

$$\varphi(s, p, q) = \mathrm{e}^{-\rho s}F(w), \quad \text{where } w = pq \tag{4.3.5}$$

and a continuation region D of the form

$$D = \{(s, p, q); pq > w_0\} \quad \text{for some } w_0 > 0.$$

Then (4.3.1)–(4.3.4) get the form

$$F(w) = \theta w - a \quad \text{for all } w \geq w_0, \tag{4.3.6}$$

$$F(w) > \theta w - a \quad \text{for all } w < w_0, \tag{4.3.7}$$

$$-\rho F(w) + (\alpha - v)wF'(w) + \frac{1}{2}\beta^2 w^2 F''(w)$$

$$+ \int_{\mathbb{R}} \{F(w + \gamma zw) - F(w) - F'(w)\gamma zw\}\nu(dz) + v(w - K_1) - K_0 \leq 0$$

$$\text{for all } w < w_0, \ v \in [0, m], \tag{4.3.8}$$

$$\sup_{v \in [0,m]} \left\{ -\rho F(w) + (\alpha - v)wF'(w) + \frac{1}{2}\beta^2 w^2 F''(w) + \int_{\mathbb{R}} \{F(w + \gamma zw) - F(w)\right.$$

$$\left. - F'(w)\gamma zw\}\nu(dz) + v(w - K_1) - K_0 \right\} = 0 \quad \text{for all } w > w_0. \tag{4.3.9}$$

From (4.3.9) and (xi) of Theorem 4.2 we get the following candidate $\hat{u}$ for the optimal control:

$$v = \hat{u}(w) = \operatorname*{Argmax}_{v \in [0,m]} \left\{ v(w(1 - F'(w)) - K_1) \right\}$$

$$= \begin{cases} m & \text{if } F'(w) < 1 - (K_1/w) \\ 0 & \text{if } F'(w) > 1 - (K_1/w) \end{cases}. \tag{4.3.10}$$

Let $F_m(w)$ be the solution of (4.3.9) with $v = m$, i.e., the solution of

$$-\rho F_m(w) + (\alpha - m)wF_m'(w) + \frac{1}{2}\beta^2 w^2 F_m''(w)$$

$$+ \int_{\mathbb{R}} \{F(w + \gamma zw) - F(w) - F'(w)\gamma zw\}\nu(dz) = K_0 + mK_1 - mw. \tag{4.3.11}$$

A solution of (4.3.11) is

$$F_m(w) = C_1 w^{\lambda_1} + C_2 w^{\lambda_2} + \frac{mw}{\rho + m - \alpha} - \frac{K_0 + mK_1}{\rho}, \tag{4.3.12}$$

where C_1, C_2 are constants and $\lambda_1 > 0, \lambda_2 < 0$ are roots of the equation

$$h(\lambda) = 0 \tag{4.3.13}$$

with

$$h(\lambda) = -\rho + (\alpha - m)\lambda + \frac{1}{2}\beta^2 \lambda(\lambda - 1) + \int_{\mathbb{R}} \{(1 + \gamma z)^{\lambda} - 1 - \lambda \gamma z\}\nu(dz). \tag{4.3.14}$$

(Note that $h(0) = -\rho < 0$ and $\lim_{|\lambda| \to \infty} h(\lambda) = \infty$.) The solution will depend on the relation between the parameters involved and we will not give a complete discussion, but only consider some special cases.

Case 1

Let us assume that

$$\alpha \le \rho, \quad K_1 = a = 0, \quad \text{and} \quad 0 < \theta < \tfrac{m}{\rho+m-\alpha}. \qquad (4.3.15)$$

It is easy to see that

$$\lambda_1 > 1 \iff \rho + m > \alpha. \qquad (4.3.16)$$

Let us try (guess) that

$$C_1 = 0 \qquad (4.3.17)$$

and that the continuation region $D = \{(s, p, q); pq > w_0\}$ is such that (see (4.3.10))

$$F'_m(w) < 1 \quad \text{for all } w > w_0. \qquad (4.3.18)$$

The intuitive motivation for trying this is the belief that it is optimal to use the maximal extraction intensity m all the time until closure, at least if θ is small enough.

These guesses lead to the following candidate for the value function $F(w)$:

$$F(w) = \begin{cases} \theta w & \text{if } 0 \le w \le w_0 \\ F_m(w) = C_2 w^{\lambda_2} + \frac{mw}{\rho+m-\alpha} - \frac{K_0}{\rho} & \text{if } w > w_0. \end{cases} \qquad (4.3.19)$$

We now use continuity and differentiability at $w = w_0$ to determine w_0 and C_2:

$$\text{(Continuity)} \quad C_2 w_0^{\lambda_2} + \frac{mw_0}{\rho+m-\alpha} - \frac{K_0}{\rho} = \theta w_0, \qquad (4.3.20)$$

$$\text{(Differentiability)} \quad C_2 \lambda_2 w_0^{\lambda_2-1} + \frac{m}{\rho+m-\alpha} = \theta. \qquad (4.3.21)$$

Easy calculations show that the unique solution of (4.3.20) and (4.3.21) is

$$w_0 = \frac{(-\lambda_2) K_0 (\rho + m - \alpha)}{(1 - \lambda_2)\rho[m - \theta(\rho + m - \alpha)]} \quad (> 0 \quad \text{by (4.3.15)}) \qquad (4.3.22)$$

and

$$C_2 = \frac{[m - \theta(\rho + m - \alpha)] w_0^{1-\lambda_2}}{(-\lambda_2)(\rho + m - \alpha)} \quad (> 0 \quad \text{by (4.3.15)}). \qquad (4.3.23)$$

It remains to verify that with these values of w_0 and C_2 the set $D = \{(s, p, q); pq > w_0\}$ and the function $F(w)$ given by (4.3.19) satisfies (4.3.6)–(4.3.9), as well as all the other conditions of Theorem 4.2.

To verify (4.3.6) we have to check that (4.3.18) holds, i.e., that

$$F'_m(w) = C_2\lambda_2 w^{\lambda_2 - 1} + \frac{m}{\rho + m - \alpha} < 1 \quad \text{for all } w > w_0.$$

Since $\lambda_2 < 0$ and we have assumed $\alpha \leq \rho$ (in (4.3.15)) this is clear. So (4.3.9) holds. If we substitute $F(w) = \theta w$ in (4.3.8) we get

$$-\rho \theta w + (\alpha - m)w\theta + mw - K_0 = w[m - \theta(\rho + m - \alpha)] - K_0.$$

We know that this is 0 for $w = w_0$ by (4.3.20) and (4.3.21). Hence it is less than 0 for $w < w_0$. So (4.3.8) holds. Condition (4.3.6) holds by definition of D and F. Finally, since $F(w_0) = \theta w_0$, $F'(w_0) = \theta$, and

$$F''(w) = F''_m(w) = C_2\lambda_2(\lambda_2 - 1)w^{\lambda_2 - 2} > 0$$

we must have $F(w) > \theta w$ for $w > w_0$. Hence (4.3.7) holds. Similarly one can verify all the other conditions of Theorem 4.2. We have proved.

Theorem 4.5. *Suppose (4.3.15) holds. Then the optimal strategy (u^*, τ^*) for problems (4.1.3) and (4.1.4) is*

$$u^* = m, \quad \tau^* = \inf\{t > 0; \, P_t Q_t \leq w_0\}, \tag{4.3.24}$$

where w_0 is given by (4.3.22). The corresponding value function is $\Phi(s, p, q) = e^{-\rho s} F(p \cdot q)$, where F is given by (4.3.19) with $\lambda_2 < 0$ as in (4.3.11) and $C_2 > 0$ as in (4.3.23).

For other values of the parameters it might be optimal not to produce at all but just wait for the best closing/sellout time. For example, we mention without proof the following cases (see Exercise 4.2):

Case 2

Assume that

$$\theta = 1 \quad \text{and} \quad \rho \leq \alpha. \tag{4.3.25}$$

Then $u^* = 0$ and

$$\Phi = \infty.$$

Case 3

Assume that

$$\theta = 1, \quad \rho > \alpha, \quad \text{and} \quad K_0 < \rho a < K_0 + \rho K_1. \tag{4.3.26}$$

Then

$$u^* = 0 \quad \text{and} \quad \tau^* = \inf\{t > 0; \, P_t Q_t \geq w_1\},$$

for some $w_1 > 0$.

4.4 Exercises

Exercise* 4.1. (a) Solve the following stochastic control problem

$$\Phi(s,x) = \sup_{u(t)\geq 0} J^{(u)}(s,x) = J^{(u^*)}(s,x),$$

where

$$J^{(u)}(s,x) = E^x\left[\int_0^{\tau_S} e^{-\delta(s+t)}\frac{u^\gamma(t)}{\gamma}dt\right].$$

Here

$$\tau_S = \tau_S(\omega) = \inf\{t > 0; X(t) \leq 0\} \quad \text{(the time of bankruptcy)}$$

and

$$dX(t) = (\mu X(t) - u(t))dt + \sigma X(t)dB(t) + \theta X(t^-)\int_{\mathbb{R}} z\bar{N}(dt, dz), \quad X_0 = x > 0$$

with $\gamma \in (0,1), \delta > 0, \mu, \sigma \neq 0, \theta$ constants, $\theta z > -1$ a.s. ν. The interpretation of this is the following. $X(t)$ represents the total wealth at time t, $u(t) = u(t,\omega) \geq 0$ represents the chosen consumption rate (the control). We want to find the consumption rate $u^*(t)$ which maximizes the expected total discounted utility of the consumption up to the time of bankruptcy, τ_S.

[Hint: Try a value function of the form

$$\phi(s,x) = Ke^{-\delta s}x^\gamma$$

for a suitable value of the constant K.]

(b) Consider the following combined stochastic control and optimal stopping problem

$$\Phi(s,x) = \sup_{u,\tau} J^{(u,\tau)}(s,x) = J^{(u^*,\tau^*)}(s,x),$$

where

$$J^{(u,\tau)}(s,x) = E^x\left[\int_0^\tau e^{-\delta(s+t)}\frac{u^\gamma(t)}{\gamma}dt + \lambda e^{-\delta(s+\tau)}X^\gamma(\tau)\right]$$

with $X(t)$ as in (a), $\lambda > 0$ a given constant.

Now the supremum is taken over all $\mathcal{F}_t$-adapted controls $u(t) \geq 0$ and all $\mathcal{F}_t$-stopping times $\tau \leq \tau_S$.

Let K be the constant found in (a). Show that:

1. If $\lambda \geq K$ then it is optimal to stop immediately.
2. If $\lambda < K$ then it is never optimal to stop.

Exercise 4.2. (a) Verify the statements in Cases 2 and 3 at the end of Sect. 4.3.

(b) What happens in the cases

Case 4: $\theta = 1$, $\rho > \alpha$, and $\rho a \leq K_0$?

Case 5: $\theta = 1$, $\rho > \alpha$, and $K_0 + \rho K_1 \leq \rho a$?

Exercise 4.3 (A Stochastic Linear Regulator Problem with Optimal Stopping).

Consider the stochastic linear regulator problem in Exercise 3.5, with the additional option of stopping, i.e., solve the problem

$$\Phi(s, x) = \inf \left\{ J^{(u,\tau)}(s, x); u \in \mathcal{U}, \tau \in \mathcal{T} \right\},$$

where

$$J^{(u,\tau)}(s, x) = E^{s,x}\left[\int_0^\tau e^{-\rho(s+t)}\left(X^2(t) + \theta u^2(t)\right) dt + \lambda e^{-\rho(s+\tau)}X^2(\tau)\chi_{\{\tau<\infty\}}\right],$$

($\rho > 0, \theta > 0, \lambda > 0$ constants), where the state process is

$$Y(t) = \begin{bmatrix} s+t \\ X(t) \end{bmatrix}; \quad t \geq 0, \quad Y(0) = \begin{bmatrix} s \\ x \end{bmatrix} = y \in \mathbb{R}^2$$

with

$$dX(t) = dX^{(u)}(t) = u(t)dt + \sigma\, dB(t) + \int_{\mathbb{R}} z\tilde{N}(dt, dz), \quad X(0) = x.$$

(a) Explain why

$$\Phi(s, x) \leq \lambda e^{-\rho s}x^2 \quad \text{for all } s, x.$$

(b) Prove that

$$\Phi(s, x) \leq e^{-\rho s}\left(x^2 + \sigma^2 + \int_{\mathbb{R}} z^2\nu(dz) \right) \quad \text{for all } s, x.$$

[Hint: What happens if we choose $\tau = \infty$?]

(c) Show that if

$$0 < \lambda \leq \frac{1}{2}\left(-\rho\theta + \sqrt{\rho^2\theta^2 + 4\theta} \right)$$

then

$$\Phi(s, x) = \lambda e^{-\rho s}x^2$$

and it is optimal to stop immediately.

[Hint: Apply the (minimum version of) Theorem 4.2a to the function $\Phi(s, x) = g(s, x) = \lambda e^{-\rho s}x^2$.]

Singular Control for Jump Diffusions

5.1 An Illustrating Example

We illustrate singular control problems by the following example, studied in [FØS2].

Example 5.1 (Optimal Consumption and Portfolio Under Proportional Transaction Costs). Consider again a financial market of the forms (3.3.1) and (3.3.2), where we have two investment possibilities:

(i) A bank account/bond where the value/price $P_1(t)$ at time t grows with interest rate r, i.e.,

$$dP_1(t) = rP_1(t)dt, \quad P_1(0) = 1.$$

(ii) A stock, with price $P_2(t)$ satisfying the equation

$$dP_2(t) = P_2(t^-) \left[\alpha\, dt + \beta\, dB(t) + \int_{\mathbb{R}} z\tilde{N}(dt, dz) \right], \quad P_2(0) = p_2 > 0.$$

Here $r > 0$, α and $\beta > 0$ are given constants, and we assume that $z > -1$ a.s. ν.

Assume that at any time t the investor can choose an adapted and càdlàg consumption rate process $c(t) = c(t, \omega) \geq 0$, taken from the bank account. Moreover, the investor can at any time transfer money from one investment to another with a transaction cost which is proportional to the size of the transaction. Let $X_1(t)$ and $X_2(t)$ denote the amounts of money invested in the bank and in the stocks, respectively. Then the evolution equations for $X_1(t)$ and $X_2(t)$ are

$$dX_1(t) = dX_1^{x_1, c, \xi}(t) = (rX_1(t) - c(t))dt - (1 + \lambda)d\xi_1(t)$$
$$+ (1 - \mu)d\xi_2(t), \quad X_1(0^-) = x_1 \in \mathbb{R},$$

$$dX_2(t) = dX_2^{x,c,\xi}(t) = X_2(t^-)\left[\alpha\,dt + \beta\,dB(t) + \int_{\mathbb{R}} z\tilde{N}(dt, dz)\right]$$
$$+ d\xi_1(t) - d\xi_2(t), \quad X_2(0^-) = x_2 \in \mathbb{R}.$$

Here $\xi = (\xi_1, \xi_2)$, where $\xi_1(t)$ and $\xi_2(t)$ represent cumulative purchase and sale, respectively, of stocks up to time t. The constants $\lambda \geq 0$, $\mu \in [0, 1]$ represent the constants of proportionality of the transaction costs.

Define the *solvency region* S to be the set of states (x_1, x_2) such that the net wealth is nonnegative, i.e., (see Fig. 5.1)

$$S = \{(x_1, x_2) \in \mathbb{R}^2,\ x_1 + (1 + \lambda)x_2 \geq 0, \quad \text{and} \quad x_1 + (1 - \mu)x_2 \geq 0\}. \quad (5.1.1)$$

We define the set $\mathcal{A}$ of *admissible* controls as the set of adapted consumption–investment policies (c, ξ) such that $c = c(t) \geq 0$ is càdlàg and $\xi = (\xi_1, \xi_2)$ where each $\xi_i(t), i = 1, 2$ is right-continuous, nondecreasing, $\xi(0^-) = 0$.

Define

$$\tau_S = \inf\{t > 0;\ (X_1(t), X_2(t)) \notin S\}.$$

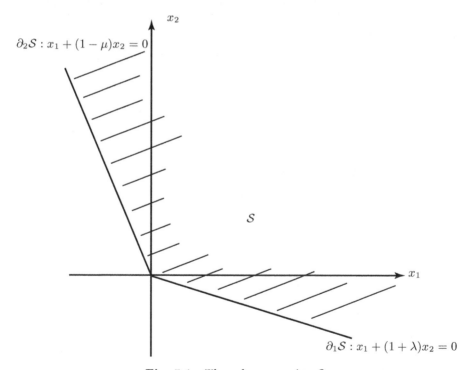

Fig. 5.1. The solvency region S

The performance criterion is defined by

$$
J^{c,\xi}(s, x_1, x_2) = E^{s,x_1,x_2} \left[\int_0^{\tau_S} e^{-\delta(s+t)} \frac{c^\gamma(t)}{\gamma} dt \right],
\tag{5.1.2}
$$

where $\delta > 0$, $\gamma \in (0,1)$ are constants. We seek $(c^*, \xi^*) \in \mathcal{A}$ and $\Phi(s, x_1, x_2)$ such that

$$
\Phi(s, x_1, x_2) = \sup_{(c,\xi) \in \mathcal{A}} J^{c,\xi}(s, x_1, x_2) = J^{c^*, \xi^*}(s, x_1, x_2).
\tag{5.1.3}
$$

This is an example of a *singular* stochastic control problem. It is called singular because the investment control measure $d\xi(t)$ is allowed to be singular with respect to Lebesgue measure dt. In fact, as we shall see, the optimal control measure $d\xi^*(t)$ turns out to be singular.

We now give a general theory of singular control of jump diffusions and return to the above example afterward.

5.2 A General Formulation

Let $\kappa = [\kappa_{ij}] : \mathbb{R}^k \to \mathbb{R}^{k \times p}$ and $\theta = [\theta_i] : \mathbb{R}^k \to \in \mathbb{R}^p$ be given continuous functions. Suppose the state $Y(t) = Y^{u,\xi}(t) \in \mathbb{R}^k$ is described by the equation

$$
dY(t) = b(Y(t), u(t))dt + \sigma(Y(t), u(t))dB(t)
$$
$$
+ \int_{\mathbb{R}^\ell} \gamma(Y(t^-), u(t^-), z)\tilde{N}(dt, dz) + \kappa(Y(t^-))d\xi(t), \quad Y(0^-) = y \in \mathbb{R}^k.
\tag{5.2.1}
$$

Here $\xi(t) \in \mathbb{R}^p$ is an adapted càdlàg finite variation process with increasing components and $\xi(0^-) = 0$. Since $d\xi(t)$ may be singular with respect to Lebesgue measure dt, we call ξ our *singular control* or our *intervention control*. The process $u(t)$ is an adapted càdlàg process with values in a given set U ($u(t)dt$ is our absolutely continuous control). Suppose we are given a performance functional $J^{u,\xi}(y)$ of the form

$$
J^{u,\xi}(y) = E^y \left[\int_0^{\tau_S} f(Y(t), u(t))dt + g(Y(\tau_S)) \cdot \mathcal{X}_{\{\tau_S < \infty\}} \right.
$$
$$
\left. + \int_0^{\tau_S} \theta^T(Y(t^-))d\xi(t) \right],
$$

where $f : \mathbb{R}^k \times U \to \mathbb{R}$, $g : \mathbb{R}^k \to \mathbb{R}$ are continuous functions and

$\tau_S = \inf\{t > 0; Y^{u,\xi}(t) \notin S\} \leq \infty$ is the time of bankruptcy,

where $S \subset \mathbb{R}^k$ is a given *solvency set*, assumed to satisfy $S \subset \overline{S^0}$.

Let $\mathcal{A}$ be a given family of *admissible controls* (u, ξ), contained in the set of (u, ξ) such that a unique strong solution $Y(t)$ of (5.2.1) exists and

$$E^y \left[\int_0^{\tau_S} |f(Y(t), u(t))| dt + |g(Y(\tau_S))| \cdot \mathcal{X}_{\{\tau_S < \infty\}} \right.$$

$$\left. + \int_0^{\tau_S} \sum_{j=1}^p |\theta_j(Y(t^-))| d\xi_j(t) \right] < \infty.$$

The problem is to find the value function $\Phi(y)$ and an optimal control $(u^*, \xi^*) \in \mathcal{A}$ such that

$$\Phi(y) = \sup_{(u,\xi) \in \mathcal{A}} J^{u,\xi}(y) = J^{u^*,\xi^*}(y).$$

Note that if we apply a Markov control $u(t) = u(Y(t)) \in U$ and $d\xi = 0$, then $Y(t) = Y^{u,0}(t)$ has the generator A^u given by

$$A^u \phi(y) = \sum_{i=1}^k b_i(y, u(y)) \frac{\partial \phi}{\partial y_i} + \frac{1}{2} \sum_{i,j=1}^k (\sigma \sigma^T)_{ij}(y, u(y)) \frac{\partial^2 \phi}{\partial y_i \partial y_j}$$

$$+ \sum_{j=1}^\ell \int_{\mathbb{R}} \left\{ \phi(y + \gamma^{(j)}(y, u(y), z_j)) \right.$$

$$\left. - \phi(y) - \nabla \phi(y)^T \gamma^{(j)}(y, u(y), z_j) \right\} \nu_j(dz_j).$$

In the following we let A^v denote this same integrodifferential operator obtained by choosing $u(y) = v$ (constant) $\in U$.

Note that we distinguish between the jumps of $Y^{u,\xi}(t)$ caused by the jump of $N(t, z)$, denoted by $\Delta_N Y(t)$, and the jump caused by the singular control ξ, denoted by $\Delta_\xi Y(t)$. Thus

$$\Delta_N Y(t) = \int_{\mathbb{R}^k} \gamma(Y(t^-), u(t^-), z) \tilde{N}(\{t\}, dz), \quad \text{while} \quad \Delta_\xi Y(t) = \kappa(Y(t^-)) \Delta \xi(t),$$

$$(5.2.2)$$

where

$$\tilde{N}(\{t\}, dz) = \tilde{N}(t, dz) - \tilde{N}(t^-, dz).$$

We let $t_1, t_2, \ldots$ denote the jumping times of $\xi(t)$ and we let

$$\Delta_\xi \phi(Y(t_n)) = \phi(Y(t_n)) - \phi(Y(t_n^-) + \Delta_N Y(t_n)) \quad (5.2.3)$$

be the increase in ϕ due to the jump $\Delta \xi(t) = \xi(t) - \xi(t^-)$ of $\xi(t)$ at $t = t_n$.
We give now a verification theorem for singular control problems.

Theorem 5.2 (Integrovariational Inequalities for Singular Control).

(a) *Suppose there exists a function $\phi \in C^2(\mathcal{S}^0) \cap C(\mathbb{R}^k)$ such that*
   (i) $A^v\phi(y) + f(y, v) \leq 0$ *for all (constant) $v \in U$ and $y \in \mathcal{S}$.*

(ii) $\displaystyle\sum_{i=1}^{k} \kappa_{ij}(y)\frac{\partial\phi}{\partial y_i}(y) + \theta_j(y) \leq 0$ *for all $y \in \mathcal{S}$, $j = 1, \ldots, p$.*

(iii) $\displaystyle E^y\left[\int_0^{\tau_S}\left\{|\sigma^T(Y(t), u(t))\nabla\phi(Y(t))|^2 + \right.\right.$

$\left.\left.\displaystyle\sum_{k=1}^{\ell}\int_{\mathbb{R}}|\phi(Y(t) + \gamma^{(k)}(Y(t), u(t), z)) - \phi(Y(t))|^2\nu_k(dz)\right\}dt\right] < \infty$

   for all $(u, \xi) \in \mathcal{A}$.
   (iv) $\displaystyle\lim_{t\to\tau_S^-}\phi(Y(t)) = g(Y(\tau_S))\mathcal{X}_{\{\tau_S<\infty\}}$ *a.s., for all $(u, \xi) \in \mathcal{A}$.*
   (v) $\{\phi^-(Y(\tau))\}_{\tau\leq\tau_S}$ *is uniformly integrable for all $(u, \xi) \in \mathcal{A}$, $y \in \mathcal{S}$.*
   Then

$$\phi(y) \geq \Phi(y) \quad \text{for all } y \in \mathcal{S}. \tag{5.2.4}$$

(b) *Define the* nonintervention region D *by*

$$D = \left\{y \in \mathcal{S};\ \max_{1\leq j\leq p}\left\{\sum_{i=1}^{k}\kappa_{ij}(y)\frac{\partial\phi}{\partial y_i}(y) + \theta_j(y)\right\} < 0\right\}. \tag{5.2.5}$$

Suppose, in addition to (i)–(v) above, that for all $y \in \overline{D}$ there exists $v = \hat{u}(y)$ such that
   (vi) $A^{\hat{u}(y)}\phi(y) + f(y, \hat{u}(y)) = 0$.
   Moreover, suppose there exists $\hat{\xi}$ such that $(\hat{u}, \hat{\xi}) \in \mathcal{A}$ and
(vii) $Y^{\hat{u},\hat{\xi}}(t) \in \overline{D}$ *for all t.*

(viii) $\displaystyle\sum_{j=1}^{p}\left\{\sum_{i=1}^{k}\kappa_{ij}(Y(t^-))\frac{\partial\phi}{\partial y_i}(Y(t^-)) + \theta_j(Y(t^-))\right\}d\hat{\xi}_j^{(c)} = 0$

   for all t, $1 \leq j \leq p$,
   where $\xi^{(c)}(t)$ is the continuous part of $\xi(t)$, i.e., the process obtained by removing the jumps of $\xi(t)$.

(ix) $\displaystyle\Delta_{\hat{\xi}}\phi(Y(t_n)) + \sum_{j=1}^{p}\theta_j(Y(t_n^-))\Delta\hat{\xi}_j(t_n) = 0$ *for all jumping times t_n of*

$\hat{\xi}(t)$
and
   (x) $\displaystyle\lim_{R\to\infty}E^y\left[\phi\left(Y^{\hat{u},\hat{\xi}}(T_R)\right)\right] = E^y\left[g\left(Y^{\hat{u},\hat{\xi}}(\tau_S)\right)\cdot\mathcal{X}_{\{\tau_S<\infty\}}\right]$,
   where
$$T_R = \min(\tau_S, R) \quad \text{for } R < \infty.$$

   Then
$$\phi(y) = \Phi(y)$$
   and
$$(\hat{u}, \hat{\xi}) \quad \text{is an optimal control.}$$

Proof. (a) Choose $(u, \xi) \in \mathcal{A}$. Then by the Itô formula for the semimartingale $Y(t) = Y^{u,\xi}(t)$ we have (see [P, Theorem II.33] and Theorem 1.24)

$$
\begin{aligned}
E^y[\phi(Y(T_R))] = \phi(y) + E^y \Bigg[&\int_0^{T_R} A^u \phi(Y(t)) dt \\
&+ \int_0^{T_R} \sum_{i=1}^k \frac{\partial \phi}{\partial y_i}(Y(t^-)) \sum_{j=1}^p \kappa_{ij}(Y(t^-)) d\xi_j^{(c)}(t) \\
&+ \sum_{0 < t_n \le T_R} \Delta_\xi \phi(Y(t_n)) \Bigg].
\end{aligned}
\tag{5.2.6}
$$

By the mean value theorem we have

$$
\begin{aligned}
\Delta_\xi \phi(Y(t_n)) &= \nabla \phi(\hat{Y}^{(n)})^T \Delta_\xi Y(t_n) \\
&= \sum_{i=1}^k \sum_{j=1}^p \frac{\partial \phi}{\partial y_i}(\hat{Y}^{(n)}) \kappa_{ij}(Y(t_n^-)) \Delta \xi_j(t_n),
\end{aligned}
\tag{5.2.7}
$$

where $\hat{Y}^{(n)}$ is some point on the straight line between $Y(t_n)$ and $Y(t_n^-) + \Delta_N Y(t_n)$ (see (5.2.3)).

Hence, by (i) and (ii) and (5.2.6), (5.2.7),

$$
\begin{aligned}
\phi(y) \ge E^y \Bigg[&\int_0^{T_R} f(Y(t), u(t)) dt + \phi(Y(T_R)) \\
&- \sum_{j=1}^p \sum_{i=1}^k \Bigg\{ \int_0^{T_R} \frac{\partial \phi}{\partial y_i}(Y(t^-)) \kappa_{ij} d\xi_j^{(c)}(t) \\
&\quad + \sum_{0 < t_n \le T_R} \frac{\partial \phi}{\partial y_i}(\hat{Y}^{(n)}) \kappa_{ij}(Y(t_n^-)) \Delta \xi_j(t_n) \Bigg\} \Bigg] \\
\ge E^y \Bigg[&\int_0^{T_R} f(Y(t), u(t)) dt + \phi(Y(T_R)) + \sum_{j=1}^p \int_0^{T_R} \theta_j(Y(t^-)) d\xi_j(t) \Bigg].
\end{aligned}
$$

Letting $R \to \infty$ and applying (iv) and (v) we obtain from (5.2.7) that

$$
\phi(y) \ge J^{u,\xi}(y).
$$

Since $(u, \xi) \in \mathcal{A}$ was arbitrary, this proves (5.2.4).

(b) Now apply the above argument to $(\hat{u}, \hat{\xi}) \in \mathcal{A}$, as given by (vi)–(xi). Then we get *equality* everywhere in (a) and we end up with

$$
\phi(y) = J^{\hat{u},\hat{\xi}}(y)
$$

and the proof is complete. □

The optimal singular control $\hat{\xi}$ described in Theorem 5.2b can be characterized as a *local time* at ∂D of the jump diffusion $Y(t)$ reflected back into $\overline{D}$ with respect to the directions given by the rows of the matrix κ.

Definition 5.3. *Let $D \subset \mathbb{R}^k$ be an open set and let $b : \mathbb{R}^k \to \mathbb{R}^k$, $\sigma : \mathbb{R}^k \to \mathbb{R}^{k \times m}$, $\gamma : \mathbb{R}^k \times \mathbb{R}^\ell \to \mathbb{R}^{k \times \ell}$, and $\kappa : \mathbb{R}^k \to \mathbb{R}^{k \times k}$ be given continuous functions. The* Skorohod problem *for reflected jump diffusions into $\overline{D}$ (with respect to the direction matrix κ) is to find a pair $(Y(t), \xi(t))$ of càdlàg and adapted processes such that the following conditions are satisfied:*

$$dY(t) = b(Y(t))dt + \sigma(Y(t))dB(t) + \int_{\mathbb{R}^\ell} \gamma(Y(t^-), z)\tilde{N}(dt, dz)$$

$$+ \kappa(Y(t^-))d\xi(t), \quad Y(0^-) = y \in \mathbb{R}^k, \tag{5.2.8}$$

$$Y(t) \in \overline{D} \quad \text{for all } t \geq 0, \tag{5.2.9}$$

$$\xi(t) \in \mathbb{R}^k \text{ has finite variation and } d\xi(t) = 0 \text{ if } Y(t) \in D. \tag{5.2.10}$$

The process $\xi(t)$ is called a local time *at the boundary ∂D for the reflected process $Y(t)$.*

The following result is an extension to jump diffusions of a well-known result for (continuous) Itô diffusions. For related results see [CEM].

Theorem 5.4. *Suppose ∂D is C^1, i.e., ∂D is locally the graph of a C^1-function. Suppose that $b(\cdot), \sigma(\cdot), \gamma(\cdot, z)$, and $\kappa(\cdot)$ are Lipschitz continuous with at most linear growth (uniformly in $z \in \mathbb{R}^\ell$). Then there exists a solution $(Y(t), \xi(t))$ to the Skorohod problem (5.2.8)–(5.2.10).*

Sketch of Proof. We adapt the approach of Anderson and Orey [AO] to the jump diffusion case. (See also the presentation in [Fre].) Let us first assume that D is the half-space

$$D = \{(y_1, \ldots, y_k) \in \mathbb{R}^k, \quad y_1 > 0\} \tag{5.2.11}$$

and that the rows $\kappa_1, \ldots, \kappa_k$ of the matrix κ are

$$\kappa_1 = (1, 0, \ldots, 0), \kappa_i \in \mathbb{R}^k \text{ arbitrary constants}, \, i = 2, \ldots, k. \tag{5.2.12}$$

Let $\mathcal{C}$ denote the set of all càdlàg functions $f : [0, \infty) \to \mathbb{R}^k$.
Define the map $\Gamma : \mathcal{C} \to \mathcal{C}$ by

$$\Gamma f(t) = \left(f_1(t) + \sup_{0 \leq s \leq t} f_1^-(s), f_2(t), \ldots, f_k(t) \right), \quad f = (f_1, \ldots, f_k) \in \mathcal{C} \tag{5.2.13}$$

and define the map $\Lambda : \mathcal{C} \to \mathcal{C}$ by

$$\Lambda(f) = (\Lambda(f)_1, 0, \ldots, 0),$$

where
$$\Lambda(f)_1(t) = \Gamma(f)_1(t) - f_1(t) = \sup_{0 \le s \le t} f_1^-(s). \tag{5.2.14}$$

Note that
$$\sup_{0 \le s \le t} |\Gamma(f)(s) - \Gamma(g)(s)| \le 2 \sup_{0 \le s \le t} |f(s) - g(s)|$$

and
$$|\Gamma(f)(s) - \Gamma(f)(t)| \le |f(s) - f(t)|, \quad f, g \in \mathcal{C}.$$

Hence we can define $Z(t)$ to be the càdlàg adapted solution of the stochastic differential equation

$$\mathrm{d}Z(t) = b(\Gamma(Z)(t))\mathrm{d}t + \sigma(\Gamma(Z)(t))\mathrm{d}B(t) + \int_{\mathbb{R}^\ell} \gamma(\Gamma(Z)(t^-), z)\tilde{N}(\mathrm{d}t, \mathrm{d}z)$$

$$Z(0) = y. \tag{5.2.15}$$

Now define
$$Y(t) := \Gamma(Z)(t) \quad \text{and} \quad \xi(t) := \Lambda(Z)(t). \tag{5.2.16}$$

Then $(Y(t), \xi(t))$ satisfies (5.2.8). This is because

$$\begin{aligned}
\mathrm{d}Y_1(t) - \kappa_1(Y(t))\mathrm{d}\xi(t) &= \mathrm{d}Y_1(t) - \mathrm{d}\xi_1(t) \\
&= \mathrm{d}(\Gamma(Z)_1(t) - \Lambda_1(Z)(t)) \\
&= \mathrm{d}(\Gamma(Z)_1(t) - \Gamma(Z)_1(t) + Z_1(t)) = \mathrm{d}Z_1(t) \\
&= b_1(Y(t))\mathrm{d}t + \sigma_1(Y(t))\mathrm{d}B(t) \\
&\quad + \int_{\mathbb{R}} \gamma_1(Y(t^-), z)\tilde{N}(\mathrm{d}t, \mathrm{d}z)
\end{aligned}$$

and, trivially,
$$\mathrm{d}Y_i(t) - \kappa_i(Y(t))\mathrm{d}\xi(t) = \mathrm{d}Y_i(t) = \mathrm{d}Z_i(t)$$

for $i = 2, \ldots, k$. It follows from the definition of Γ that

$$Y(t) = (\Gamma(Z)_1(t), Z_2(t), \ldots, Z_k(t)) \in \overline{D}$$

since
$$\Gamma(Z)_1(t) = Z_1(t) + \sup_{0 \le s \le t} Z_1^-(s) \ge 0 \quad \text{for all } t.$$

Moreover, by (5.2.14) we see that $\xi_1(t) = \Lambda(Z)_1(t)$ is nondecreasing and hence $\xi(t)$ has finite variation. Moreover, if $Y(t) = \Gamma(Z)(t) \in D$ then

$$\Gamma(Z)_1(t) = Z_1(t) + \sup_{0 \le s \le t} Z_1^-(s) > 0$$

and hence
$$Z_1(t) > \min\left(\inf_{0 \le s \le t} Z(s), 0\right).$$

Therefore either $Z_1(t) > 0$ or $Z_1(t) > \inf_{0 \leq s \leq t} Z(s)$. In either case the process $\xi_1(t) = \sup_{0 \leq s \leq t} Z_1^-(s)$ is constant in an interval $[t, t + \epsilon)$ for some $\epsilon > 0$, by right continuity of $Z(t)$. Therefore $d\xi(t) = 0$ if $Y(t) \in D$, as claimed. Hence $(Y(t), \xi(t))$ defined by (5.2.16) solves the Skorohod equation (5.2.8)–(5.2.10) in the case when D and κ are given by (5.2.11) and (5.2.12).

The proof in the general case follows by mapping neighborhoods of $x \in \partial D$ into neighborhoods of the boundary point 0 of a half-space and applying the argument above. □

5.3 Application to Portfolio Optimization with Transaction Costs

We now apply this theorem to Example 5.1. In this case our state process is

$$
dY(t) = \begin{bmatrix} dt \\ dX_1(t) \\ dX_2(t) \end{bmatrix} = \begin{bmatrix} 1 \\ rX_1(t) - c(t) \\ \alpha X_2(t) \end{bmatrix} dt + \begin{bmatrix} 0 \\ 0 \\ \beta X_2(t) \end{bmatrix} dB(t)
$$

$$
+ \begin{bmatrix} 0 \\ 0 \\ X_2(t^-) \int_{\mathbb{R}} z \tilde{N}(dt, dz) \end{bmatrix} + \begin{bmatrix} 0 & 0 \\ -(1 + \lambda) & (1 - \mu) \\ 1 & -1 \end{bmatrix} \begin{bmatrix} d\xi_1(t) \\ d\xi_2(t) \end{bmatrix}.
$$

The generator of $Y(t)$ when there are no interventions is

$$
A^c \phi(y) = \frac{\partial \phi}{\partial s} + (rx_1 - c) \frac{\partial \phi}{\partial x_1} + \alpha x_2 \frac{\partial \phi}{\partial x_2} + \frac{1}{2} \beta^2 x_2^2 \frac{\partial^2 \phi}{\partial x_2^2}
$$

$$
+ \int_{\mathbb{R}} \left\{ \phi(s, x_1, x_2 + x_2 z) - \phi(s, x_1, x_2) - x_2 z \frac{\partial \phi}{\partial x_2}(s, x_1, x_2) \right\} \nu(dz).
$$

Or, if

$$
\phi(s, x_1, x_2) = e^{-\delta s} \psi(x_1, x_2)
$$

we have

$$
A^c \phi(s, x_1, x_2) = e^{-\delta s} A_0^c \psi(x_1, x_2),
$$

where

$$
A_0^c \psi(x_1, x_2) = -\delta \psi + (rx_1 - c) \frac{\partial \psi}{\partial x_1} + \alpha x_2 \frac{\partial \psi}{\partial x_2} + \frac{1}{2} \beta^2 x_2^2 \frac{\partial^2 \psi}{\partial x_2^2}
$$

$$
+ \int_{\mathbb{R}} \left\{ \psi(x_1, x_2 + x_2 z) - \psi(x_1, x_2) - x_2 z \frac{\partial \psi}{\partial x_2}(x_1, x_2) \right\} \nu(dz).
$$

Here
$$\theta = g = 0, \quad u(t) = c(t), \quad f(y,u) = f(s,x_1,x_2,c) = e^{-\delta s}\frac{c^\gamma}{\gamma}.$$

Condition (ii) of Theorem 5.2 gets the form
$$-(1+\lambda)\frac{\partial\psi}{\partial x_1} + \frac{\partial\psi}{\partial x_2} \leq 0$$

and
$$(1-\mu)\frac{\partial\psi}{\partial x_1} - \frac{\partial\psi}{\partial x_2} \leq 0.$$

The nonintervention region D in (5.2.5) therefore becomes
$$D = \left\{ (s,x_1,x_2) \in \mathcal{S}; -(1+\lambda)\frac{\partial\psi}{\partial x_1} + \frac{\partial\psi}{\partial x_2} < 0 \text{ and } (1-\mu)\frac{\partial\psi}{\partial x_1} - \frac{\partial\psi}{\partial x_2} < 0 \right\}.$$

Conditions (i) and (vi) become, respectively,
$$A_0^c\psi(x_1,x_2) + \frac{c^\gamma}{\gamma} \leq 0 \quad \text{for all } c \geq 0,$$

$$A_0^{\hat{c}}\psi(x_1,x_2) + \frac{\hat{c}^\gamma}{\gamma} = 0 \quad \text{on } D.$$

Using these integrovariational inequalities together with the remaining conditions of Theorem 5.2 it is possible to prove the following about the optimal consumption rate $c^*(t)$ and the optimal portfolio $\xi^*(t)$.

Theorem 5.5 ([FØS2]). *Suppose*
$$\delta > \gamma\alpha - \frac{1}{2}\beta^2\gamma(1-\gamma) - \gamma\|\nu\| + \int_{-1}^{\infty}\{(1+z)\gamma - 1\}\nu(dz),$$

where
$$\|\nu\| = \nu((-1,\infty)) < \infty.$$

Then $\Phi \in C^2(\mathcal{S}^0) \cap C(\overline{\mathcal{S}})$ and
$$c^*(x_1,x_2) = \left(\frac{\partial\Phi}{\partial x_1}\right)^{1/(\gamma-1)}$$

is optimal and there exist $\hat{\theta}_1,\hat{\theta}_2 \in [0,\pi/2]$ with $\hat{\theta}_1 < \hat{\theta}_2$ such that
$$D = \{re^{i\theta}; \hat{\theta}_1 < \theta < \hat{\theta}_2\} \quad (i = \sqrt{-1})$$

is the nonintervention region and the optimal intervention strategy (portfolio) $\xi^*(t)$ is the local time of the process $(X_1(t), X_2(t))$ reflected back into D in the direction parallel to $\partial_1\mathcal{S}$ at $\theta = \hat{\theta}_1$ and in the direction parallel to $\partial_2\mathcal{S}$ at $\theta = \hat{\theta}_2$. See Fig. 5.2.

For proofs and more details we refer to [FØS2].

Remark 5.6. For other applications of singular control theory of jump diffusions see [Ma].

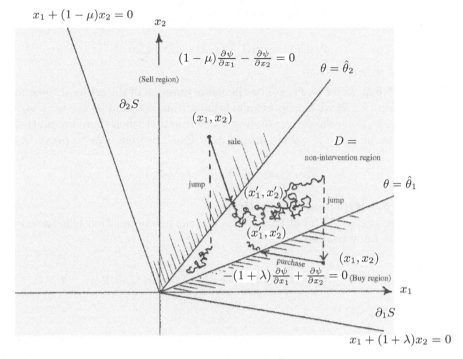

$$x_1 + (1-\mu)x_2 = 0$$

$$(1-\mu)\frac{\partial\psi}{\partial x_1} - \frac{\partial\psi}{\partial x_2} = 0$$

$\theta = \hat{\theta}_2$

(Sell region)

$\partial_2 S$

(x_1, x_2)

sale

$D =$ non-intervention region

jump

(x_1', x_2')

jump

$\theta = \hat{\theta}_1$

(x_1', x_2')

purchase

(x_1, x_2)

$-(1+\lambda)\frac{\partial\psi}{\partial x_1} + \frac{\partial\psi}{\partial x_2} = 0$ (Buy region)

$\partial_1 S$

$x_1 + (1+\lambda)x_2 = 0$

ℓ = purchase amount, m = sale amount (at transaction)
$x_1' = x_1 - (1+\lambda)\ell + (1-\mu)m$ (new value of x_1 after transaction)
$x_2' = x_2 + \ell - m$ (new value of x_2 after transaction)

Fig. 5.2. The optimal portfolio $\xi^*(t)$

5.4 Exercises

Exercise* 5.1 (Optimal Dividend Policy Under Proportional Transaction Costs).
Suppose the cash flow $X(t) = X^{(\xi)}(t)$ of a firm at time t is given by (with $\alpha, \sigma, \beta, \lambda > 0$ constants)

$$dX(t) = \alpha\, dt + \sigma\, dB(t) + \beta\int_{\mathbb{R}} z\tilde{N}(dt, dz) - (1+\lambda)d\xi(t), \quad X(0^-) = x > 0,$$

where $\xi(t)$ is an increasing, adapted càdlàg process representing the total dividend taken out up to time t (our control process). We assume that $\beta z \leq 0$ for a.a. z (ν). Let

$$\tau_S = \inf\{t > 0; X^{(\xi)}(t) \leq 0\}$$

be the time of bankruptcy of the firm and let

$$J^\xi(s, x) = E^{s,x}\left[\int_0^{\tau_S} e^{-\rho(s+t)}d\xi(t)\right], \quad \rho > 0\, \text{constant}$$

be the expected total discounted amount taken out up to bankruptcy time.

Find $\Phi(s,x)$ and a dividend policy ξ^* such that

$$\Phi(s,x) = \sup_\xi J^{(\xi)}(s,x) = J^{\xi^*}(s,x).$$

Exercise* 5.2. Let $\Phi(s,x_1,x_2)$ be the value function of the optimal consumption problem (5.1.2) with proportional transaction costs and let $\Phi_0(s,x_1,x_2) = Ke^{-\delta s}(x_1+x_2)^\gamma$ be the corresponding value function when there are no transaction costs, i.e., $\mu = \lambda = 0$ (Example 3.2). Use Theorem 5.2a to prove that

$$\Phi(s,x_1,x_2) \le Ke^{-\delta s}(x_1 + x_2)^\gamma.$$

Exercise 5.3 (Optimal Harvesting).

Suppose the size $X(t)$ at time t of a certain fish population is modeled by a geometric Lévy process, i.e.,

$$dX(t) = dX^{(\xi)}(t) = X(t^-)\left[\mu\,dt + \int_{\mathbb{R}} z\tilde{N}(dt,dz)\right] - d\xi(t), \quad t > 0,$$

$$X(0^-) = x > 0,$$

where $\mu > 0$ is a constant, $z > -1$ a.s. $\nu(dz)$ and $\xi(t)$ is an increasing adapted process giving the amount harvested from the population from time 0 up to time t. We assume that $\xi(t)$ is right-continuous. Consider the *optimal harvesting problem*

$$\Phi(s,x) = \sup_\xi J^{(\xi)}(s,x),$$

where

$$J^{(\xi)}(s,x) = E^{s,x}\left[\int_0^{\tau_s} \theta e^{-\rho(s+t)}d\xi(t)\right],$$

with $\theta > 0$, $\rho > 0$ constants and

$$\tau_S = \inf\{t > 0; X^{(\xi)}(t) \le 0\} \quad \text{(extinction time)}.$$

If we interpret θ as the price per unit harvested, then $J^{(\xi)}(s,x)$ represents the expected total discounted value of the harvested amount up to extinction time.

(a) Write down the integrovariational inequalities (i), (ii), (vi), and (ix) of Theorem 5.2 in this case, with the state process

$$Y(t) = \begin{bmatrix} s+t \\ X(t) \end{bmatrix}, \quad t \ge 0, \quad Y(0) = y = \begin{bmatrix} s \\ x \end{bmatrix} \in \mathbb{R}^+ \times \mathbb{R}^+.$$

(b) Suppose $\mu \leq \rho$.

Show that in this case it is optimal to harvest all the population immediately, i.e., it is optimal to choose the harvesting strategy $\hat{\xi}$ defined by

$$\hat{\xi}(t) = x \text{ for all } t \geq 0$$

(sometimes called the "take the money and run" strategy).
This gives the value function

$$\Phi(s, x) = \theta e^{-\rho s} x.$$

(c) Suppose $\mu > \rho$.

Show that in this case $\Phi(s, x) = \infty$.

6

Impulse Control of Jump Diffusions

6.1 A General Formulation and a Verification Theorem

Suppose that – if there are no interventions – the state $Y(t) \in \mathbb{R}^k$ of the system we consider is a jump diffusion of the form

$$dY(t) = b(Y(t))dt + \sigma(Y(t))dB(t) + \int_{\mathbb{R}^\ell} \gamma(Y(t^-), z)\tilde{N}(dt, dz), \qquad (6.1.1)$$

$$Y(0) = y \in \mathbb{R}^k,$$

where $b : \mathbb{R}^k \to \mathbb{R}^k$, $\sigma : \mathbb{R}^k \to \mathbb{R}^{k \times m}$, and $\gamma : \mathbb{R}^k \times \mathbb{R}^\ell \to \mathbb{R}^{k \times \ell}$ are given functions satisfying the conditions for the existence and uniqueness of a solution $Y(t)$ (see Theorem 1.19).

The generator A of $Y(t)$ is

$$A\phi(y) = \sum_{i=1}^k b_i(y)\frac{\partial \phi}{\partial y_i} + \frac{1}{2}\sum_{i,j=1}^k (\sigma\sigma^T)_{ij}(y)\frac{\partial^2 \phi}{\partial y_i \partial y_j}$$

$$+ \sum_{j=1}^\ell \int_{\mathbb{R}} \{\phi(y + \gamma^{(j)}(y, z_j)) - \phi(y) - \nabla\phi(y) \cdot \gamma^{(j)}(y, z_j)\}\nu_j(dz_j).$$

Now suppose that at any time t and any state y we are free to intervene and give the system an impulse $\zeta \in \mathcal{Z} \subset \mathbb{R}^p$, where $\mathcal{Z}$ is a given set (the set of admissible impulse values). Suppose the result of giving the impulse ζ when the state is y is that the state jumps immediately from $y = Y(t^-)$ to $Y(t) = \Gamma(y, \zeta) \in \mathbb{R}^k$, where $\Gamma : \mathbb{R}^k \times \mathcal{Z} \to \mathbb{R}^k$ is a given function.

An *impulse control* for this system is a double (possibly finite) sequence

$$v = (\tau_1, \tau_2, \ldots, \tau_j, \ldots; \zeta_1, \zeta_2, \ldots, \zeta_j, \ldots)_{j \le M}, \qquad M \le \infty,$$

where $0 \le \tau_1 \le \tau_2 \le \cdots$ are $\mathcal{F}_t$-stopping times (the *intervention times*) and $\zeta_1, \zeta_2, \ldots$ are the corresponding *impulses* at these times. We assume that ζ_j is $\mathcal{F}_{\tau_j}$-measurable for all j.

If $v = (\tau_1, \tau_2, \ldots; \zeta_1, \zeta_2, \ldots)$ is an impulse control, the corresponding state process $Y^{(v)}(t)$ is defined by

$$Y^{(v)}(0^-) = y \quad \text{and} \quad Y^{(v)}(t) = Y(t), \quad 0 < t \leq \tau_1, \tag{6.1.2}$$

$$Y^{(v)}(\tau_j) = \Gamma(\check{Y}^{(v)}(\tau_j^-), \zeta_j), \quad j = 1, 2, \ldots \tag{6.1.3}$$

(If $\tau_1 = 0$ we put $Y^{(v)}(\tau_1) = \Gamma(Y^{(v)}(0^-), \zeta_1) = \Gamma(y, \zeta_1)$.)

$$dY^{(v)}(t) = b(Y^{(v)}(t))dt + \sigma(Y^{(v)}(t))dB(t)$$
$$+ \int_{\mathbb{R}^\ell} \gamma(Y^{(v)}(t), z)\tilde{N}(dt, dz) \quad \text{for } \tau_j < t < \tau_{j+1} \wedge \tau^*. \tag{6.1.4}$$

If $\tau_{j+1} = \tau_j$, then $Y^{(v)}(t)$ jumps from $\check{Y}^{(v)}(\tau_j^-)$ to $\Gamma(\Gamma(\check{Y}^{(v)}(\tau_j^-), \zeta_j), \zeta_{j+1})$,

where

$$\check{Y}^{(v)}(\tau_j^-) = Y^{(v)}(\tau_j^-) + \Delta_N Y(\tau_j), \tag{6.1.5}$$

$\Delta_N Y^{(v)}(t)$ is as in (5.2.2) the jump of $Y^{(v)}$ stemming from the jump of the random measure $N(t, \cdot)$ only and

$$\tau^* = \tau^*(\omega) = \lim_{R \to \infty} (\inf\{t > 0; |Y^{(v)}(t)| \geq R\}) \leq \infty \tag{6.1.6}$$

is the explosion time of $Y^{(v)}(t)$. Note that here we must distinguish between the (possible) jump of $Y^{(v)}(\tau_j)$ stemming from the random measure N, denoted by $\Delta_N Y^{(v)}(\tau_j)$ and the jump caused by the intervention v, given by

$$\Delta_v Y^{(v)}(\tau_j) := \Gamma(\check{Y}^{(v)}(\tau_j^-), \zeta) - \check{Y}^{(v)}(\tau_j^-). \tag{6.1.7}$$

Let $\mathcal{S} \subset \mathbb{R}^k$ be a fixed open set (the solvency region). Define

$$\tau_{\mathcal{S}} = \inf\{t \in (0, \tau^*); Y^{(v)}(t) \notin \mathcal{S}\}. \tag{6.1.8}$$

Suppose we are given a continuous *profit function* $f : \mathcal{S} \to \mathbb{R}$ and a continuous *bequest function* $g : \mathbb{R}^k \to \mathbb{R}$. Moreover, suppose the profit/utility of making an intervention with impulse $\zeta \in \mathcal{Z}$ when the state is y is $K(y, \zeta)$, where $K : \mathcal{S} \times \mathcal{Z} \to \mathbb{R}$ is a given continuous function.

We assume we are given a set $\mathcal{V}$ of *admissible impulse controls* which is included in the set of $v = (\tau_1, \tau_2, \ldots; \zeta_1, \zeta_2, \ldots)$ such that a unique solution $Y^{(v)}$ of (6.1.2)–(6.1.4) exists and

$$\tau^* = \infty \quad \text{a.s.} \tag{6.1.9}$$

and (if $M = \infty$)

$$\lim_{j \to \infty} \tau_j = \tau_{\mathcal{S}}. \tag{6.1.10}$$

We also assume that

$$E^y \left[\int_0^{\tau_{\mathcal{S}}} \left| f(Y^{(v)}(t)) \right| dt \right] < \infty \quad \text{for all } y \in \mathbb{R}^k, v \in \mathcal{V}, \tag{6.1.11}$$

$$E\left[g^{-}(Y^{(v)}(\tau_{\mathcal{S}}))\mathcal{X}_{\{\tau_{\mathcal{S}}<\infty\}}\right] < \infty \quad \text{for all } y \in \mathbb{R}^{k}, v \in \mathcal{V}, \tag{6.1.12}$$

and

$$E\left[\sum_{\tau_{j}\leq\tau_{\mathcal{S}}} K^{-}(\check{Y}^{(v)}(\tau_{j}^{-}),\zeta_{j})\right] < \infty \quad \text{for all } y \in \mathbb{R}^{k}, v \in \mathcal{V}. \tag{6.1.13}$$

Now define the performance criterion

$$J^{(v)}(y) = E^{y}\left[\int_{0}^{\tau_{\mathcal{S}}} f(Y^{(v)}(t))\mathrm{d}t + g(Y^{(v)}(\tau_{\mathcal{S}}))\mathcal{X}_{\{\tau_{\mathcal{S}}<\infty\}}\right.$$
$$\left. + \sum_{\tau_{j}\leq\tau_{\mathcal{S}}} K(\check{Y}^{(v)}(\tau_{j}^{-}),\zeta_{j})\right].$$

The *impulse control* problem is the following.
    Find $\Phi(y)$ and $v^{*} \in \mathcal{V}$ such that

$$\Phi(y) = \sup\{J^{(v)}(y); v \in \mathcal{V}\} = J^{(v^{*})}(y). \tag{6.1.14}$$

The following concept is crucial.

Definition 6.1. *Let $\mathcal{H}$ be the space of all measurable functions $h : \mathcal{S} \to \mathbb{R}$. The* intervention operator *$\mathcal{M} : \mathcal{H} \to \mathcal{H}$ is defined by*

$$\mathcal{M}h(y) = \sup\{h(\Gamma(y,\zeta)) + K(y,\zeta); \zeta \in \mathcal{Z} \quad \text{and} \quad \Gamma(y,\zeta) \in \mathcal{S}\}. \tag{6.1.15}$$

As in Chap. 2 we put

$$\mathcal{T} = \{\tau; \tau \text{ stopping time}, 0 \leq \tau \leq \tau_{\mathcal{S}} \text{ a.s.}\}.$$

We can now state the main result of this chapter, a verification theorem for impulse control problems.

Theorem 6.2 (Quasi-Integrovariational Inequalities for Impulse Control).

(a) *Suppose we can find $\phi : \bar{\mathcal{S}} \to \mathbb{R}$ such that*
    (i) $\phi \in C^{1}(\mathcal{S}) \cap C(\bar{\mathcal{S}})$.
    (ii) $\phi \geq \mathcal{M}\phi$ *on* $\mathcal{S}$.
        Define

$$D = \{y \in \mathcal{S}; \phi(y) > \mathcal{M}\phi(y)\} \quad \text{(the continuation region)}.$$

    Assume
    (iii) $E^{y}\left[\int_{0}^{\tau_{\mathcal{S}}} \mathcal{X}_{\partial D}(Y^{(v)}(t))\mathrm{d}t\right] = 0 \quad \text{for all } y \in \mathcal{S}, v \in \mathcal{V}$.
    (iv) ∂D *is a Lipschitz surface.*

(v) $\phi \in C^2(\mathcal{S} \setminus \partial D)$ with locally bounded derivatives near ∂D.

(vi) $A\phi + f \leq 0$ on $\mathcal{S} \setminus \partial D$.

(vii) $Y^{(v)}(\tau_{\mathcal{S}}) \in \partial \mathcal{S}$ a.s. on $\{\tau_{\mathcal{S}} < \infty\}$ and
$$\phi(Y^{(v)}(t)) \rightarrow g(Y^{(v)}(\tau_{\mathcal{S}})) \cdot \chi_{\{\tau_{\mathcal{S}} < \infty\}} \text{ as } t \rightarrow \tau_{\mathcal{S}}^- \text{ a.s., for all } y \in \mathcal{S},$$
$v \in \mathcal{V}$.

(viii) $\{\phi^-(Y^{(v)}(\tau)); \tau \in \mathcal{T}\}$ is uniformly integrable, for all $y \in \mathcal{S}$, $v \in \mathcal{V}$.

(ix) $E^y \left[|\phi(Y^{(v)}(\tau))| + \int_0^{\tau_{\mathcal{S}}} |A\phi(Y^{(v)}(t))| dt \right] < \infty$ for all $\tau \in \mathcal{T}, v \in \mathcal{V}$, $y \in \mathcal{S}$.

Then
$$\phi(y) \geq \Phi(y) \quad \text{for all } y \in \mathcal{S}. \tag{6.1.16}$$

(b) Suppose in addition that

(x) $A\phi + f = 0$ in D.

(xi) $\hat{\zeta}(y) \in \text{Argmax}\{\phi(\Gamma(y, \cdot)) + K(y, \cdot)\} \in \mathcal{Z}$ exists for all $y \in \mathcal{S}$ and $\hat{\zeta}(\cdot)$ is a Borel measurable selection.

Put $\hat{\tau}_0 = 0$ and define $\hat{v} = (\hat{\tau}_1, \hat{\tau}_2, \ldots; \hat{\zeta}_1, \hat{\zeta}_2, \ldots)$ inductively by
$\hat{\tau}_{j+1} = \inf\{t > \hat{\tau}_j; Y^{(\hat{v}_j)}(t) \notin D\} \wedge \tau_{\mathcal{S}}$ and $\hat{\zeta}_{j+1} = \hat{\zeta}(Y^{(\hat{v}_j)}(\hat{\tau}_{j+1}^-))$
if $\hat{\tau}_{j+1} < \tau_{\mathcal{S}}$, where $Y^{(\hat{v}_j)}$ is the result of applying
$\hat{v}_j := (\hat{\tau}_1, \ldots, \hat{\tau}_j; \hat{\zeta}_1, \ldots, \hat{\zeta}_j)$ to Y.
Suppose

(xii) $\hat{v} \in \mathcal{V}$ and $\{\phi(Y^{(\hat{v})}(\tau)); \tau \in \mathcal{T}\}$ is uniformly integrable.

Then
$$\phi(y) = \Phi(y) \quad \text{and} \quad \hat{v} \text{ is an optimal impulse control.} \tag{6.1.17}$$

Sketch of Proof. (a) By Theorem 2.1 and (iii)–(v), we may assume that $\phi \in C^2(\mathcal{S}) \cap C(\bar{\mathcal{S}})$. Choose $v = (\tau_1, \tau_2, \ldots; \zeta_1, \zeta_2, \ldots) \in \mathcal{V}$ and set $\tau_0 = 0$. By another approximation argument we may assume that we can apply the Dynkin formula to the stopping times τ_j. Then for $j = 0, 1, 2, \ldots$, with $Y = Y^{(v)}$

$$E^y[\phi(Y(\tau_j))] - E^y[\phi(\check{Y}(\tau_{j+1}^-))] = -E^y \left[\int_{\tau_j}^{\tau_{j+1}} A\phi(Y(t)) dt \right], \tag{6.1.18}$$

where $\check{Y}(\tau_{j+1}^-) = Y(\tau_{j+1}^-) + \Delta_N Y(\tau_{j+1})$, as before. Summing this from $j = 0$ to $j = m$ we get

$$\phi(y) + \sum_{j=1}^m E^y[\phi(Y(\tau_j)) - \phi(\check{Y}(\tau_j^-))] - E^y[\phi(\check{Y}(\tau_{m+1}^-))]$$
$$= -E^y \left[\int_0^{\tau_{m+1}} A\phi(Y(t)) dt \right] \geq E^y \left[\int_0^{\tau_{m+1}} f(Y(t)) dt \right]. \tag{6.1.19}$$

Now

$$\phi(Y(\tau_j)) = \phi(\Gamma(\check{Y}(\tau_j^-), \zeta_j))$$
$$\leq \mathcal{M}\phi(\check{Y}(\tau_j^-)) - K(\check{Y}(\tau_j^-), \zeta_j) \quad \text{if } \tau_j < \tau_S \text{ by (6.1.15)}$$

and

$$\phi(Y(\tau_j)) = \phi(\check{Y}(\tau_j^-)) \quad \text{if } \tau_j = \tau_S \text{ by (vii)}.$$

Therefore

$$\mathcal{M}\phi(\check{Y}(\tau_j^-)) - \phi(\check{Y}(\tau_j^-)) \geq \phi(Y(\tau_j)) - \phi(\check{Y}(\tau_j^-)) + K(\check{Y}(\tau_j^-), \zeta_j)$$

and

$$\phi(y) + \sum_{j=1}^{m} E^y[\{\mathcal{M}\phi(\check{Y}(\tau_j^-)) - \phi(\check{Y}(\tau_j^-))\} \cdot \mathcal{X}_{\{\tau_j < \tau_S\}}]$$

$$\geq E^y\left[\int_0^{\tau_{m+1}} f(Y(t))\mathrm{d}t + \phi(\check{Y}(\tau_{m+1}^-)) + \sum_{j=1}^{m} K(\check{Y}(\tau_j^-), \zeta_j)\right].$$

Letting $m \to M$ we get

$$\phi(y) \geq E^y\left[\int_0^{\tau_S} f(Y(t))\mathrm{d}t + g(Y(\tau_S))\mathcal{X}_{\{\tau_S < \infty\}} + \sum_{j=1}^{M} K(\check{Y}(\tau_j^-), \zeta_j)\right] = J^{(v)}(y).$$

$$(6.1.20)$$

Hence $\phi(y) \geq \Phi(y)$.

(b) Next assume (x)–(xii) also hold. Apply the above argument to $\hat{v} = (\hat{\tau}_1, \hat{\tau}_2, \dots; \hat{\zeta}_1, \hat{\zeta}_2, \dots)$. Then by (x) we get *equality* in (6.1.19) and by our choice of $\zeta_j = \hat{\zeta}_j$ we have *equality* in (6.1.20). Hence

$$\phi(y) = J^{(\hat{v})}(y),$$

which combined with (a) completes the proof.   □

Remark 6.3. In the case of a pure diffusion process, the same verification theorem holds, just skip condition (ix).

6.2 Examples

Example 6.4 (Optimal Stream of Dividends Under Transaction Costs). This example is an extension to the jump diffusion case of a problem studied in [J-PS]. Suppose that if we make no interventions the amount $X(t)$ available (cash flow) is given by

$$\mathrm{d}X(t) = \mu\,\mathrm{d}t + \sigma\,\mathrm{d}B(t) + \theta\int_{\mathbb{R}} z\tilde{N}(\mathrm{d}t, \mathrm{d}z), \quad X(0) = x > 0, \qquad (6.2.1)$$

where $\mu, \sigma > 0$, $\theta \geq 0$ are constants and we assume that $z \leq 0$ a.s. ν. Suppose that at any time t we are free to take out an amount $\zeta > 0$ from $X(t)$ by applying the transaction cost

$$k(\zeta) = c + \lambda\zeta, \tag{6.2.2}$$

where $c > 0$, $\lambda \geq 0$ are constants. The constant c is called the *fixed* part and the quantity $\lambda\zeta$ is called the *proportional* part, respectively, of the transaction cost. The resulting cash flow $X^{(v)}(t)$ is given by

$$X^{(v)}(t) = X(t) \quad \text{if } 0 \leq t < \tau_1, \tag{6.2.3}$$

$$X^{(v)}(\tau_j) = \Gamma(X^{(v)}(\tau_j^-) + \Delta_N X(\tau_j), \zeta_j) = \check{X}^{(v)}(\tau_j^-) - (1+\lambda)\zeta_j - c, \tag{6.2.4}$$

and

$$dX^{(v)}(t) = \mu\,dt + \sigma\,dB(t) + \theta \int_{\mathbb{R}} z\tilde{N}(dt, dz) \text{ if } \tau_j \leq t < \tau_{j+1}. \tag{6.2.5}$$

Put

$$\tau_S = \inf\{t > 0; X^{(v)}(t) \leq 0\} \quad \text{(time of bankruptcy)} \tag{6.2.6}$$

and

$$J^{(v)}(s, x) = E^{s,x}\left[\sum_{\tau_j \leq \tau_S} e^{-\rho(s+\tau_j)}\zeta_j\right], \tag{6.2.7}$$

where $\rho > 0$ is constant (the discounting exponent).

We seek $\Phi(s, x)$ and $v^* = (\tau_1^*, \tau_2^*, \ldots; \zeta_1^*, \zeta_2^*, \ldots) \in \mathcal{V}$ such that

$$\Phi(s, x) = \sup_{v \in \mathcal{V}} J^{(v)}(s, x) = J^{(v^*)}(s, x), \tag{6.2.8}$$

where $\mathcal{V}$ is the set of impulse controls s.t. $X^{(v)}(t) \geq 0$ for all $t \leq \tau_S$. This is a problem of the type (6.1.14), with

$$Y^{(v)}(t) = \begin{bmatrix} s+t \\ X^{(v)}(t) \end{bmatrix}, \quad t \geq 0, \quad Y^{(v)}(0^-) = \begin{bmatrix} s \\ x \end{bmatrix} = y,$$

$$\Gamma(y, \zeta) = \Gamma(s, x, \zeta) = \begin{bmatrix} s \\ x - c - (1+\lambda)\zeta \end{bmatrix},$$

$$K(y, \zeta) = K(s, x, \zeta) = e^{-\rho s}\zeta, \quad f = g = 0,$$

and

$$\mathcal{S} = \{(s, x); x > 0\}.$$

As a candidate for the value function Φ we try

$$\phi(s, x) = e^{-\rho s}\psi(x). \tag{6.2.9}$$

Then

$$\mathcal{M}\psi(x) = \sup\left\{\psi(x - c - (1 + \lambda)\zeta) + \zeta,\ 0 < \zeta < \frac{x - c}{1 + \lambda}\right\}.$$

We now guess that the continuation region has the form

$$D = \{(s, x); 0 < x < x^*\} \quad \text{for some } x^* > 0. \tag{6.2.10}$$

Then (x) of Theorem 6.2 gives

$$-\rho\psi(x) + \mu\psi'(x) + \frac{1}{2}\sigma^2\psi''(x) + \int_{\mathbb{R}}\{\psi(x + \theta z) - \psi(x) - \psi'(x)\theta z\}\nu(\mathrm{d}z) = 0.$$

To solve this equation we try a function of the form

$$\psi(x) = e^{rx}$$

for some constant $r \in \mathbb{R}$. Then r must solve the equation

$$h(r) := -\rho + \mu r + \frac{1}{2}\sigma^2 r^2 + \int_{\mathbb{R}}\{e^{r\theta z} - 1 - r\theta z\}\nu(\mathrm{d}z) = 0. \tag{6.2.11}$$

Since $h(0) = -\rho < 0$ and $\lim_{|r|\to\infty} h(r) = \infty$, we see that there exist two solutions r_1, r_2 of $h(r) = 0$ such that

$$r_2 < 0 < r_1.$$

Moreover, since $e^{r\theta z} - 1 - r\theta z \geq 0$ for all r, z we have

$$|r_2| > r_1.$$

With such a choice of r_1, r_2 we try

$$\psi(x) = A_1 e^{r_1 x} + A_2 e^{r_2 x}, \quad A_i \text{ constants}.$$

Since

$$\psi(0) = 0 \quad \text{we have} \quad A_1 + A_2 = 0$$

so we write $A_1 = A = -A_2 > 0$ and

$$\psi(x) = A(e^{r_1 x} - e^{r_2 x}), \quad 0 < x < x^*.$$

Define

$$\psi_0(x) = A(e^{r_1 x} - e^{r_2 x}) \quad \text{for all } x > 0. \tag{6.2.12}$$

To study $\mathcal{M}\psi$ we first consider

$$g(\zeta) := \psi_0(x - c - (1 + \lambda)\zeta) + \zeta, \quad \zeta > 0.$$

The first-order condition for a maximum point $\hat{\zeta} = \hat{\zeta}(x)$ for $g(\zeta)$ is that

$$\psi_0'(x - c - (1 + \lambda)\hat{\zeta}) = \frac{1}{1 + \lambda}.$$

Now

$$\psi_0'(x) > 0 \quad \text{for all } x \text{ and}$$

$$\psi_0''(x) < 0 \quad \text{iff} \quad x < \tilde{x} := \frac{2(\ln |r_2| - \ln r_1)}{r_1 - r_2}.$$

Therefore the equation $\psi_0'(x) = 1/(1 + \lambda)$ has exactly two solutions $x = \underline{x}$ and $x = \bar{x}$ where

$$0 < \underline{x} < \tilde{x} < \bar{x}$$

(provided that $\psi_0'(\tilde{x}) < 1/(1 + \lambda) < \psi_0'(0)$). See Fig. 6.1.

Choose

$$x^* = \bar{x} \quad \text{and put} \quad \hat{x} = \underline{x}. \tag{6.2.13}$$

If we require that $\psi(x) = \mathcal{M}\psi_0(x)$ for $x \ge x^*$ we get

$$\psi(x) = \psi_0(\hat{x}) + \hat{\zeta}(x) \quad \text{for} \quad x \ge x^*,$$

where

$$x - c - (1 + \lambda)\hat{\zeta}(x) = \hat{x},$$

i.e.,

$$\hat{\zeta}(x) = \frac{x - \hat{x} - c}{1 + \lambda} \quad \text{for} \quad x \ge x^*. \tag{6.2.14}$$

Hence we propose that ψ has the form

$$\psi(x) = \begin{cases} \psi_0(x) = A(e^{r_1 x} - e^{r_2 x}), & 0 < x < x^*, \\ \psi_0(\hat{x}) + \frac{x - \hat{x} - c}{1 + \lambda}, & x \ge x^*. \end{cases} \tag{6.2.15}$$

Now choose A such that ψ is continuous at $x = x^*$. This gives

$$A = (1 + \lambda)^{-1} \left[e^{r_1 x^*} - e^{r_2 x^*} - e^{r_1 \hat{x}} + e^{r_2 \hat{x}} \right]^{-1} (x^* - \hat{x} - c). \tag{6.2.16}$$

By our choice of x^* we then have that ψ is also differentiable at $x = x^*$.

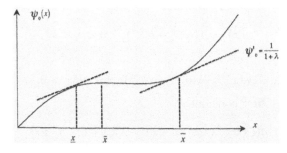

Fig. 6.1. The function $\psi_0(x)$

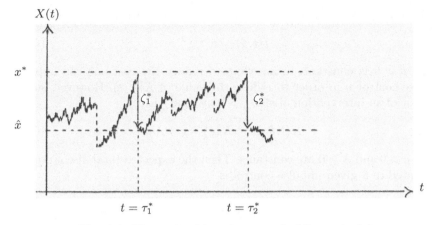

Fig. 6.2. The optimal impulse control of Example 6.4

We can now check that, with these values of $x^*, \hat{x}$, and A, our choice of $\phi(s, x) = e^{-\rho s}\psi(x)$ satisfies all the requirements of Theorem 6.2, provided that some conditions on the parameters are satisfied. We leave this verification to the reader.

Thus the solution of the impulse control problem (6.2.8) can be described as follows. As long as $X(t) < x^*$ we do nothing. If $X(t)$ reaches the value x^* then immediately we make an intervention to bring $X(t)$ down to the value $\hat{x}$. See Fig. 6.2.

Example 6.5. As another illustration of how to apply Theorem 6.2 we consider the following example, which is a jump diffusion version of the example in [Ø2] studied in connection with questions involving vanishing fixed costs. Variations of this problem have been studied by many authors (see, e.g., [HST, J-P, MØ, ØS, ØUZ, V]). One possible economic interpretation is that the given process represents the exchange rate of a given currency and the impulses represent the interventions taken in order to keep the exchange rate in a given "target zone." See, e.g., [J-P, MØ].

Suppose that without interventions the system has the form

$$Y(t) = \begin{bmatrix} s+t \\ X(t) \end{bmatrix} \in \mathbb{R}^2, \quad Y(0) = y = (s, x), \tag{6.2.17}$$

where $X(t) = x + B(t) + \int_0^t \int_{\mathbb{R}} z\tilde{N}(ds, dz)$ and $B(0) = 0$. We assume that $z \leq 0$ a.s. z. Suppose that we are only allowed to give the system impulses ζ with values in $\mathcal{Z} := (0, \infty)$ and that if we apply an impulse control $v = (\tau_1, \tau_2, \ldots; \zeta_1, \zeta_2, \ldots)$ to $Y(t)$ it gets the form

$$Y^{(v)}(t) = \begin{bmatrix} s+t \\ X(t) - \sum_{\tau_k \leq t} \zeta_k \end{bmatrix} = \begin{bmatrix} s+t \\ X^{(v)}(t) \end{bmatrix}. \tag{6.2.18}$$

Suppose that the *cost rate* $f(t, \xi)$ if $X^{(v)}(t) = \xi$ at time t is given by

$$f(t, \xi) = e^{-\rho t} \xi^2, \tag{6.2.19}$$

where $\rho > 0$ is constant. In an effort to reduce the cost one can apply the impulse control v in order to reduce the value of $X^{(v)}(t)$. However, suppose the cost of an intervention of size $\zeta > 0$ at time t is

$$K(t, \xi, \zeta) = K(\zeta) = c + \lambda \zeta, \tag{6.2.20}$$

where $c > 0$ and $\lambda \geq 0$ are constants. Then the expected total discounted cost associated to a given impulse control is

$$J^{(v)}(s, x) = E^x \left[\int_0^\infty e^{-\rho(s+t)} (X^{(v)}(t))^2 dt + \sum_{k=1}^N e^{-\rho(s+\tau_k)} (c + \lambda \zeta_k) \right]. \tag{6.2.21}$$

We seek $\Phi(s, x)$ and $v^* = (\tau_1^*, \tau_2^*, \ldots; \zeta_1^*, \zeta_2^*, \ldots)$ such that

$$\Phi(s, x) = \inf_v J^{(v)}(s, x) = J^{(v^*)}(s, x). \tag{6.2.22}$$

This is an impulse control problem of the type described above, except that it is a minimum problem rather than a maximum problem. Theorem 6.2 still applies, with the corresponding changes.

Note that it is not optimal to move $X(t)$ downward if $X(t)$ is already below 0. Hence we may restrict ourselves to consider impulse controls $v = (\tau_1, \tau_2, \ldots; \zeta_1, \zeta_2, \ldots)$ such that

$$\sum_{j=1}^{\tau_k} \zeta_j \leq X(\tau_k) \quad \text{for all } k. \tag{6.2.23}$$

We let $\mathcal{V}$ denote the set of such impulse controls.

We guess that the optimal strategy is to wait until the level of $X(t)$ reaches an (unknown) value $x^* > 0$. At this time, τ_1, we intervene and give $X(t)$ an impulse ζ_1, which brings it down to a lower value $\hat{x} > 0$. Then we do nothing until the next time, τ_2, that $X(t)$ reaches the level x^*, etc. This suggests that the continuation region D in Theorem 6.2 has the form

$$D = \{(s, x), \ x < x^*\}. \tag{6.2.24}$$

See Fig. 6.3.

Let us try a value function φ of the form

$$\varphi(s, x) = e^{-\rho s} \psi(x), \tag{6.2.25}$$

where ψ remains to be determined.

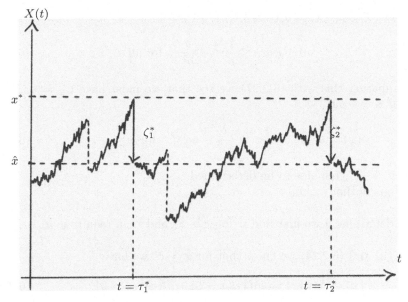

Fig. 6.3. The optimal impulse control of Example 6.5

Condition (x) of Theorem 6.2 gives that for $x < x^*$ we should have

$$A\varphi + f = e^{-\rho s}\left(-\rho\psi(x) + \frac{1}{2}\psi''(x) + \int_{\mathbb{R}}\{\psi(x+z) - \psi(x) - z\psi'(x)\}\nu(dz)\right)$$
$$+ e^{-\rho s}x^2 = 0.$$

So for $x < x^*$ we let ψ be a solution $h(x)$ of the equation

$$\int_{\mathbb{R}}\{h(x+z) - h(x) - zh'(x)\}\nu(dz) + \frac{1}{2}h''(x) - \rho h(x) + x^2 = 0. \quad (6.2.26)$$

We see that any function $h(x)$ of the form

$$h(x) = C_1 e^{r_1 x} + C_2 e^{r_2 x} + \frac{1}{\rho}x^2 + \frac{1 + \int_{\mathbb{R}} z^2\nu(dz)}{\rho^2}, \quad (6.2.27)$$

where C_1, C_2 are arbitrary constants, is a solution of (6.2.26), provided that $r_1 > 0, r_2 < 0$ are roots of the equation

$$K(r) := \int_{\mathbb{R}}\{e^{rz} - 1 - rz\}\nu(dz) + \frac{1}{2}r^2 - \rho = 0.$$

Note that if we make no interventions at all, then the cost is

$$J^{(v)}(s, x) = e^{-\rho s}E^x\left[\int_0^\infty e^{-\rho t}(X(t))^2 dt\right]$$
$$= e^{-\rho s}\int_0^\infty e^{-\rho t}(x^2 + tb)dt = e^{-\rho s}\left(\frac{1}{\rho}x^2 + \frac{b}{\rho^2}\right), \quad (6.2.28)$$

where $b = 1 + \int_{\mathbb{R}} z^2 \nu(dz)$. Hence we must have

$$0 \le \psi(x) \le \frac{1}{\rho}x^2 + \frac{b}{\rho^2} \quad \text{for all } x. \tag{6.2.29}$$

Comparing this with (6.2.27) we see that we must have $C_2 = 0$. Hence $C_1 \le 0$. So we put

$$\psi(x) = \psi_0(x) := \frac{1}{\rho}x^2 + \frac{b}{\rho^2} - ae^{r_1 x} \quad \text{for } x \le x^*, \tag{6.2.30}$$

where $a = -C_1$ remains to be determined.
We guess that $a > 0$.

To determine a we first find ψ for $x > x^*$ and then require ψ to be C^1 at $x = x^*$.
By (ii) and (6.2.24) we know that for $x > x^*$ we have

$$\psi(x) = \mathcal{M}\psi(x) := \inf\{\psi(x - \zeta) + c + \lambda\zeta, \zeta > 0\}. \tag{6.2.31}$$

The first-order condition for a minimum $\hat{\zeta} = \hat{\zeta}(x)$ of the function

$$G(\zeta) := \psi(x - \zeta) + c + \lambda\zeta, \quad \zeta > 0$$

is

$$\psi'(x - \hat{\zeta}) = \lambda.$$

Suppose there is a unique point $\hat{x} \in (0, x^*)$ such that

$$\psi'(\hat{x}) = \lambda. \tag{6.2.32}$$

Then

$$\hat{x} = x - \hat{\zeta}(x), \text{ i.e., } \hat{\zeta}(x) = x - \hat{x}$$

and from (6.2.31) we deduce that

$$\psi(x) = \psi_0(\hat{x}) + c + \lambda(x - \hat{x}) \quad \text{for } x \ge x^*.$$

In particular,

$$\psi'(x^*) = \lambda \tag{6.2.33}$$

and

$$\psi(x^*) = \psi_0(\hat{x}) + c + \lambda(x^* - \hat{x}). \tag{6.2.34}$$

To summarize we put

$$\psi(x) = \begin{cases} \frac{1}{\rho}x^2 + \frac{b}{\rho^2} - ae^{r_1 x}, & \text{for } x \le x^*, \\ \psi_0(\hat{x}) + c + \lambda(x - \hat{x}), & \text{for } x > x^*, \end{cases} \tag{6.2.35}$$

where $\hat{x}, x^*$, and a are determined by (6.2.32)–(6.2.34), i.e.,

$$ar_1 e^{r_1 \hat{x}} = \frac{2}{\rho} \hat{x} - \lambda \quad \text{(i.e., } \psi'(\hat{x}) = \lambda), \tag{6.2.36}$$

$$ar_1 e^{r_1 x^*} = \frac{2}{\rho} x^* - \lambda \quad \text{(i.e., } \psi'(x^*) = \lambda), \tag{6.2.37}$$

$$ae^{r_1 x^*} - ae^{r_1 \hat{x}} = \frac{1}{\rho}((x^*)^2 - (\hat{x})^2) - c - \lambda(x^* - \hat{x}). \tag{6.2.38}$$

One can now prove (see [Ø2, Theorem 2.5]).

For each $c > 0$ there exists $a = a^*(c) > 0$, $\hat{x} = \hat{x}(c) > 0$, and $x^* = x^*(c) > \hat{x}$ such that (6.2.36)–(6.2.38) hold. With this choice of a, $\hat{x}$, and x^*, the function $\varphi(s, x) = e^{-\rho s} \psi(x)$ with ψ given by (6.2.35) coincides with the value function Φ defined in (6.2.22). Moreover, the optimal impulse control $v^* = (\tau_1^*, \tau_2^*, \ldots; \zeta_1^*, \zeta_2^*, \ldots)$ is to do nothing while $X(t) < x^*$, then move $X(t)$ from x^* down to $\hat{x}$ (i.e., apply $\zeta_1^* = x^* - \hat{x}$) at the first time τ_1^* when $X(t)$ reaches a value $\geq x^*$, then wait until the next time, τ_2^*, $X(t)$ again reaches the value x^*, etc.

Remark 6.6. In [Ø2] this result is used to study how the value function $\Phi(s, x) = \Phi_c(s, x)$ depends on the fixed part $c > 0$ of the intervention cost. It is proved that the function

$$c \to \Phi_c(s, x)$$

is continuous but not differentiable at $c = 0$. In fact, we have

$$\frac{\partial}{\partial c} \Phi_c(s, x) \to \infty \text{ as } c \to 0^+.$$

Subsequently this high c-sensitivity of the value function for c close to 0 was proved for other processes as well. See [ØUZ].

Remark 6.7. For applications of impulse control theory in inventory control see, e.g., [S, S2] and the references therein.

6.3 Exercises

Exercise* 6.1. Solve the impulse control problem

$$\Phi(s, x) = \inf_v J^{(v)}(s, x) = J^{(v^*)}(s, x),$$

where

$$J^{(v)}(s, x) = E\left[\int_0^\infty e^{-\rho(s+t)}(X^{(v)}(t))^2 dt + \sum_{k=1}^N e^{-\rho(s+\tau_k)}(c + \lambda|\zeta_k|)\right].$$

The inf is taken over all impulse controls $v = (\tau_1, \tau_2, \ldots; \zeta_1, \zeta_2, \ldots)$ with $\zeta_i \in \mathbb{R}$ and the corresponding process $X^{(v)}(t)$ is given by

$$dX^{(v)}(t) = dB(t) + \int_{\mathbb{R}} \theta(X^{(v)}(t), z)\tilde{N}(dt, dz), \quad \tau_k < t < \tau_{k+1},$$

$$X^{(v)}(\tau_{k+1}) = \check{X}^{(v)}(\tau_{k+1}^-) + \zeta_{k+1}, \quad k = 0, 1, 2, \ldots,$$

where $B(0) = 0$, $x \in \mathbb{R}$, and we assume that there exists $\xi > 0$ such that

$$\theta(x, z) = 0 \quad \text{for a.a. } z \text{ if } |x| < \xi, \tag{6.3.1}$$

$$\theta(x, z) \geq 0 \text{ if } x \geq \xi \text{ and } \theta(x, z) \leq 0 \text{ if } x \leq -\xi \quad \text{for a.a. } z, \tag{6.3.2}$$

Exercise* 6.2 (Optimal Stream of Dividends with Transaction Costs from a Geometric Lévy Process).
For $v = (\tau_1, \tau_2, \ldots; \zeta_1, \zeta_2, \ldots)$ with $\zeta_i \in \mathbb{R}_+$ define $X^{(v)}(t)$ by

$$dX^{(v)}(t) = \mu X^{(v)}(t)dt + \sigma X^{(v)}(t)dB(t)$$

$$+ \theta X^{(v)}(t^-) \int_{\mathbb{R}} z\tilde{N}(ds, dz), \quad \tau_i \leq t \leq \tau_{i+1},$$

$$X^{(v)}(\tau_{i+1}) = \check{X}^{(v)}(\tau_{i+1}^-) - (1 + \lambda)\zeta_{i+1} - c, \quad i = 0, 1, 2, \ldots,$$

$$X^{(v)}(0^-) = x > 0,$$

where $\mu, \sigma \neq 0$, $\theta, \lambda \geq 0$, and $c > 0$ are constants (see (6.1.5)), $-1 \leq \theta z \leq 0$ a.s. ν.

Find Φ and v^* such that

$$\Phi(s, x) = \sup_v J^{(v)}(s, x) = J^{(v^*)}(s, x).$$

Here

$$J^{(v)}(s, x) = E^x \left[\sum_{\tau_k < \tau_S} e^{-\rho(s + \tau_k)} \zeta_k \right] \quad (\rho > 0 \text{ constant})$$

is the expected discounted total dividend up to time τ_S, where

$$\tau_S = \tau_S(\omega) = \inf\{t > 0; X^{(v)}(t) \leq 0\}$$

is the time of bankruptcy. (See also Exercise 7.2.)

Exercise* 6.3 (Optimal Forest Management (Inspired by Y. Willassen [W])).
Suppose the biomass of a forest at time t is given by

$$X(t) = x + \mu t + \sigma B(t) + \theta \int_{\mathbb{R}} z\tilde{N}(t, dz),$$

where $\mu > 0$, $\sigma > 0$, $\theta > 0$ are constants and we assume that $z \leq 0$ a.s. ν. At times $0 \leq \tau_1 < \tau_2 < \cdots$ we decide to cut down the forest and replant it, with the cost

$$c + \lambda \check{X}(\tau_k^-), \quad \text{with} \quad \check{X}(\tau_k^-) = X(\tau_k^-) + \Delta_N X(\tau_k),$$

where $c > 0$ and $\lambda \in [0, 1)$ are constants and $\Delta_N X(t)$ is the (possible) jump in X at t coming from the jump in $N(t, \cdot)$ only, not from the intervention.

Find the sequence of stopping times $v = (\tau_1, \tau_2, \ldots)$ which maximizes the expected total discounted net profit $J^{(v)}(s, x)$ given by

$$J^{(v)}(s, x) = E^x \left[\sum_{k=1}^{\infty} e^{-\rho(s+\tau_k)} (\check{X}(\tau_k^-) - c - \lambda \check{X}(\tau_k^-)) \right],$$

where $\rho > 0$ is a given discounting exponent.

7

Approximating Impulse Control by Iterated Optimal Stopping

7.1 Iterative Scheme

In general it is not possible to reduce impulse control to optimal stopping, because the choice of the first intervention time τ_1 and the first impulse ζ_1 will influence the next and so on. However, if we allow only (up to) a fixed finite number n of interventions, then the corresponding impulse control problem can be solved by solving iteratively n optimal stopping problems. Moreover, if we restrict the number of interventions to (at most) n in a given impulse control problem, then the value function of this restricted problem will converge to the value function of the original problem as $n \to \infty$. Thus it is possible to reduce a given impulse control problem to a sequence of iterated optimal stopping problems. This is useful both for theoretical purposes and numerical applications.

We now make this more precise.

Using the notation of Chap. 6 consider the impulse control problem

$$\Phi(y) = \sup\{J^{(v)}(y); v \in \mathcal{V}\} = J^{(v^*)}(y), \quad y \in \mathcal{S}, \qquad (7.1.1)$$

where, with $\tau_{\mathcal{S}} = \tau_{\mathcal{S}}^{(v)} = \inf\{t > 0; Y^{(v)}(t) \notin \mathcal{S}\}$,

$$J^{(v)}(y)$$
$$= E^y \left[\int_0^{\tau_{\mathcal{S}}} f(Y^{(v)}(t))\mathrm{d}t + g(Y^{(v)}(\tau_{\mathcal{S}}))\chi_{\{\tau_{\mathcal{S}}<\infty\}} + \sum_{\tau_j < \tau_{\mathcal{S}}} K(\check{Y}^{(v)}(\tau_j^-), \zeta_j) \right].$$
$$(7.1.2)$$

Here $\mathcal{V}$ denotes the set of admissible controls $v = (\tau_1, \tau_2, \ldots; \zeta_1, \zeta_2, \ldots)$. See (6.1.9)–(6.1.13).

For $n = 1, 2, \ldots$ let $\mathcal{V}_n$ denote the set of all $v \in \mathcal{V}$ such that $v = (\tau_1, \tau_2, \ldots, \tau_k; \zeta_1, \zeta_2, \ldots, \zeta_k)$ with $k \leq n$. In other words, $\mathcal{V}_n$ is the set of all admissible controls *with at most n interventions*. Then

$$\mathcal{V}_n \subseteq \mathcal{V}_{n+1} \subseteq \mathcal{V} \quad \text{for all } n. \tag{7.1.3}$$

Define

$$\Phi_n(y) = \sup\{J^{(v)}(y); v \in \mathcal{V}_n\}, \quad n = 1, 2, \ldots \tag{7.1.4}$$

Then $\Phi_n(y) \leq \Phi_{n+1}(y) \leq \Phi(y)$ because $\mathcal{V}_n \subseteq \mathcal{V}_{n+1} \subseteq \mathcal{V}$. Moreover, we have:

Lemma 7.1. *Suppose $g \geq 0$. Then*

$$\lim_{n\to\infty} \Phi_n(y) = \Phi(y) \quad \text{for all } y \in \mathcal{S}.$$

Proof. We have already seen that

$$\lim_{n\to\infty} \Phi_n(y) \leq \Phi(y).$$

To get the opposite inequality let us first assume $\Phi(y) < \infty$. Then for each $\varepsilon > 0$ there exists $v = (\tau_1, \tau_2, \ldots; \zeta_1, \zeta_2, \ldots) \in \mathcal{V}$ such that

$$
J^{(v)}(y) = E^y \Bigg[\int_0^{\tau_S^{(v)}} f(Y^{(v)}(t))dt + g(Y^{(v)}(\tau_S^{(v)}))\chi_{\{\tau_S^{(v)} < \infty\}}
$$
$$
+ \sum_{\tau_j < \tau_S^{(v)}} K(\check{Y}^{(v)}(\tau_j^-), \zeta_j) \Bigg] \geq \Phi(y) - \varepsilon. \tag{7.1.5}
$$

For $n = 1, 2, \ldots$ define $v_n = (\tau_1, \tau_2, \ldots, \tau_n, \tau_S; \zeta_1, \zeta_2, \ldots, \zeta_n)$, i.e., v_n is obtained by truncating the v sequence after n steps. Then

$$Y^{(v_n)}(t) = Y^{(v)}(t) \quad \text{for all } t \leq \tau_n. \tag{7.1.6}$$

Since $\tau_j \to \tau_S$ a.s. when $j \to \infty$, we get by assumptions (6.1.11) and (6.1.13) that there exists n such that

$$
E^y \Bigg[\int_{\tau_n}^{\tau_S^{(v)}} \big| f(Y^{(v)}(t)) \big| dt + \int_{\tau_n}^{\tau_S^{(v_n)}} \big| f(Y^{(v_n)}(t)) \big| dt \Bigg] < \varepsilon \tag{7.1.7}
$$

and

$$
E^y \Bigg[\sum_{\substack{j > n \\ \tau_j < \tau_S^{(v)}}} K^-(\check{Y}^{(v)}(\tau_j^-), \zeta_j) \Bigg] < \varepsilon. \tag{7.1.8}
$$

Moreover, by (7.1.6) we have

$$\chi_{\{\tau_S^{(v)} < \infty\}} \leq \liminf_{n \to \infty} \chi_{\{\tau_S^{(v_n)} < \infty\}}. \tag{7.1.9}$$

Combining (7.1.6)–(7.1.9) we get

$$\liminf_{n \to \infty} J^{(v_n)}(y) = \liminf_{n \to \infty} \left\{ E^y \left[\int_0^{\tau_n} + \int_{\tau_n}^{\tau_S^{(v_n)}} f(Y^{(v_n)}(t)) dt \right] \right.$$

$$\left. + E^y \left[g(Y^{(v_n)}(\tau_S^{(v_n)})) \chi_{\{\tau_S^{(v_n)} < \infty\}} \right] + E^y \left[\sum_{j=1}^n K(\check{Y}^{(v_n)}(\tau_j^-), \zeta_j) \right] \right\}$$

$$\geq J^{(v)}(y) - 2\varepsilon + \liminf_{n \to \infty} E^y \left[g(Y^{(v_n)}(\tau_S^{(v_n)})) \chi_{\{\tau_S^{(v_n)} < \infty\}} \right.$$

$$\left. - g(Y^{(v)}(\tau_S^{(v)})) \chi_{\{\tau_S^{(v)} < \infty\}} \right] \geq J^{(v)}(y) - 2\varepsilon$$

$$+ E^y \left[\liminf_{n \to \infty} (g(Y^{(v_n)}(\tau_S^{(v_n)})) - g(Y^{(v)}(\tau_S^{(v)}))) \chi_{\{\tau_S^{(v_n)} < \infty\}} \right]$$

$$= J^{(v)}(y) - 2\varepsilon.$$

Hence by (7.1.5)

$$\liminf_{n \to \infty} \Phi_n(y) \geq \liminf_{n \to \infty} J^{(v_n)}(y) \geq \Phi(y) - 3\varepsilon.$$

Since $\varepsilon > 0$ was arbitrary this proves Lemma 7.1 in the case when $\Phi(y) < \infty$. If $\Phi(y) = \infty$ the proof is similar, except that now we use that for each $M < \infty$ there exists $v \in \mathcal{V}$ such that $J^{(v)}(y) \geq M$. Choosing v_n as before and using (7.1.6)–(7.1.9) with $\varepsilon = 1$ we get $J^{(v_n)}(y) \geq M - 2$. Since M was arbitrary this shows that

$$\lim_{n \to \infty} \Phi_n(y) \geq \lim_{n \to \infty} J^{(v_n)}(y) = \infty. \qquad \square$$

Let

$$\mathcal{M}h(y) = \sup_{\zeta \in \mathcal{Z}} \{h(\Gamma(y, \zeta)) + K(y, \zeta)\}, \quad h \in \mathcal{H}, \ y \in \mathbb{R}^k \tag{7.1.10}$$

be the intervention operator (Definition 6.1). The iterative procedure is the following.

Let $Y(t) = Y^{(0)}(t)$ be the process (6.1.1) without interventions. Define

$$\varphi_0(y) = E^y \left[\int_0^{\tau_S} f(Y(t)) dt + g(Y(\tau_S)) \chi_{\{\tau_S < \infty\}} \right] \tag{7.1.11}$$

and inductively, for $j = 1, 2, \ldots, n$,

$$\varphi_j(y) = \sup_{\tau \in \mathcal{T}} E^y \left[\int_0^\tau f(Y(t)) \mathrm{d}t + \mathcal{M}\varphi_{j-1}(Y(\tau))\chi_{\{\tau_S < \infty\}} \right], \qquad (7.1.12)$$

where, as before $\mathcal{T}$ denotes the set of stopping times $\tau \leq \tau_S$, with

$$\tau_S = \inf\{t > 0; Y(t) \notin \mathcal{S}\}.$$

Let $\mathcal{P}(\mathbb{R}^k)$ denote the set of functions $h : \mathbb{R}^k \to \mathbb{R}$ of at most *polynomial growth*, i.e., with the property that there exists constants C and m (depending on h) such that

$$|h(y)| \leq C(1 + |y|^m) \quad \text{for all } y \in \mathbb{R}^k. \qquad (7.1.13)$$

The main result of this chapter is the following.

Theorem 7.2. *Suppose*

$$f, g \text{ and } \mathcal{M}\varphi_{j-1} \in \mathcal{P}(\mathbb{R}^k) \quad \text{for } j = 1, 2, \ldots, n. \qquad (7.1.14)$$

Then

$$\varphi_n = \Phi_n.$$

To prove this we need a *dynamic programming principle* (or *Bellman* principle). This principle is due to Krylov [K, Theorems 9 and 11, p. 134] when there are no jumps. The proof of the dynamic programming principle for jump processes can be found in Iskikawa [Isk, Sect. 4].

Lemma 7.3. *Suppose $G \in \mathcal{P}(\mathbb{R}^k)$. Define*

$$\psi(y) = \sup_{\tau \in \mathcal{T}} E^y \left[\int_0^\tau f(Y(s)) \mathrm{d}s + G(Y(\tau))\chi_{\{\tau_S < \infty\}} \right]. \qquad (7.1.15)$$

(a) *Then for all finite stopping times β we have*

$$\psi(y) = \sup_{\tau \in \mathcal{T}} E^y \left[\int_0^{\tau \wedge \beta} f(Y(t)) \mathrm{d}t + G(Y(\tau))\chi_{\{\tau \leq \beta\}} + \psi(Y(\beta))\chi_{\{\tau > \beta\}} \right].$$
$$(7.1.16)$$

(b) *For $\varepsilon > 0$ define*

$$D^{(\varepsilon)} = \{y \in \mathcal{S}; \psi(y) > G(y) + \varepsilon\}$$

and put

$$\tau^{(\varepsilon)} = \inf\{t > 0; Y(t) \notin D^{(\varepsilon)}\}.$$

Then if β is a finite stopping time such that $\beta \leq \tau^{(\varepsilon)}$ for some $\varepsilon > 0$ we have

$$\psi(y) = E^y \left[\int_0^\beta f(Y(t)) \mathrm{d}t + \psi(Y(\beta)) \right]. \qquad (7.1.17)$$

Corollary 7.4. (a) *For each given $\tau \in T$ the process*

$$U(t) := \int_0^{t \wedge \tau} f(Y(s))ds + \psi(Y(t \wedge \tau)), \quad t \geq 0$$

is a supermartingale. In particular, if τ_1, τ_2 are finite stopping times, $\tau_1 \leq \tau_2 \leq \tau_S$, then

$$E^y[\psi(Y(\tau_1))] \geq E^y \left[\int_{\tau_1}^{\tau_2} f(Y(s))ds + \psi(Y(\tau_2)) \right].$$

(b) *For $\varepsilon > 0$ let $\tau^{(\varepsilon)}$ be as in Lemma 7.3a and let β_1, β_2 be finite stopping times, $\beta_1 \leq \beta_2 \leq \tau^{(\varepsilon)}$. Then*

$$E^y[\psi(Y(\beta_1))] = E^y \left[\int_{\beta_1}^{\beta_2} f(Y(t))dt + \psi(Y(\beta_2)) \right].$$

Proof of Corollary 7.4. (a) Define

$$R(t) = \left(Y(t \wedge \tau), \int_0^{t \wedge \tau} f(Y(r))dr \right) \in \mathbb{R}^{k+1}$$

and put

$$H(y, u) = \psi(y) + u, \quad (y, u) \in \mathbb{R}^k \times \mathbb{R}.$$

Then if $t > s$ we have by the Markov property

$$E^y[U(t)|\mathcal{F}_s] = E^y[H(R(t))|\mathcal{F}_s] = E^{R(s)}[H(R(t-s))].$$

Then by (7.1.16) applied to $\beta = (t-s) \wedge \tau$ and $(y, u) = R(s)$

$$E^{R(s)}[H(R(t-s))] = E^y \left[\psi(Y((t-s) \wedge \tau)) + u + \int_0^{(t-s) \wedge \tau} f(Y(r))dr \right]$$

$$= E^y \left[\psi(Y(\beta)) + u + \int_0^\beta f(Y(r))dr \right]$$

$$\leq u + \sup_{\tau \geq \beta} E^y \left[\int_0^{\tau \wedge \beta} f(Y(r))dr + \psi(Y(\tau \wedge \beta)) \right]$$

$$\leq u + \psi(y) = H(R(s)) = U(s).$$

Hence $E^y[U(t)|\mathcal{F}_s] \geq U(s)$ for all $t > s$. This proves that $U(t)$ is a supermartingale. The second statement follows from Doob's optional sampling theorem.

(b) By Lemma 7.3b we have

$$\psi(y) = E^y \left[\int_0^{\beta_1} f(Y(t))dt + \psi(Y(\beta_1)) \right]$$

$$+ E^y \left[\int_{\beta_1}^{\beta_2} f(Y(t))dt + \psi(Y(\beta_2)) - \psi(Y(\beta_1)) \right]$$

$$= \psi(y) + E^y \left[\int_{\beta_1}^{\beta_2} f(Y(t))dt + \psi(Y(\beta_2)) - \psi(Y(\beta_1)) \right],$$

from which (b) follows. $\square$

Proof of Theorem 7.2. Choose $v_n = (\tau_1, \tau_2, \ldots, \tau_n; \zeta_1, \zeta_2, \ldots, \zeta_n)$ with $\tau_n \leq \tau_S$ and set $\tau_{n+1} = \tau_S$. Let $Y(t) = Y^{(v_n)}(t)$. Then by Corollary 7.4a we have

$$E^y[\varphi_{n-j}(Y(\tau_j))] \geq E^y \left[\int_{\tau_j}^{\tau_{j+1}} f(Y(t))dt + \varphi_{n-j}(\check{Y}(\tau_{j+1}^-)) \right]. \qquad (7.1.18)$$

Choosing $\tau = 0$ in (7.1.12) we obtain that

$$\varphi_{n-j} \geq \mathcal{M}\varphi_{n-j-1} \quad \text{if } n-j \geq 1. \qquad (7.1.19)$$

Moreover, from the definition of $\mathcal{M}$ it follows that

$$\mathcal{M}\varphi_{n-j-1}(\check{Y}(\tau_{j+1}^-)) \geq \varphi_{n-j-1}(Y(\tau_{j+1})) + K(\check{Y}(\tau_{j+1}^-), \zeta_{j+1}). \qquad (7.1.20)$$

Combining (7.1.18)–(7.1.20) we get

$$E^y[\varphi_{n-j}(Y(\tau_j))]$$

$$\geq E^y \left[\int_{\tau_j}^{\tau_{j+1}} f(Y(t))dt + \varphi_{n-j-1}(Y(\tau_{j+1})) + K(\check{Y}(\tau_{j+1}^-), \zeta_{j+1}) \right]$$

$$(7.1.21)$$

for $j = 0, 1, \ldots, n-1$, where we have put $\tau_0 = 0$. Summing (7.1.21) from $j = 0$ to $j = n-1$ we get

$$\sum_{j=0}^{n-1} E^y[\varphi_{n-j}(Y(\tau_j)) - \varphi_{n-j-1}(Y(\tau_{j+1}))]$$

$$\geq E^y \left[\int_0^{\tau_n} f(Y(t))dt + \sum_{j=1}^{n} K(\check{Y}(\tau_j^-), \zeta_j) \right]$$

or

$$E^y[\varphi_n(Y(0)) - \varphi_0(Y(\tau_n))] \geq E^y \left[\int_0^{\tau_n} f(Y(t))dt + \sum_{j=1}^n K(\check{Y}(\tau_j^-), \zeta_j) \right].$$
(7.1.22)

Now

$$E^y[\varphi_n(Y(0))] = \varphi_n(y)$$
(7.1.23)

and by the strong Markov property

$$E^y[\varphi_0(Y(\tau_n))] = E^y \left[E^{Y(\tau_n)} \left[\int_0^{\tau_S} f(Y(t))dt + g(Y(\tau_S))\chi_{\{\tau_S < \infty\}} \right] \right]$$

$$= E^y \left[\int_{\tau_n}^{\tau_S} f(Y(t))dt + g(Y(\tau_S))\chi_{\{\tau_S < \infty\}} \right].$$
(7.1.24)

Combining (7.1.22)–(7.1.24) we obtain

$$\varphi_n(y) \geq E^y \left[\int_0^{\tau_S} f(Y(t))dt + g(Y(\tau_S))\chi_{\{\tau_S < \infty\}} + \sum_{j=1}^n K(\check{Y}(\tau_j^-), \zeta_j) \right]$$

$$= J^{(v)}(y).$$
(7.1.25)

Since $v \in \mathcal{V}_n$ was arbitrary we conclude that

$$\varphi_n(y) \geq \Phi_n(y).$$
(7.1.26)

To get the opposite inequality choose $\varepsilon > 0$ and define an increasing sequence of stopping times $0 = \hat{\tau}_0 < \hat{\tau}_1 < \cdots < \hat{\tau}_n$ as follows.

For $j = 1, 2, \ldots, n$ let

$$D_j^{(\varepsilon)} = \{y; \varphi_j(y) > \mathcal{M}\varphi_{j-1}(y) + \varepsilon\}.$$
(7.1.27)

Define

$$\hat{\tau}_1 = \inf\{t > 0; Y^{(0)}(t) \notin D_n^{(\varepsilon)}\},$$
(7.1.28)

where $Y^{(0)}(t) = Y(t)$ is the process without interventions. Then choose $\hat{\zeta}_1 = \bar{\zeta}_1(Y(\hat{\tau}_1^-))$, where $\bar{\zeta}_1 = \bar{\zeta}_1(y) \in \mathcal{Z}$ is ε-optimal for φ_{n-1}, in the sense that

$$\mathcal{M}\varphi_{n-1}(y) \leq \varphi_{n-1}(\Gamma(y, \bar{\zeta}_1)) + K(y, \bar{\zeta}_1) + \varepsilon.$$
(7.1.29)

Inductively, if $0 = \hat{\tau}_0, \ldots, \hat{\tau}_j; \hat{\zeta}_1, \ldots, \hat{\zeta}_j$ have been chosen, where $j \leq n - 1$, we let $Y^{(j)}(t)$ be the process obtained by applying $\hat{v}_j = (\hat{\tau}_1, \ldots, \hat{\tau}_j; \hat{\zeta}_1, \ldots, \hat{\zeta}_j)$ to $Y(t)$. Define

$$\hat{\tau}_{j+1} = \inf\{t > \hat{\tau}_j; Y^{(j)}(t) \notin D^{(\varepsilon)}_{n-j}\} \tag{7.1.30}$$

and choose $\hat{\zeta}_{j+1} = \bar{\zeta}_{j+1}(Y^{(j)}(\hat{\tau}^-_{j+1}))$, where $\bar{\zeta}_{j+1} = \bar{\zeta}_{j+1}(y) \in \mathcal{Z}$ is ε-optimal for φ_{n-j-1}, in the sense that

$$\mathcal{M}\varphi_{n-j-1}(y) \leq \varphi_{n-j-1}(\Gamma(y, \bar{\zeta}_{j+1})) + K(y, \bar{\zeta}_{j+1}) + \varepsilon. \tag{7.1.31}$$

Finally define

$$\hat{v} = (\hat{\tau}_1, \ldots, \hat{\tau}_n; \hat{\zeta}_1, \ldots, \hat{\zeta}_n) \in \mathcal{V}_n.$$

Now apply the argument (7.1.18)–(7.1.25) to $\hat{v}$.

By Corollary 7.4b we have

$$E^y[\varphi_{n-j}(Y(\hat{\tau}_j))] \leq E^y\left[\int_{\hat{\tau}_j}^{\hat{\tau}_{j+1}} f(Y(t))dt + \varphi_{n-j}(\check{Y}(\hat{\tau}^-_{j+1}))\right]. \tag{7.1.32}$$

Since $\check{Y}(\hat{\tau}^-_{j+1}) \notin D^{(\varepsilon)}_{n-j}$ we deduce that

$$\varphi_{n-j}(\check{Y}(\hat{\tau}^-_{j+1})) \leq \mathcal{M}\varphi_{n-j-1}(\check{Y}(\hat{\tau}^-_{j+1})) + \varepsilon \tag{7.1.33}$$

and by (7.1.31), we get, with $Y = Y^{(\hat{v})}$,

$$\mathcal{M}\varphi_{n-j-1}(\check{Y}(\hat{\tau}^-_{j+1})) \leq \varphi_{n-j-1}(Y(\hat{\tau}_{j+1})) + K(\check{Y}(\hat{\tau}^-_{j+1}), \hat{\zeta}_{j+1}) + \varepsilon. \tag{7.1.34}$$

Combining (7.1.32)–(7.1.34) we obtain

$$E^y[\varphi_{n-j}(Y(\hat{\tau}))]$$
$$\leq E^y\left[\int_{\hat{\tau}_j}^{\hat{\tau}_{j+1}} f(Y(t))dt + \varphi_{n-j-1}(Y(\hat{\tau}_{j+1})) + K(\check{Y}(\hat{\tau}^-_{j+1}), \hat{\zeta}_{j+1})\right] + 2\varepsilon. \tag{7.1.35}$$

Now sum (7.1.35) from $j = 0$ to $j = n - 1$. The result is

$$\varphi_n(y) \leq E^y\left[\int_0^{\hat{\tau}_n} f(Y(t))dt + \varphi_0(Y(\hat{\tau}_n)) + \sum_{j=1}^n K(\check{Y}(\hat{\tau}^-_j), \hat{\zeta}_j)\right] + 2n\varepsilon.$$

Therefore by (7.1.24),

$$\varphi_n(y) \leq E^y\left[\int_0^{\hat{\tau}_s} f(Y(t))dt + g(Y(\hat{\tau}_s))\chi_{\{\hat{\tau}_s < \infty\}} + \sum_{j=1}^n K(\check{Y}(\hat{\tau}^-_j), \hat{\zeta}_j)\right] + 2n\varepsilon$$

$$= J^{(\hat{v})}(y) + 2n\varepsilon.$$

Since ε was arbitrary we deduce that

$$\varphi_n(y) \le \sup\{J^{(v)}(y); v \in \mathcal{V}_n\} = \Phi_n(y).$$

Combined with (7.1.26) this proves Theorem 7.2. □

Remark 7.5. Note that the proof of Theorem 7.2 actually also gives a $2n\varepsilon$-optimal impulse control $\hat{v} = (\hat{\tau}_1, \ldots, \hat{\tau}_n; \hat{\zeta}_1, \ldots, \hat{\zeta}_n)$. It is defined inductively by (7.1.27)–(7.1.31).

In particular, if it is possible to choose $\hat{\zeta}_j = \zeta_j^*$ to be optimal (i.e., (7.1.31) holds with $\varepsilon = 0$), then $v^* = (\hat{\tau}_1, \ldots, \hat{\tau}_n; \zeta_1^*, \ldots, \zeta_n^*)$ will be an *optimal impulse* control for Φ_n given by the following procedure.

For $j = 1, 2, \ldots, n$ let

$$D_j = \{y; \varphi_j(y) > M\varphi_{j-1}(y)\}. \tag{7.1.36}$$

Define

$$\hat{\tau}_1 = \inf\{t > 0; Y^{(0)}(t) \notin D_n\} \tag{7.1.37}$$

and

$$\hat{\zeta}_1 = \bar{\zeta}_1(Y^{(0)}(\hat{\tau}_1^-)), \tag{7.1.38}$$

where $\bar{\zeta}_1 = \bar{\zeta}_1(y)$ is a Borel measurable function such that

$$M\varphi_{n-1}(y) = \varphi_{n-1}(\Gamma(y, \bar{\zeta}_1)) + K(y, \bar{\zeta}_1), \quad y \in \mathcal{S}. \tag{7.1.39}$$

Then if $(\hat{\tau}_1, \ldots, \hat{\tau}_j; \hat{\zeta}_1, \ldots, \hat{\zeta}_j)$ is defined, put

$$\hat{\tau}_{j+1} = \inf\{t > \hat{\tau}_j; Y^{(j)}(t) \notin D_{n-j}\} \tag{7.1.40}$$

and

$$\hat{\zeta}_{j+1} = \bar{\zeta}_{j+1}(Y^{(j)}(\hat{\tau}_{j+1}^-)), \tag{7.1.41}$$

where $\bar{\zeta}_{j+1} = \bar{\zeta}_{j+1}(y)$ is a Borel measurable function such that

$$M\varphi_{n-(j+1)}(y) = \varphi_{n-(j+1)}(\Gamma(y, \bar{\zeta}_{j+1})) + K(y, \bar{\zeta}_{j+1}), \quad y \in \mathcal{S}, j+1 \le n. \tag{7.1.42}$$

As before $Y^{(j)}(t)$ denotes the result of applying the impulse control $\hat{v}_j = (\hat{\tau}_1, \ldots, \hat{\tau}_j; \hat{\zeta}_1, \ldots, \hat{\zeta}_j)$ to Y.

Corollary 7.6. *Assume that* $g \ge 0$ *and that* $f, g, M\varphi_n \in \mathcal{P}(\mathbb{R}^k)$ *for* $n = 0, 1, 2, \ldots$, *where* φ_n *is as defined in (7.1.11) and (7.1.12). Then*

$$\varphi_n(y) \uparrow \Phi(y) \text{ as } n \to \infty, \quad \text{for all } y.$$

Corollary 7.7. *Suppose* $g \ge 0$ *and* $f, g, M\varphi_n \in \mathcal{P}(\mathbb{R}^k)$ *for* $n = 0, 1, 2, \ldots$
Then Φ *is a solution of the following nonlinear optimal stopping problem*

$$\Phi(y) = \sup_{\tau \in \mathcal{T}} E^y \left[\int_0^\tau f(Y(t))dt + M\Phi(Y(\tau))\chi_{\{\tau < \infty\}} \right]. \tag{7.1.43}$$

Proof. Let $\{\varphi_n\}_{n=0}^{\infty}$ be as in (7.1.11) and (7.1.12). Then $\varphi_n \uparrow \Phi$ as $n \to \infty$, by Corollary 7.6. Therefore

$$\varphi_n(y) \leq \sup_{\tau \in \mathcal{T}} E^y \left[\int_0^{\tau} f(Y(t))dt + \mathcal{M}\Phi(Y(\tau))\chi_{\{\tau<\infty\}} \right]$$

for all n and hence

$$\Phi(y) \leq \sup_{\tau \in \mathcal{T}} E^y \left[\int_0^{\tau} f(Y(t))dt + \mathcal{M}\Phi(Y(\tau))\chi_{\{\tau<\infty\}} \right].$$

To get the opposite inequality, choose $\varepsilon > 0$ and let $\hat{\tau} \in \mathcal{T}$ be a stopping time such that

$$E^y \left[\int_0^{\hat{\tau}} f(Y(t))dt + \mathcal{M}\Phi(Y(\hat{\tau}))\chi_{\{\hat{\tau}<\infty\}} \right]$$

$$\geq \sup_{\tau \in \mathcal{T}} E^y \left[\int_0^{\tau} f(Y(t))dt + \mathcal{M}\Phi(Y(\tau))\chi_{\{\tau<\infty\}} \right] - \varepsilon.$$

Then by monotone convergence

$$\Phi(y) = \lim_{n\to\infty} \varphi_n(y) \geq \lim_{n\to\infty} E^y \left[\int_0^{\hat{\tau}} f(Y(t))dt + \mathcal{M}\varphi_{n-1}(Y(\hat{\tau}))\chi_{\{\hat{\tau}<\infty\}} \right]$$

$$= E^y \left[\int_0^{\hat{\tau}} f(Y(t))dt + \mathcal{M}\Phi(Y(\hat{\tau}))\chi_{\{\hat{\tau}<\infty\}} \right]$$

$$\geq \sup_{\tau \in \mathcal{T}} E^y \left[\int_0^{\tau} f(Y(t))dt + \mathcal{M}\Phi(Y(\tau))\chi_{\{\tau<\infty\}} \right] - \varepsilon.$$

$$(7.1.44)$$

Combining the results above we get the following general picture of the optimal impulse control $(\hat{\tau}_1, \ldots, \hat{\tau}_n; \hat{\zeta}_1, \ldots, \hat{\zeta}_n)$ for Φ_n (see (7.1.11), (7.1.12), and (7.1.36)–(7.1.42)). Define

$$D = \{y; \Phi(y) > \mathcal{M}\Phi(y)\} \tag{7.1.45}$$

and

$$D_n = \{y; \varphi_n(y) > \mathcal{M}\varphi_{n-1}(y)\}, \quad n = 1, 2, \ldots \tag{7.1.46}$$

Make the first intervention the first time $t = \hat{\tau}_1$ that $Y(t) \notin D_n$. Then give the system the impulse $\hat{\zeta}_1$ according to (7.1.38). Now we have only $n - 1$ interventions left, so we wait until $Y(t)$ exits from the set D_{n-1} before making the next intervention, and so on. The last intervention time $\hat{\tau}_n$ is the first time after $\hat{\tau}_{n-1}$ that $Y(t) \notin D_1$. See Fig. 7.1.

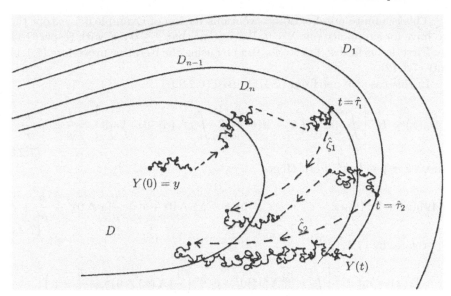

Fig. 7.1. Optimal impulse control for Φ_n

7.2 Examples

Example 7.8. Consider the impulse control problems

$$\Phi_n(s,x) = \inf_{v \in \mathcal{V}_n} J^{(v)}(s,x) = J^{(v_n^*)}(s,x), \quad n = 1,2,\ldots, \qquad (7.2.1)$$

$$\Phi(s,x) = \inf_{v \in \mathcal{V}} J^{(v)}(s,x) = J^{(v^*)}(s,x), \qquad (7.2.2)$$

where

$$J^{(v)}(s,x) = E^x \left[\int_0^\infty e^{-\rho(s+t)}(X^{(v)}(t))^2 dt + c \sum_j e^{-\rho(s+\tau_j)} \right],$$

$c > 0$ and $\rho > 0$ are constants, and

$$X^{(v)}(t) = x + B(t) + \int_0^t \int_{\mathbb{R}} z \tilde{N}(ds,dz) - \sum_j \zeta_j \chi_{\{\tau_j \le t\}}, \quad B(0) = 0,$$

when

$$v = (\tau_1, \tau_2, \ldots, \tau_n; \zeta_1, \zeta_2, \ldots, \zeta_n) \in \mathcal{V}_n$$

or

$$v = (\tau_1, \tau_2, \ldots; \zeta_1, \zeta_2, \ldots) \in \mathcal{V}, \quad \zeta_j > 0.$$

Put

$$X(t) = x + B(t) + \int_0^t \int_{\mathbb{R}} z \tilde{N}(ds,dz), \quad t \ge 0.$$

This is a finite number of interventions version of Example 6.5, except that we have for simplicity put $\lambda = 0$. We assume that $z \leq 0$ a.s. with respect to ν.

First, let us find $\Psi_n(x) := \Phi_n(0, x)$ by using the iterative procedure (7.1.11) and (7.1.12).

In this case we get from (7.1.11) (see (6.2.29))

$$\Psi_0(x) = E^x \left[\int_0^\infty e^{-\rho t} \left(x + B(t) + \int_0^t \int_{\mathbb{R}} z \tilde{N}(\mathrm{d}s, \mathrm{d}z) \right)^2 \mathrm{d}t \right] = \frac{1}{\rho} x^2 + \frac{b}{\rho^2},$$

(7.2.3)

where $b = 1 + \int_{\mathbb{R}} z^2 \nu(\mathrm{d}z)$. Hence

$$\mathcal{M}\Psi_0(x) = \inf \{ \Psi_0(x - \zeta) + c; \zeta > 0 \} = \Psi_0(x \wedge 0) + c = \frac{1}{\rho} (x \wedge 0)^2 + \frac{b}{\rho^2} + c.$$

(7.2.4)

Therefore, by (7.1.12)

$$\Psi_1(x) = \inf_{\tau \geq 0} E^x \left[\int_0^\tau e^{-\rho t} X^2(t) \mathrm{d}t + e^{-\rho \tau} \left(\frac{1}{\rho} (X(\tau) \wedge 0)^2 + \frac{b}{\rho^2} + c \right) \right].$$

(7.2.5)

To solve this optimal stopping problem we consider the three basic associated variational inequalities

$$-\rho \psi_1(x) + \frac{1}{2} \psi_1''(x) + \int_{\mathbb{R}} \{ \psi_1(x + z) - \psi_1(x) - z \psi_1'(x) \} \nu(\mathrm{d}z) + x^2 \geq 0 \quad \text{for all } x,$$

(7.2.6)

$$\psi_1(x) \leq \frac{1}{\rho} (x \wedge 0)^2 + \frac{b}{\rho^2} + c \quad \text{for all } x,$$

(7.2.7)

$$-\rho \psi_1(x) + \frac{1}{2} \psi_1''(x) + \int_{\mathbb{R}} \{ \psi_1(x + z) - \psi_1(x) - z \psi_1'(x) \} \nu(\mathrm{d}z) + x^2 = 0$$

$$\text{on } D_1 := \left\{ x; \psi_1(x) < \frac{1}{\rho} (x \wedge 0)^2 + \frac{b}{\rho^2} + c \right\}.$$

If we guess that D_1 has the form

$$D_1 = (-\infty, x_1^*) \quad \text{for some } x_1^* > 0$$

then we are led to the following candidate for ψ_1 (see Example 6.5)

$$\psi_1(x) = \begin{cases} \frac{1}{\rho} x^2 + \frac{b}{\rho^2} - a_1 \exp(rx), & x < x_1^*, \\ \frac{b}{\rho^2} + c, & x \geq x_1^*, \end{cases}$$

(7.2.8)

where a_1 is a constant to determine and $r = r_1 > 0$ is a root of the equation

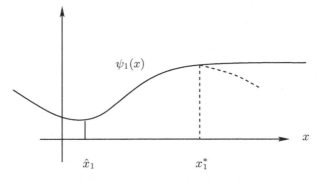

Fig. 7.2. The graph of $\psi_1(x)$

$$K(r) := \int_{\mathbb{R}} \{e^{rz} - 1 - rz\}\nu(dz) + \frac{1}{2}r^2 - \rho = 0.$$

See Fig. 7.2.

If we require ψ_1 to be continuous and differentiable at $x = x_1^*$, we get the following two equations to determine x_1^* and a_1:

$$\frac{1}{\rho}(x_1^*)^2 + \frac{b}{\rho^2} - a_1 \exp(rx_1^*) = \frac{b}{\rho^2} + c, \qquad (7.2.9)$$

$$\frac{2}{\rho}x_1^* - a_1 r \exp(rx_1^*) = 0. \qquad (7.2.10)$$

Combining these two equations we get

$$x_1^* = \frac{1}{r}\left(1 + \sqrt{1 + c\rho r^2}\right) \qquad (7.2.11)$$

and

$$a_1 = \frac{2x_1^*}{\rho r} \exp(-rx_1^*). \qquad (7.2.12)$$

If we choose these values of x_1^* and a_1, we can now verify that, under some additional assumptions on ν, the candidate ψ_1 given by (7.2.8) satisfies all the conditions of the verification theorem and we conclude that

$$\psi_1 = \Psi_1.$$

We now repeat the procedure to find Ψ_2.

First note that

$$\mathcal{M}\Psi_1(x) = \inf\{\Psi_1(x - \zeta) + c; \zeta > 0\} = \Psi_1(x \wedge \hat{x}_1) + c$$
$$= \frac{1}{\rho}(x \wedge \hat{x}_1)^2 + \frac{b}{\rho^2} + c - a_1 \exp(r(x \wedge \hat{x}_1)), \qquad (7.2.13)$$

where
$$\hat{x}_1 \in \operatorname{Argmin} \Psi_1(x).$$

Hence

$$\Psi_2(x) = \inf_{\tau \geq 0} E^x \Bigg[\int_0^\tau e^{-\rho t} X^2(t) dt$$

$$+ e^{-\rho\tau} \left(\frac{1}{\rho}(X(\tau) \wedge \hat{x}_1)^2 + \frac{b}{\rho^2} + c - a_1 \exp(r(X(\tau) \wedge \hat{x}_1)) \right) \Bigg].$$

The same procedure as above leads us to the candidate

$$\psi_2(x) = \begin{cases} \dfrac{1}{\rho}x^2 + \dfrac{b}{\rho^2} - a_2 \exp(rx), & x < x_2^*, \\[2mm] \dfrac{1}{\rho}\hat{x}_1^2 + \dfrac{b}{\rho^2} + c - a_1 \exp(r\hat{x}_1), & x \geq x_2^*, \end{cases} \tag{7.2.14}$$

where x_2^* and a_2 are given by

$$x_2^* = \frac{1}{r}\left(1 + \sqrt{1 + r^2(\hat{x}_1^2 + \rho(c - a_1 \exp(r\hat{x}_1)))} \right), \tag{7.2.15}$$

$$a_2 = \frac{2x_2^*}{\rho r} \exp(-rx_2^*). \tag{7.2.16}$$

If we choose these values of x_2^* and $a_2 > 0$ we can verify that the candidate ψ_2 given by (7.2.14) coincides with Ψ_2. Note that $x_2^* < x_1^*$ and $a_2 > a_1$.

Continuing this inductively we get a sequence of functions ψ_n of the form

$$\psi_n(x) = \begin{cases} \dfrac{1}{\rho}x^2 + \dfrac{b}{\rho^2} - a_n \exp(rx), & x < x_n^*, \\[2mm] \dfrac{1}{\rho}\hat{x}_{n-1}^2 + \dfrac{b}{\rho^2} + c - a_{n-1} \exp(r\hat{x}_{n-1}), & x \geq x_n^*, \end{cases} \tag{7.2.17}$$

where x_n^* and a_n are given by

$$x_n^* = \frac{1}{r}\left(1 + \sqrt{1 + r^2(\hat{x}_{n-1}^2 + \rho(c - a_{n-1} \exp(r\hat{x}_{n-1})))} \right) \tag{7.2.18}$$

with
$$\hat{x}_{n-1} \in \operatorname{Argmin} \Psi_{n-1}(x),$$

and
$$a_n = \frac{2x_n^*}{\rho r} \exp(-rx_n^*). \tag{7.2.19}$$

We find that $x_n^* < x_{n-1}^*$ and $a_n > a_{n-1}$. Using the verification theorem we conclude that $\psi_n = \Psi_n$. In the limiting case when there is no bound on the number of interventions the corresponding value function will be

$$\Psi(x) = \begin{cases} \dfrac{1}{\rho}x^2 + \dfrac{b}{\rho^2} - a\exp(rx), & x < x^*, \\[3mm] \dfrac{1}{\rho}\hat{x}^2 + \dfrac{b}{\rho^2} + c - a\exp(r\hat{x}), & x \ge x^*, \end{cases} \qquad (7.2.20)$$

where x^* and a solve the coupled system of equations

$$x^* = \frac{1}{r}\left(1 + \sqrt{1 + r^2(\hat{x}^2 + \rho(c - a\exp(r\hat{x})))}\right), \qquad (7.2.21)$$

$$a = \frac{2x^*}{\rho r}\exp(-rx^*), \qquad (7.2.22)$$

where

$$\hat{x} \in \operatorname{Argmin}\Psi(x).$$

It is easy to see that this system has a unique solution $x^* > 0$, $a > 0$, and that $x^* < x_n^*$ and $a > a_n$ for all n. The situation in the case $n = 3$ is shown in Fig. 7.3. Note that with only three interventions allowed the optimal strategy is first to wait until the first time τ_1^* when $X(t) \ge x_3^*$, then move $X(t)$ down to $\hat{x}_2$, next wait until the first time $\tau_2^* > \tau_1^*$ when $X(t) \ge x_2^*$, then move $X(t)$ down to $\hat{x}_1$ and finally wait until the first time $t = \tau_3^* > \tau_2^*$ when $X(t) \ge x_1^*$ before making the last intervention, which is moving $X(t)$ down to $\hat{x}_0 = 0$.

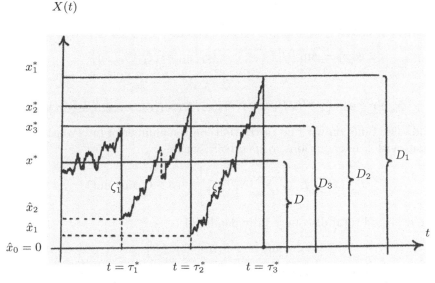

Fig. 7.3. The optimal impulse control for Ψ_3 (Example 7.8)

7.3 Exercises

Exercise* 7.1 (Optimal Forest Management Revisited). Using the notation of Exercise 6.3 let

$$\Phi(x) = \sup \left\{ J^{(v)}(x); v = (\tau_1, \tau_2, \ldots) \right\}$$

be the value function when there are no restrictions on the number of interventions. For $n = 1, 2, \ldots$ let

$$\Phi_n(x) = \sup \left\{ J^{(v)}(x); v = (\tau_1, \tau_2, \ldots, \tau_k); k \le n \right\}$$

be the value function when up to n interventions are allowed. Use Theorem 7.2 to find Φ_1 and Φ_2.

Exercise* 7.2 (Optimal Stream of Dividends with Transaction Costs from a Geometric Lévy Process). This is an addition to Exercise 6.2. Suppose that we at times $0 \le \tau_1 \le \tau_2 \le \cdots$ decide to take out dividends of sizes $\zeta_1, \zeta_2, \ldots \in (0, \infty)$ from an economic quantity growing like a geometric Lévy process. If we let $X^{(v)}(t)$ denote the size at time t of this quantity when the dividend policy $v = (\tau_1, \tau_2, \ldots; \zeta_1, \zeta_2, \ldots)$ is applied, we assume that $X^{(v)}(t)$ is described by (see (6.2.4))

$$\mathrm{d}X^{(v)}(t) =$$

$$\mu X^{(v)}(t)\mathrm{d}t + \sigma X^{(v)}(t)\mathrm{d}B(t) + \theta X^{(v)}(t^-) \int_{\mathbb{R}} z\tilde{N}(\mathrm{d}t, \mathrm{d}z), \quad \tau_i \le t < \tau_{i+1},$$

$$X^{(v)}(\tau_{i+1}) = \check{X}^{(v)}(\tau_{i+1}^-) - (1 + \lambda)\zeta_{i+1} - c, \quad i = 0, 1, 2, \ldots \quad \text{(see (6.1.5))},$$

where $\mu, \sigma \ne 0$, $\theta, \lambda \ge 0$, and $c > 0$ are constants, $-1 \le \theta z \le 0$, a.s. (ν). Let

$$\Phi(x) = \sup \left\{ J^{(v)}(x); v = (\tau_1, \tau_2, \ldots; \zeta_1, \zeta_2, \ldots) \right\}$$

and

$$\Phi_n(x) = \sup \left\{ J^{(v)}(x); v = (\tau_1, \tau_2, \ldots, \tau_k; \zeta_1, \zeta_2, \ldots, \zeta_k); k \le n \right\}$$

be the value function with no restrictions on the number of interventions and with at most n interventions, respectively, where

$$J^{(v)}(x) = E^x \left[\sum_{\tau_k < \tau_S} \mathrm{e}^{-\rho\tau_k} \zeta_k \right] \quad (\rho > 0 \text{ constant})$$

is the expected total discounted dividend and

$$\tau_S = \inf \left\{ t > 0; X^{(v)}(t) \le 0 \right\}$$

is the time of bankruptcy. Show that

$$\Phi(x) = \Phi_n(x) = \Phi_1(x) \quad \text{for all } n.$$

Thus in this case we achieve the optimal result with just one intervention.

8

Combined Stochastic Control and Impulse Control of Jump Diffusions

8.1 A Verification Theorem

Consider the general situation in Chap. 6, except that now we assume that we, in addition, are free at any state $y \in \mathbb{R}^k$ to choose a Markov control $u(y) \in U$, where U is a given closed convex set in $\mathbb{R}^p$. Let $\mathcal{U}$ be a given family of such Markov controls u. If, as before, $v = (\tau_1, \tau_2, \ldots; \zeta_1, \zeta_2, \ldots) \in \mathcal{V}$ denotes a given impulse control we call $w := (u, v)$ a *combined control*. If $w = (u, v)$ is applied, we assume that the corresponding state $Y(t) = Y^{(w)}(t)$ at time t is given by (see (3.1.1) and (6.1.2)–(6.1.5))

$$Y(0^-) = y,$$
$$dY(t) = b(Y(t), u(t))dt + \sigma(Y(t), u(t))dB(t)$$
$$+ \int_{\mathbb{R}^\ell} \gamma(Y(t^-), u(t^-), z)\tilde{N}(dt, dz), \quad \tau_j < t < \tau_{j+1}, \qquad (8.1.1)$$
$$Y(\tau_{j+1}) = \Gamma(\check{Y}(\tau_{j+1}^-), \zeta_{j+1}), \quad j = 0, 1, 2, \ldots \qquad (8.1.2)$$

where $u(t) = u(Y(t))$ and $b : \mathbb{R}^k \times U \to \mathbb{R}^k$, $\sigma : \mathbb{R}^k \times U \to \mathbb{R}^{k \times m}$, and $\gamma : \mathbb{R}^k \times U \times \mathbb{R}^\ell \to \mathbb{R}^{k \times \ell}$ are given continuous functions, $\tau_0 = 0$.

As before we let our "universe" S be a fixed Borel set in $\mathbb{R}^k$ such that $S \subset \bar{S}^0$ and we define

$$\tau^* = \lim_{R \to \infty} \inf\{t > 0; |Y^{(w)}(t)| \geq R\} \leq \infty \qquad (8.1.3)$$

and

$$\tau_S = \inf\{t \in (0, \tau^*(\omega)); Y^{(w)}(t, \omega) \notin S\}. \qquad (8.1.4)$$

If $Y^{(w)}(t, \omega) \in S$ for all $t < \tau^*$ we put $\tau_S = \tau^*$.

We assume that we are given a set $\mathcal{W} = \mathcal{U} \times \mathcal{V}$ of *admissible* combined controls $w = (u, v)$ which is included in the set of combined controls $w = (u, v)$ such that a unique strong solution $Y^{(w)}(t)$ of (8.1.1) and (8.1.2) exists and

$$\tau^* = \infty \quad \text{and} \quad \lim_{j \to \infty} \tau_j = \tau_S \text{ a.s. } Q^y \text{ for all } y \in \mathbb{R}^k. \qquad (8.1.5)$$

Define the *performance* or *total expected profit/utility* of $w = (u, v) \in \mathcal{W}$, $v = (\tau_1, \tau_2, \ldots; \zeta_1, \zeta_2, \ldots)$, by

$$J^{(w)}(y) = E^y \left[\int_0^{\tau_S} f(Y^{(w)}(t), u(t)) dt + g(Y^{(w)}(\tau_S)) \mathcal{X}_{\{\tau_S < \infty\}} \right.$$

$$\left. + \sum_{\tau_j \leq \tau_S} K(\check{Y}^{(w)}(\tau_j^-), \zeta_j) \right], \qquad (8.1.6)$$

where $f : \mathcal{S} \times U \to \mathbb{R}$, $g : \mathbb{R}^k \to \mathbb{R}$, and $K : \bar{\mathcal{S}} \times \mathcal{Z} \to \mathbb{R}$ are given functions satisfying the conditions similar to (6.1.11)–(6.1.13).

The *combined stochastic control and impulse control problem* is the following.

Find the value function $\Phi(y)$ and an optimal control $w^* = (u^*, v^*) \in \mathcal{W}$ such that

$$\Phi(y) = \sup\{J^{(w)}(y); w \in \mathcal{W}\} = J^{(w^*)}(y). \qquad (8.1.7)$$

We now state a verification theorem for this problem. It is a combination of the HJB equation of control theory and the QVI for impulse control.

Define

$$L^\alpha h(y) = \sum_{i=1}^k b_i(y, \alpha) \frac{\partial h}{\partial y_i} + \frac{1}{2} \sum_{i,j=1}^k (\sigma \sigma^T)_{ij}(y, \alpha) \frac{\partial^2 h}{\partial y_i \partial y_j}$$

$$+ \int_{\mathbb{R}} \sum_{j=1}^\ell \{h(y + \gamma^{(j)}(y, \alpha, z)) - h(y) - \nabla h(y).\gamma^{(j)}(y, \alpha, z)\} \nu_j(dz_j)$$

$$(8.1.8)$$

for each $\alpha \in U$ and for each twice differentiable function h. This is the generator of $Y^{(w)}(t)$ if we apply the constant control α and no (impulse) interventions.

As in Chap. 6 we let

$$\mathcal{M}h(y) = \sup\{h(\Gamma(y, \zeta)) + K(y, \zeta); \zeta \in \mathcal{Z} \quad \text{and} \quad \Gamma(y, \zeta) \in \mathcal{S}\} \qquad (8.1.9)$$

be the intervention operator.

Then the verification theorem is the following (compare with Theorems 3.1 and 6.2).

Theorem 8.1 (HJBQVI Verification Theorem for Combined Stochastic and Impulse Control).

(a) *Suppose we can find a function* $\phi : \bar{\mathcal{S}} \to \mathbb{R}$ *such that*
   (i) $\phi \in C^1(\mathcal{S}^0) \cap C(\bar{\mathcal{S}})$
   (ii) $\phi \geq \mathcal{M}\phi$ *on* $\mathcal{S}$

Define

$$D = \{y \in \mathcal{S}; \phi(y) > M\phi(y)\} \quad \text{(the continuation region)}. \quad (8.1.10)$$

Suppose that $Y^{(w)}(t)$ spends 0 time on ∂D a.s., i.e.,

(iii) $E^y \left[\displaystyle\int_0^{\tau_S} \mathcal{X}_{\partial D}(Y^{(w)}(t))dt \right] = 0$ *for all $y \in \mathcal{S}$, $w \in \mathcal{W}$*

and suppose that
(iv) *∂D is a Lipschitz surface*
(v) *$\phi \in C^2(\mathcal{S}^0 \setminus \partial D)$ and the second-order derivatives of ϕ are locally bounded near ∂D*
(vi) *$L^\alpha \phi(y) + f(y, \alpha) \leq 0$ for all $\alpha \in U$, $y \in \mathcal{S}^0 \setminus \partial D$*
(vii) *$Y^{(w)}(\tau_S) \in \partial \mathcal{S}$ a.s. on $\{\tau_S < \infty\}$ and*
$$\phi(Y^{(w)}(t)) \to g(Y^{(w)}(\tau_S)) \cdot \mathcal{X}_{\{\tau_S < \infty\}} \text{ as } t \to \tau_S^- \text{ a.s. } Q^y, \text{ for all } y \in \mathcal{S},$$
$w \in \mathcal{W}$
(viii) *The family $\{\phi^-(Y^{(w)}(\tau)); \tau \in \mathcal{T}\}$ is uniformly Q^y-integrable for all $y \in \mathcal{S}$, $w \in \mathcal{W}$*
(ix) $E^y \left[|\phi(Y(\tau))| + \displaystyle\int_0^{\tau_S} |A\phi(Y(t))|dt \right] < \infty$ *for all $\tau \in \mathcal{T}$, $w \in \mathcal{W}$, $y \in \mathcal{S}$.*

Then

$$\phi(y) \geq \Phi(y) \quad \text{for all } y \in \mathcal{S}.$$

(b) *Suppose in addition that*
(x) *There exists a function $\hat{u} : \mathcal{S} \to \mathbb{R}$ such that*

$$L^{\hat{u}(y)}\phi(y) + f(y, \hat{u}(y)) = 0 \quad \text{for all } y \in D$$

and
(xi)
$$\bar{\zeta}(y) \in \text{Argmax}\{\phi(\Gamma(y, \cdot)) + K(y, \cdot)\}$$

exists for all $y \in \mathcal{S}$ and $\bar{\zeta}(\cdot)$ is a Borel measurable selection.
Define an impulse control $\hat{v} = (\hat{\tau}_1, \hat{\tau}_2, \ldots; \hat{\zeta}_1, \hat{\zeta}_2, \ldots)$ as follows.
Put $\hat{\tau}_0 = 0$ and inductively

$$\hat{\tau}_{k+1} = \inf\{t > \hat{\tau}_k; Y^{(k)}(t) \notin D\},$$

$$\hat{\zeta}_{k+1} = \bar{\zeta}(Y^{(k)}(\hat{\tau}_{k+1}^-)), \quad k = 0, 1, \ldots,$$

where $Y^{(k)}(t)$ is the result of applying the combined control

$$\widehat{w}_k := (\hat{u}, (\hat{\tau}_1, \ldots, \hat{\tau}_k; \hat{\zeta}_1, \ldots, \hat{\zeta})).$$

Put $\widehat{w} = (\hat{u}, \hat{v})$. Suppose $\widehat{w} \in \mathcal{W}$ and that
(xii) *$\{\phi(Y^{(\widehat{w})}(\tau)); \tau \in \mathcal{T}, \tau \leq \tau_D\}$ is Q^y-uniformly integrable for all $y \in \mathcal{S}$.*

Then

$$\phi(y) = \Phi(y) \quad \text{for all } y \in \mathcal{S}$$

and

$$\widehat{w} \in \mathcal{W} \quad \text{is an optimal combined control.}$$

Proof. The proof is similar to the proof of Theorem 6.2 and is omitted. □

8.2 Examples

Example 8.2 (Optimal Combined Control of the Exchange Rate).
This example is a simplification of a model studied in [MØ].

Suppose a government has two means of influencing the foreign exchange rate of its own currency:

1. At all times the government can choose the domestic interest rate r.
2. At times selected by the government it can intervene in the foreign exchange market by buying or selling large amounts of foreign currency.

Let $r(t)$ denote the interest rate chosen and let $\tau_1, \tau_2, \ldots$ be the (stopping) times when it is decided to intervene, with corresponding amounts $\zeta_1, \zeta_2, \ldots$ If $\zeta > 0$ the government buys foreign currency, if $\zeta < 0$ it sells. Let $v = (\tau_1, \tau_2, \ldots; \zeta_1, \zeta_2, \ldots)$ be corresponding impulse control.

If the combined control $w = (r, v)$ is applied, we assume that the corresponding exchange rate $X(t)$ (measured in the number of domestic monetary units it takes to buy one average foreign monetary unit) is given by

$$X(t) = x - \int_0^t F(r(s) - \bar{r}(s))\mathrm{d}s + \sigma B(t) + \sum_{j:\tau_j \leq t} \gamma(\zeta_j), \quad t \geq 0, \qquad (8.2.1)$$

where $\sigma > 0$ is a constant, $\bar{r}(s)$ is the (average) foreign interest rate, and $F : \mathbb{R} \to \mathbb{R}$ and $\gamma : \mathbb{R} \to \mathbb{R}$ are known functions which give the effects on the exchange rate by the interest rate differential $r(s) - \bar{r}(s)$ and the amount ζ_j, respectively.

The total expected cost of applying the combined control $w = (r, v)$ is assumed to be of the form

$$J^{(w)}(s, x) = E^x \left[\int_s^T e^{-\rho t} \{M(X(t) - m) + N(r(t) - \bar{r}(t))\}\mathrm{d}t + \sum_{j:\tau_j \leq T} L(\zeta_j) e^{-\rho \tau_j} \right],$$
$$(8.2.2)$$

where $M(X(t) - m)$ and $N(r(t) - \bar{r}(t))$ give the costs incurred by the difference $X(t) - m$ between $X(t)$ and an optimal value m, and by the difference $r(t) - \bar{r}(t)$ between the domestic and the average foreign interest rate $\bar{r}(t)$, respectively. The cost of buying/selling the amount ζ_j is $L(\zeta_j)$ and $\rho > 0$ is a constant discounting exponent.

The problem is to find $\Phi(s, x)$ and $w^* = (r^*, v^*)$ such that

$$\Phi(s, x) = \inf_w J^{(w)}(s, x) = J^{(w^*)}(s, x). \qquad (8.2.3)$$

Since this is a minimum problem, the corresponding HJBQVIs of Theorem 8.1 are changed to minima also and they get the form

$$\min \left(\inf_{r \in \mathbb{R}} \left\{ e^{-\rho s} \left(M(x - m) + N(r - \bar{r}(s)) \right) + \frac{\partial \phi}{\partial s} - F(r - \bar{r}(s)) \frac{\partial \phi}{\partial x} \right. \right.$$
$$\left. \left. + \frac{1}{2} \sigma^2 \frac{\partial^2 \phi}{\partial x^2} \right\}, \mathcal{M}\phi(s, x) - \phi(s, x) \right) = 0, \qquad (8.2.4)$$

where

$$\mathcal{M}\phi(s, x) = \inf_{\zeta \in \mathbb{R}} \{ \phi(s, x + \gamma(\zeta)) + e^{-\rho s} L(\zeta) \}. \qquad (8.2.5)$$

In general this is difficult to solve for ϕ, even for simple choices of the functions M, N, and F. A detailed discussion on a special case can be found in [MØ].

Example 8.3 (Optimal Consumption and Portfolio with Both Fixed and Proportional Transaction Costs (1)). This application is studied in [ØS].

Suppose there are two investment possibilities, say a bank account and a stock. Let $X_1(t)$ and $X_2(t)$ denote the amount of money invested in these two assets, respectively, at time t. In the absence of consumption and transactions suppose that

$$dX_1(t) = rX_1(t)dt \qquad (8.2.6)$$

and

$$dX_2(t) = X_2(t^-) \left(\mu\, dt + \sigma\, dB(t) + \theta \int_{\mathbb{R}} z \tilde{N}(dt, dz) \right), \qquad (8.2.7)$$

where r, μ, θ, and $\sigma \neq 0$ are constants and

$$\mu > r > 0, \quad \theta z \geq -1 \text{ a.s. } \nu. \qquad (8.2.8)$$

Suppose that at any time t the investor is free to choose a *consumption rate* $u(t) \geq 0$. This consumption is automatically drawn from the bank account holding with no extra costs. In addition the investor may at any time transfer money from the bank to the stock and conversely. Suppose that such a transaction of size ζ incurs a transaction cost given by

$$c + \lambda |\zeta|, \qquad (8.2.9)$$

where $c > 0$ and $\lambda \in [0, 1)$ are constants. (If $\zeta > 0$ we buy stocks and if $\zeta < 0$ we sell stocks.) Thus the control of the investor consists of a combination of a stochastic control $u(t)$ and an impulse control $v = (\tau_1, \tau_2, \ldots; \zeta_1, \zeta_2, \ldots)$, where

$\tau_1, \tau_2, \ldots$ are the chosen transaction times and $\zeta_1, \zeta_2, \ldots$ are corresponding transaction amounts.

If such a combined control $w = (u, v)$ is applied, the corresponding system $(X_1(t), X_2(t)) = (X_1^{(w)}(t), X_2^{(w)}(t))$ gets the form

$$dX_1(t) = (rX_1(t) - u(t))dt, \quad \tau_i < t < \tau_{i+1}, \tag{8.2.10}$$

$$dX_2(t) = X_2(t^-)\left(\mu \, dt + \sigma \, dB(t) + \theta \int_{\mathbb{R}} z\tilde{N}(dt, dz)\right), \quad \tau_i < t < \tau_{i+1}, \tag{8.2.11}$$

$$X_1(\tau_{i+1}) = X_1(\tau_{i+1}^-) - \zeta_{i+1} - c - \lambda|\zeta_{i+1}|, \tag{8.2.12}$$

$$X_2(\tau_{i+1}) = \check{X}_2(\tau_{i+1}^-) + \zeta_{i+1}. \tag{8.2.13}$$

If we do not allow any negative amounts held in the bank account or in the stock, the *solvency region* $\mathcal{S}$ is given by

$$\mathcal{S} = [0, \infty) \times [0, \infty). \tag{8.2.14}$$

We call $w = (u, v)$ *admissible* if $X^{(w)}(t) := (X_1^{(w)}(t), X_2^{(w)}(t))$ exists for all t. The set of admissible controls is denoted by $\mathcal{W}$. Let

$$\tau_{\mathcal{S}} = \inf\{t > 0; X^{(w)}(t) \notin \mathcal{S}\}$$

be the bankruptcy time. The investor's objective is to maximize

$$J^{(w)}(y) = E^y\left[\int_0^{\tau_{\mathcal{S}}} e^{-\rho(s+t)}\frac{u^\gamma(t)}{\gamma} dt\right], \tag{8.2.15}$$

where $\rho > 0$ and $\gamma \in (0, 1)$ are constants and E^y with $y = (s, x_1, x_2)$ denotes the expectation when $X_1(0^-) = x_1 \geq 0$ and $X_2(0^-) = x_2 \geq 0$. Thus we seek the value function $\Phi(y)$ and an optimal control $w^* = (u^*, v^*) \in \mathcal{W}$ such that

$$\Phi(y) = \sup_{w \in \mathcal{W}} J^{(w)}(y) = J^{(w^*)}(y). \tag{8.2.16}$$

This problem may be regarded as a generalization of optimal consumption and portfolio problems studied by Merton [M] and Davis and Norman [DN] (see also Shreve and Soner [SS]). Merton [M] considers the case with no jumps and no-transaction costs ($c = \lambda = \theta = 0$). The problem then reduces to an ordinary stochastic control problem and it is optimal to keep the positions $(X_1(t), X_2(t))$ on the line $y = (\pi^*/(1 - \pi^*))x$ in the (x, y)-plane at all times (the Merton line), where $\pi^* = (\mu - r)/(1 - \gamma)\sigma^2$ (see Example 3.2).

Davis and Norman [DN] and Shreve and Soner [SS] consider the case when the cost is proportional ($\lambda > 0$), with no fixed component ($c = 0$) and

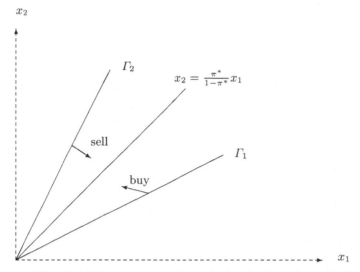

Fig. 8.1. The no-transaction cone (no fixed cost: $c = 0$), $\theta = 0$

no jumps ($\theta = 0$). In this case the problem can be formulated as a singular stochastic control problem and under some conditions it is proved that there exists a *no-transaction cone NT* bounded by two straight lines Γ_1 and Γ_2 such that it is optimal to make no transactions if $(X_1(t), X_2(t)) \in NT$ and make transactions corresponding to local time at $\partial(NT)$, resulting in reflections back to NT every time $(X_1(t), X_2(t)) \in \partial(NT)$. See Fig. 8.1. These results have subsequently been extended to jump diffusion markets by [FØS2] (see Example 5.1).

In the general combined control case numerical results indicate (see [CØS]) that the optimal control $w^* = (u^*, v^*)$ has the following form.

There exist two pairs of lines, $(\Gamma_1, \hat{\Gamma}_1)$ and $(\Gamma_2, \hat{\Gamma}_2)$ from the origin such that the following is optimal. Make no transactions (only consume at the rate $u^*(t)$) while $(X_1(t), X_2(t))$ belongs to the region D bounded by the outer curves Γ_1, Γ_2, and if $(X_1(t), X_2(t))$ hits $\partial D = \Gamma_1 \cup \Gamma_2$ then sell or buy so as to bring $(X_1(t), X_2(t))$ to the curve $\hat{\Gamma}_1$ or $\hat{\Gamma}_2$. See Fig. 8.2.

Note that if we *sell* stocks ($\zeta < 0$) then $(X_1(t), X_2(t)) = (x_1, x_2)$ moves to a point $(x_1', x_2') = (x_1'(\zeta), x_2'(\zeta))$ on the line

$$x_1' + (1 - \lambda)x_2' = x_1 + (1 - \lambda)x_2 - c, \qquad (8.2.17)$$

i.e., (x_1', x_2') lies on the straight line through (x_1, x_2) with slope $-1/(1 - \lambda)$.

Similarly, if we *buy* stocks ($\zeta > 0$) then the new position $(x_1', x_2') = (x_1'(\zeta), x_2'(\zeta))$ is on the line

$$x_1' + (1 + \lambda)x_2' = x_1 + (1 + \lambda)x_2 - c, \qquad (8.2.18)$$

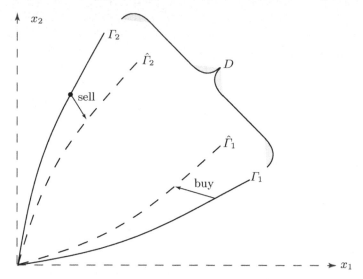

Fig. 8.2. The no-transaction region D $(c > 0)$

i.e., (x_1', x_2') lies on the straight line through (x_1, x_2) with slope $-1/(1 + \lambda)$. If there are no interventions then the process

$$Y(t) = \begin{bmatrix} s + t \\ X_1(t) \\ X_2(t) \end{bmatrix} \tag{8.2.19}$$

has the generator

$$
\begin{aligned}
L^u \phi(s, x_1, x_2) = {} & \frac{\partial \phi}{\partial s} + (rx_1 - u)\frac{\partial \phi}{\partial x_1} + \mu x_2 \frac{\partial \phi}{\partial x_2} + \frac{1}{2}\sigma^2 x_2^2 \frac{\partial^2 \phi}{\partial x_2^2} \\
& + \int_{\mathbb{R}} \left\{ \phi(s, x_1, x_2 + x_2 \theta z) - \phi(s, x_1, x_2) \right. \tag{8.2.20} \\
& \left. - x_2 \theta z \frac{\partial \phi}{\partial x_2}(s, x_1, x_2) \right\} \nu(dz).
\end{aligned}
$$

Therefore, if we put $\phi(s, x_1, x_2) = \mathrm{e}^{-\rho s}\psi(x_1, x_2)$ the corresponding HJBQVI is

$$
\begin{aligned}
\max \Bigg(& \sup_{u \geq 0} \left\{ \frac{u^\gamma}{\gamma} - \rho \psi(x_1, x_2) + (rx_1 - u)\frac{\partial \psi}{\partial x_1} + \mu x_2 \frac{\partial \psi}{\partial x_2} + \frac{1}{2}\sigma^2 x_2^2 \frac{\partial^2 \psi}{\partial x_2^2} \right. \\
& + \int_{\mathbb{R}} \left\{ \psi(x_1, x_2 + x_2 \theta z) - \psi(x_1, x_2) - x_2 \theta z \frac{\partial \psi}{\partial x_2}(x_1, x_2) \right\} \nu(dz) \Bigg\} \\
& \psi(x_1, x_2) - \mathcal{M}\psi(x_1, x_2) \Bigg) = 0 \quad \text{for all } (x_1, x_2) \in \mathcal{S}, \tag{8.2.21}
\end{aligned}
$$

where (see (8.2.17) and (8.2.18))

$$\mathcal{M}\psi(x_1, x_2) = \sup\{\psi(x_1'(\zeta), x_2'(\zeta)); \zeta \in \mathbb{R} \setminus \{0\}, (x_1'(\zeta), x_2'(\zeta)) \in \mathcal{S}\}. \quad (8.2.22)$$

See Example 9.12 for a further discussion of this.

8.3 Iterative Methods

In Chap. 7 we saw that an impulse control problem can be regarded as a limit of iterated optimal stopping problems. A similar result holds for combined control problems. More precisely, a combined stochastic control and impulse control problem can be regarded as a limit of iterated combined stochastic control and optimal stopping problems.

We now describe this in more detail. The presentation is similar to the approach in Chap. 7.

For $n = 1, 2, \ldots$ let $\mathcal{W}_n$ denote the set of all admissible combined controls $w = (u, v) \in \mathcal{W}$ with $v \in \mathcal{V}_n$, where $\mathcal{V}_n$ is the set of impulse controls $v = (\tau_1, \ldots, \tau_k; \zeta_1, \zeta_2, \ldots, \zeta_k)$ with *at most n interventions* (i.e., $k \leq n$). Then

$$\mathcal{W}_n \subseteq \mathcal{W}_{n+1} \subseteq \mathcal{W} \quad \text{for all } n. \quad (8.3.1)$$

Define, with $J^{(w)}(y)$ as in (8.1.6),

$$\Phi_n(y) = \sup\{J^{(w)}(y); w \in \mathcal{W}_n\}, \quad n = 1, 2, \ldots \quad (8.3.2)$$

Then

$$\Phi_n(y) \leq \Phi_{n+1}(y) \leq \Phi(y) \quad \text{because } \mathcal{W}_n \subseteq \mathcal{W}_{n+1} \subseteq \mathcal{W}.$$

Moreover, we have:

Lemma 8.4. *Suppose $g \geq 0$. Then*

$$\lim_{n \to \infty} \Phi_n(y) = \Phi(y) \quad \text{for all } y \in \mathcal{S}.$$

Proof. The proof is similar to the proof of Lemma 7.1 and is omitted.   □

The iterative procedure is the following.

Let $Y(t) = Y^{(u,0)}(t)$ be the process in (8.1.1) obtained by using the control u and no interventions. Define

$$\phi_0(y) = \sup_{u \in \mathcal{U}} E^y\left[\int_0^{\tau_S} f(Y(t), u(t))dt + g(Y(\tau_S))\chi_{\{\tau_S < \infty\}}\right] \quad (8.3.3)$$

and inductively, for $j = 1, 2, \ldots, n$,

$$\phi_j(y) = \sup_{u \in \mathcal{U}, \tau \in \mathcal{T}} E^y \left[\int_0^\tau f(Y(t), u(t)) dt + \mathcal{M}\phi_{j-1}(Y(\tau)) \chi_{\{\tau < \infty\}} \right]. \quad (8.3.4)$$

As in (7.1.13) we let $\mathcal{P}(\mathbb{R}^k)$ denote the set of functions $h : \mathbb{R}^k \to \mathbb{R}$ with at most polynomial growth. Then we have as in Chap. 7:

Theorem 8.5. *Suppose*

$$f, g \text{ and } \mathcal{M}\phi_{j-1} \in \mathcal{P}(\mathbb{R}^k) \quad (8.3.5)$$

for $j = 1, 2, \ldots, n$. *Then*

$$\phi_n = \Phi_n.$$

Proof. The proof is basically the same as the proof of Theorem 7.2 and is left to the reader. □

Similarly we obtain combined control versions of the rest of the results of Chap. 7, with obvious modifications. We omit the details.

8.4 Exercises

Exercise* 8.1. Let $\Phi(s, x_1, x_2)$ be the value function of the optimal consumption problem (8.2.16) with fixed and proportional transaction costs and let $\Phi_0(s, x_1, x_2) = K e^{-\rho s} (x_1 + x_2)^\gamma$ be the corresponding value function in the case when there are no-transaction costs, i.e., $c = \lambda = 0$ (see Example 3.2). Use Theorem 8.1a to prove that

$$\Phi(s, x_1, x_2) \le K e^{-\rho s} (x_1 + x_2)^\gamma.$$

Exercise* 8.2 (A Combined Impulse Linear Regulator Problem). (Compare with Exercises 3.5 and 4.3.) Suppose the state process is

$$Y(t) = \begin{bmatrix} s + t \\ X(t) \end{bmatrix}, \ t \ge 0, \quad Y(0) = \begin{bmatrix} s \\ x \end{bmatrix} = y \in \mathbb{R}^2,$$

with

$$dX(t) = dX^{(w)}(t) \text{ given by}$$

$$dX(t) = u(t) dt + \sigma \, dB(t) + \int_{\mathbb{R}} z \tilde{N}(dt, dz), \quad \tau_i < t < \tau_{i+1},$$

$$X(\tau_{i+1}) = X(\tau_{i+1}^-) + \Delta_N X(\tau_{i+1}) + \zeta_{i+1},$$

where $w = (u, v)$ is a combined control, $v = (\tau_1, \tau_2, \ldots; \zeta_1, \zeta_2, \ldots)$ with $\tau_i \leq T$ (constant), σ constant, and $\zeta_{i+1} \in \mathbb{R}\backslash\{0\}$.

Write down and study the HJBQVI for the combined control problem

$$\Phi(y) = \inf_{w \in \mathcal{W}} J^{(w)}(y),$$

where

$$J^{(w)}(y) = E^{(s,x)} \left[\int_0^{T-s} (X^2(t) + u^2(t))\mathrm{d}t + \sum_{\tau_i \leq T-s} c \right],$$

and $c > 0$ is a given constant.

This models the situation where one is trying to keep $X(t)$ close to 0 with a minimum of cost of the two controls, represented by the rate $u^2(t)$ and the intervention cost c.

9

Viscosity Solutions

The main results of Chaps. 2–6 and 8 and are all *verification theorems*. Any function ϕ which satisfies the given requirements is necessarily the value function Φ of the corresponding problem. These requirements are made as weak as possible in order to include as many cases as possible. For example, except for the singular control case, we do not require the function ϕ to be C^2 everywhere (only outside ∂D), because except for that case, Φ will usually not be C^2 everywhere. On the other hand, all the above-mentioned verification theorems require ϕ to be C^1 everywhere, because this is often the case for Φ. This C^1 assumption on Φ is usually called the "high contact" – or "smooth fit" – principle. As we have seen in the examples and exercises this principle is very convenient, because it provides us with extra information needed to find Φ and the continuation region D.

However, it is important to know that in general Φ need not be C^1. In fact, it need not even be continuous everywhere. Nevertheless, it turns out that Φ does satisfy the corresponding verification theorems, provided that we interpret these equations in an appropriate weak sense. More precisely, they should be interpreted in the sense of *viscosity solutions*. This weak solution concept was first introduced by Crandall and Lions to handle the HJB equations of stochastic control and later extended by them and others to more general equations. See [CIL, FS, BCa] and the references therein.

In case the equation in question is a *linear* partial differential operator, the viscosity solution is the same as the well-known distribution solution. See [Is2]. However, the nice feature of the viscosity solution is that it also applies to the nonlinear equations appearing in control theory.

We now proceed to define viscosity solutions. We will do this in two steps.

First we consider the viscosity solutions of the variational inequalities appearing in the optimal stopping problems of Chap. 2. Then we proceed to discuss more general equations.

9.1 Viscosity Solutions of Variational Inequalities

Consider the optimal stopping problem of Chap. 2. The state $Y(t)$ is given by

$$dY(t) = b(Y(t))dt + \sigma(Y(t))dB(t) + \int_{\mathbb{R}^\ell} \gamma(Y(t^-), z)\bar{N}(dt, dz), \quad Y(0) = y \in \mathbb{R}^k \tag{9.1.1}$$

and the performance criterion is

$$J^\tau(y) = E^y\left[\int_0^\tau f(Y(t))dt + g(Y(\tau))\chi_{\{\tau < \infty\}}\right], \quad \tau \in \mathcal{T}, \tag{9.1.2}$$

with f and g as in (2.1.2) and (2.1.3). The associated variational inequality of the optimal stopping problem

$$\Phi(y) = \sup\{J^\tau(y); \tau \in \mathcal{T}\} \tag{9.1.3}$$

is (see (4.2.13))

$$\max(L\phi(y) + f(y), g(y) - \phi(y)) = 0, \quad y \in \mathcal{S}^0. \tag{9.1.4}$$

In addition we have the boundary requirement

$$\phi(y) = g(y), \quad y \in \partial\mathcal{S} \tag{9.1.5}$$

(see Theorem 2.2).

We will prove that, under some conditions, the function $\phi = \Phi$ is the unique viscosity solution of (9.1.4) and (9.1.5). First we give the definition of a viscosity solution of such equations.

Definition 9.1. *Let $\phi \in C(\bar{\mathcal{S}})$.*

(i) We say that ϕ is a viscosity subsolution of (9.1.4) and (9.1.5) if (9.1.5) holds and for all $h \in C^2(\mathbb{R}^k)$ and all $y_0 \in \mathcal{S}^0$ such that

$$h \geq \phi \quad on\ \mathcal{S} \quad and \quad h(y_0) = \phi(y_0)$$

we have

$$\max(Lh(y_0) + f(y_0), g(y_0) - \phi(y_0)) \geq 0. \tag{9.1.6}$$

(ii) ϕ is a viscosity supersolution of (9.1.4) and (9.1.5) if (9.1.5) holds and for all $h \in C^2(\mathbb{R}^k)$ and all $y_0 \in \mathcal{S}$ such that

$$h \leq \phi \quad on\ \mathcal{S} \quad and \quad h(y_0) = \phi(y_0)$$

we have

$$\max(Lh(y_0) + f(y_0), g(y_0) - \phi(y_0)) \leq 0. \tag{9.1.7}$$

(iii) ϕ is a viscosity solution of (9.1.4) and (9.1.5) if ϕ is both a viscosity subsolution and a viscosity supersolution of (9.1.4) and (9.1.5).

Theorem 9.2 ([ØR]). *Assume that the set $\partial \mathcal{S}$ is regular for the process $Y(t)$, i.e.,*

$$\tau_{\mathcal{S}^0} = \tau_{\mathcal{S}^0}(y) := \inf\{t > 0; Y(t) \notin \mathcal{S}^0\} = 0 \qquad a.s. \ Q^y \ for \ all \ y \in \partial \mathcal{S}. \tag{9.1.8}$$

Moreover, assume that the value function Φ of the optimal stopping problem (9.1.3) is continuous on $\bar{\mathcal{S}}$. Then Φ is a viscosity solution of (9.1.4) and (9.1.5).

Proof. First note that (9.1.5) follows directly from (9.1.8). So it remains to consider (9.1.4).

We first prove that Φ is a subsolution. To this end suppose $h \in C^2(\mathbb{R}^k)$ and $y_0 \in \mathcal{S}$ with $h \geq \Phi$ on $\mathcal{S}$ and $h(y_0) = \Phi(y_0)$. As before let

$$D = \{y \in \mathcal{S}; \Phi(y) > g(y)\}. \tag{9.1.9}$$

Then if $y_0 \notin D$ we have $\Phi(y_0) = g(y_0)$ and hence (9.1.6) holds trivially.

Next, assume $y_0 \in D$. Then by the dynamic programming principle (Lemma 7.3b) we have

$$\Phi(y_0) = E^{y_0}\left[\int_0^\tau f(Y(t))dt + \Phi(Y(\tau))\right] \tag{9.1.10}$$

for all bounded stopping times $\tau \leq \tau_D = \inf\{t > 0; Y(t) \notin D\}$. Choose $h_m \in C_0^2(\mathbb{R}^k)$ such that $h_m \to h$ and $Lh_m \to Lh$ pointwise dominatedly on $\mathbb{R}^k$. Then by combining (9.1.10) with the Dynkin formula we get

$$\Phi(y_0) = E^{y_0}\left[\int_0^\tau f(Y(t))dt + \Phi(Y(\tau))\right]$$

$$\leq E^{y_0}\left[\int_0^\tau f(Y(t))dt + h(Y(\tau))\right]$$

$$= \lim_{m\to\infty} E^{y_0}\left[\int_0^\tau f(Y(t))dt + h_m(Y(\tau))\right]$$

$$= h(y_0) + \lim_{m\to\infty} E^{y_0}\left[\int_0^\tau (Lh_m(Y(t)) + f(Y(t)))dt\right].$$

Hence

$$\lim_{m\to\infty} E^{y_0}\left[\int_0^\tau (Lh_m(Y(t)) + f(Y(t)))dt\right] \geq 0.$$

In particular, if we choose $\tau = \beta_j := \inf\{t > 0; |Y(t) - y_0| \geq 1/j\} \wedge 1/j \wedge \tau_D$, we get

$$E^{y_0}\left[\int_0^{\beta_j} (Lh(Y(t)) + f(Y(t)))dt\right]$$

$$= \lim_{m\to\infty} E^{y_0}\left[\int_0^{\beta_j} (Lh_m(Y(t)) + f(Y(t)))dt\right] \geq 0. \tag{9.1.11}$$

If we divide (9.1.11) by $E^{y_0}[\beta_j]$ and let $j \to \infty$ we get, by right continuity,

$$Lh(y_0) + f(y_0) \geq 0.$$

Hence (9.1.6) holds and we have proved that Φ is a viscosity subsolution.

Finally we show that Φ is a viscosity supersolution. So we assume that $h \in C^2(\mathbb{R}^k)$ and $y_0 \in \mathcal{S}$ are such that $h \leq \Phi$ on $\mathcal{S}$ and $h(y_0) = \Phi(y_0)$. Then by the dynamic programming principle (Lemma 7.3a) we have

$$\Phi(y_0) \geq E^{y_0}\left[\int_0^\tau f(Y(t))\mathrm{d}t + \Phi(Y(\tau))\right] \tag{9.1.12}$$

for all stopping times $\tau \leq \tau_{\mathcal{S}^0}$. Hence, by the Dynkin formula, with h_m and β_j as above,

$$\Phi(y_0) \geq E^{y_0}\left[\int_0^{\beta_j} f(Y(t))\mathrm{d}t + \phi(Y(\tau))\right]$$

$$\geq \lim_{m \to \infty} E^{y_0}\left[\int_0^{\beta_j} f(Y(t))\mathrm{d}t + h(Y(\tau))\right]$$

$$= h(y_0) + \lim_{m \to \infty} E^{y_0}\left[\int_0^{\beta_j} (Lh(Y(t))\mathrm{d}t + f(Y(t)))\mathrm{d}t\right]$$

$$= h(y_0) + E^{y_0}\left[\int_0^{\beta_j} (Lh(Y(t))\mathrm{d}t + f(Y(t)))\mathrm{d}t\right] \quad \text{for all } j.$$

Hence

$$E^{y_0}\left[\int_0^{\beta_j} (Lh(Y(t)) + f(Y(t)))\mathrm{d}t\right] \leq 0$$

and by dividing by $E^{y_0}[\beta_j]$ and letting $j \to \infty$ we get, by right continuity,

$$Lh(y_0) + f(y_0) \leq 0.$$

Hence (9.1.7) holds and we have proved that Φ is also a viscosity supersolution.
□

9.1.1 Uniqueness

One important application of the viscosity solution concept is that it can be used as a verification method. In order to verify that a given function ϕ is indeed the value function Φ it suffices to verify that the function is a viscosity solution of the corresponding variational inequality. For this method to work, however, it is necessary that we know that Φ is the *unique* viscosity solution. Therefore the question of uniqueness is crucial.

In general we need not have uniqueness. The following simple example illustrates this.

Example 9.3. Let $Y(t) = B(t) \in \mathbb{R}$ and choose $f = 0$, $\mathcal{S} = \mathbb{R}$, and

$$g(y) = \frac{y^2}{1 + y^2}, \quad y \in \mathbb{R}. \tag{9.1.13}$$

Then the value function Φ of the optimal stopping problem

$$\Phi(y) = \sup_{\tau \in \mathcal{T}} E^y[g(B(\tau))] \tag{9.1.14}$$

is easily seen to be $\Phi(y) \equiv 1$. The corresponding VI is

$$\max\left(\frac{1}{2}\phi''(y), g(y) - \phi(y)\right) = 0 \tag{9.1.15}$$

and this equation is trivially satisfied by all constant functions

$$\phi(y) \equiv a$$

for any $a \geq 1$.

Theorem 9.4 (Uniqueness). *Suppose that*

$$\tau_{\mathcal{S}^0} < \infty \quad a.s. \ P^y \ for \ all \ y \in \mathcal{S}^0. \tag{9.1.16}$$

Let $\phi \in C(\bar{\mathcal{S}})$ be a viscosity solution of (9.1.4) and (9.1.5) with the property that

$$\begin{array}{c} the \ family \ \{\phi(Y(\tau)); \tau \ stopping \ time, \ \tau \leq \tau_{\mathcal{S}^0}\} \\ is \ P^y\text{-}uniformly \ integrable, \ for \ all \ y \in \mathcal{S}^0. \end{array} \tag{9.1.17}$$

Then

$$\phi(y) = \Phi(y) \quad for \ all \ y \in \bar{\mathcal{S}}.$$

Proof. We refer the reader to [ØR] for the proof in the case where there are no jumps. □

9.2 The Value Function is Not Always $\mathcal{C}^1$

Example 9.5. We now give an example of an optimal stopping problem where the value function Φ is not C^1 everywhere. In this case Theorem 2.2 cannot be used to find Φ. However, we can use Theorem 9.4. The example is taken from [ØR]:
Define

$$k(x) = \begin{cases} 1 & \text{for } x \leq 0 \\ 1 - cx & \text{for } 0 < x < a, \\ 1 - ca & \text{for } x \geq a \end{cases} \tag{9.2.1}$$

where c and a are constants to be specified more closely later. Consider the optimal stopping problem

$$\Phi(s,x) = \sup_{\tau \in \mathcal{T}} E^{(s,x)}\left[e^{-\rho(s+\tau)}k(B(\tau))\right], \qquad (9.2.2)$$

where $B(t)$ is one-dimensional Brownian motion, $B(0) = x \in \mathbb{R} = \mathcal{S}$, and $\rho > 0$ is a constant. The corresponding variational inequality is (see (9.1.4))

$$\max\left(\frac{\partial \phi}{\partial s} + \frac{1}{2}\frac{\partial^2 \phi}{\partial x^2}, e^{-\rho s}k(x) - \phi(s,x)\right) = 0. \qquad (9.2.3)$$

If we try a solution of the form

$$\phi(s,x) = e^{-\rho s}\psi(x) \qquad (9.2.4)$$

for some function ψ, then (9.2.3) becomes

$$\max\left(-\rho\psi(x) + \frac{1}{2}\psi''(x), k(x) - \psi(x)\right) = 0. \qquad (9.2.5)$$

Let us guess that the continuation region D has the form

$$D = \{(s,x); 0 < x < x_1\} \qquad (9.2.6)$$

for some $x_1 > a$. Then (9.2.5) can be split into the three equations

$$-\rho\psi(x) + \frac{1}{2}\psi''(x) = 0 \qquad ; \quad 0 < x < x_1 \qquad (9.2.7)$$
$$\psi(x) = 1 \qquad ; \quad x \le 0$$
$$\psi(x) = 1 - ca \quad ; \quad x \ge x_1.$$

The general solution of (9.2.7) is

$$\psi(x) = C_1 e^{\sqrt{2\rho}\,x} + C_2 e^{-\sqrt{2\rho}\,x}, \quad 0 < x < x_1,$$

where C_1, C_2 are arbitrary constants. If we require ψ to be continuous at $x = 0$ and at $x = x_1$ we get the two equations

$$C_1 + C_2 = 1, \qquad (9.2.8)$$
$$C_1 e^{\sqrt{2\rho}\,x_1} + C_2 e^{-\sqrt{2\rho}\,x_1} = 1 - ca \qquad (9.2.9)$$

in the three unknowns C_1, C_2, and x_1. If we also guess that ψ will be C^1 at $x = x_1$ we get the third equation

$$C_1\sqrt{2\rho}\,e^{\sqrt{2\rho}\,x_1} - C_2\sqrt{2\rho}\,e^{-\sqrt{2\rho}\,x_1} = 0. \qquad (9.2.10)$$

If we assume that

$$ca < 1 \quad \text{and} \quad \sqrt{2\rho} < \frac{1}{a} \ln\left(\frac{1-ca}{1-\sqrt{ca(2-ca)}}\right) \tag{9.2.11}$$

then the three equations (9.2.8), (9.2.9), and (9.2.10) have the unique solution

$$C_1 = \frac{1}{2}\left(1 - \sqrt{ca(2-ca)}\right) > 0, \quad C_2 = 1 - C_1 > 0 \tag{9.2.12}$$

and

$$x_1 = \frac{1}{\sqrt{2\rho}} \ln\left(\frac{1-ca}{2C_1}\right) > a. \tag{9.2.13}$$

With these values of C_1, C_2, and x_1 we put

$$\psi(x) = \begin{cases} 1 & \text{if } x \leq 0 \\ C_1 e^{\sqrt{2\rho}\,x} + C_2 e^{-\sqrt{2\rho}\,x} & \text{if } 0 < x < x_1. \\ 1 - ca & \text{if } x_1 \leq x \end{cases} \tag{9.2.14}$$

See Fig. 9.1.

We claim that ψ is a viscosity solution of (9.2.5).

(i) First we verify that ψ is a viscosity *subsolution*: let $h \in C^2(\mathbb{R})$, $h \geq \psi$, and $h(x_0) = \psi(x_0)$. Then if $x_0 \leq 0$ or $x_0 \geq x_1$ we have $k(x_0) - \psi(x_0) = 0$. And if $0 < x_0 < x_1$ then $h - \psi$ is C^2 at $x = x_0$ and has a local minimum at x_0 so

$$h''(x_0) - \psi''(x_0) \geq 0.$$

Therefore

$$-\rho h(x_0) + \frac{1}{2}h''(x_0) \geq -\rho\psi(x_0) + \frac{1}{2}\psi''(x_0) = 0.$$

This proves that

$$\max\left(-\rho h(x_0) + \frac{1}{2}h''(x_0), k(x_0) - \psi(x_0)\right) \geq 0,$$

so ψ is a viscosity subsolution of (9.2.5).

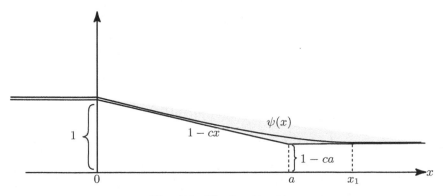

Fig. 9.1. The function ψ

(ii) Second, we prove that ψ is a viscosity *super*solution. So let $h \in C^2(\mathbb{R})$, $h \leq \psi$, and $h(x_0) = \psi(x_0)$. Note that we always have

$$k(x_0) - \psi(x_0) \leq 0$$

so in order to prove that

$$\max\left(-\rho h(x_0) + \frac{1}{2}h''(x_0), k(x_0) - \psi(x_0)\right) \leq 0$$

it suffices to prove that

$$-\rho h(x_0) + \frac{1}{2}h''(x_0) \leq 0.$$

At any point x_0 where ψ is C^2 this follows in the same way as in (i) above. So it remains only to consider the two cases $x_0 = 0$ and $x_0 = x_1$. If $x_0 = 0$ then no such h exists, so the conclusion trivially holds. If $x_0 = x_1$ then the function $h - \psi$ has a local maximum at $x = x_0$ and it is C^2 to the left of x_0 so

$$\lim_{x \to x_0^-} h''(x) - \psi''(x) \leq 0,$$

i.e., $h''(x_0) - \psi''(x_0^-) \leq 0$.
This gives

$$-\rho h(x_0) + \frac{1}{2}h''(x_0) \leq -\rho \psi(x_0) + \frac{1}{2}\psi''(x_0^-) = 0,$$

and the proof is complete.
We have proved.
Suppose (9.2.11) holds. Then the value function $\Phi(s, x)$ of problem (9.2.2) is given by

$$\Phi(s, x) = e^{-\rho s}\psi(x)$$

with ψ as in (9.2.14), C_1, C_2, and x_1 as in (9.2.12) and (9.2.13). Note in particular that $\psi(x)$ is not C^1 at $x = 0$.

9.3 Viscosity Solutions of HJBQVI

We now turn to the general combined stochastic control and impulse control problem from Chap. 8. Thus the state $Y(t) = Y^{(w)}(t)$ is

$$Y(0^-) = y \in \mathbb{R}^k,$$
$$dY(t) = b(Y(t), u(t))dt + \sigma(Y(t), u(t))dB(t)$$
$$+ \int_{\mathbb{R}^\ell} \gamma(Y(t^-), u(t^-), z)\tilde{N}(dt, dz), \quad \tau_i < t < \tau_{i+1}, \qquad (9.3.1)$$
$$Y(\tau_{i+1}) = \Gamma(\check{Y}(\tau_{i+1}^-), \zeta_{i+1}), \quad i = 0, 1, 2, \ldots,$$

where $w = (u, v) \in \mathcal{W}, u \in \mathcal{U}, v = (\tau_1, \tau_2, \ldots; \zeta_1, \zeta_2, \ldots) \in \mathcal{V}$.

The performance is given by

$$J^{(w)}(y) = E^y\left[\int_0^{\tau_S} f(Y(t), u(t))dt + g(Y(\tau_S))\chi_{\{\tau_S < \infty\}} + \sum_j K(\breve{Y}(\tau_j^-), \zeta_j)\right]$$
(9.3.2)

and we want to find the value function Φ defined by

$$\Phi(y) = \sup_{w \in W} J^{(w)}(y).$$
(9.3.3)

To simplify the presentation we will from now on assume that

$$S \text{ is an open set, i.e., } S = S^0,$$
(9.3.4)

and that ∂S is *regular* for $Y^{(w)}(t)$ for all $w \in W$, i.e.,

$$\tau_S = \tau_S(y) = \inf\{t > 0; Y^{(w)}(t) \notin S\} = 0 \quad \text{for all } y \in \partial S \text{ and all } w \in W.$$
(9.3.5)

These conditions (9.3.4) and (9.3.5) exclude cases where Φ also satisfies certain HJBQVIs on ∂S (see, e.g., [ØS]), but it is often easy to see how to extend the results to such situations.

We need to make the following two assumptions on the set of admissible controls W:

(1) If $w = (u, v)$, with $v = (\tau_1, \tau_2, \ldots; \zeta_1, \zeta_2, \ldots)$, belongs to W and $\hat{\zeta}$ is any point in Z, then $w := (u, \hat{v})$ belongs to W also, when $\hat{v} := (0, \tau_1, \tau_2, \ldots; \hat{\zeta}, \zeta_1, \zeta_2, \ldots)$.

(2) If α is any constant in U, then the combined control $w := (\alpha, 0)$ (no interventions, just the constant control α) belongs to W.

Theorem 8.1 associates Φ to the HJBQVI

$$\max\left(\sup_{\alpha \in U}\{L^\alpha \Phi(y) + f(y, \alpha)\}, \mathcal{M}\Phi(y) - \Phi(y)\right) = 0, \quad y \in S$$
(9.3.6)

with boundary values

$$\Phi(y) = g(y), \quad y \in \partial S,$$
(9.3.7)

where

$$L^\alpha \Phi(y) = \sum_{i=1}^k b_i(y, \alpha)\frac{\partial \Phi}{\partial y_i} + \frac{1}{2}\sum_{i,j=1}^k (\sigma\sigma^T)_{ij}(y, \alpha)\frac{\partial^2 \Phi}{\partial y_i \partial y_j}$$

$$+ \sum_{j=1}^\ell \int_{\mathbb{R}}\left\{\Phi\left(y + \gamma^{(j)}(y, \alpha, z_j)\right) - \Phi(y)\right.$$

$$\left. - \nabla\Phi(y) \cdot \gamma^{(j)}(y, \alpha, z_j)\right\}\nu_j(dz_j)$$
(9.3.8)

and

$$M\Phi(y) = \sup \left\{ \Phi(\Gamma(y,\zeta)) + K(y,\zeta); \zeta \in \mathcal{Z}, \Gamma(y,\zeta) \in \mathcal{S} \right\}. \qquad (9.3.9)$$

Unfortunately, as we have seen already for optimal stopping problems, the value function Φ need not be C^1 everywhere – in general not even continuous! So (9.3.6) is not well defined, if we interpret the equation in the usual sense. However, it turns out that if we interpret (9.3.6) in the *weak sense of viscosity* then Φ does indeed solve the equation. In fact, under some assumptions Φ is the *unique* viscosity solution of (9.3.6) and (9.3.7) (see Theorem 9.11). This result is an important supplement to Theorem 8.1.

We now define the concept of viscosity solutions of general HJBQVIs of type (9.3.6) and (9.3.7).

Definition 9.6. *Let $\varphi \in C(\bar{\mathcal{S}})$.*

(i) We say that φ is a viscosity subsolution *of*

$$\max \left(\sup_{\alpha \in U} \{L^\alpha \varphi(y) + f(y,\alpha)\}, \, M\varphi(y) - \varphi(y) \right) = 0, \quad y \in \mathcal{S}, \qquad (9.3.10)$$

$$\varphi(y) = g(y), \quad y \in \partial\mathcal{S} \qquad (9.3.11)$$

if (9.3.11) holds and for every $h \in C^2(\mathbb{R}^k)$ and every $y_0 \in \mathcal{S}$ such that $h \geq \varphi$ on $\mathcal{S}$ and $h(y_0) = \varphi(y_0)$ we have

$$\max \left(\sup_{\alpha \in U} \{L^\alpha h(y_0) + f(y_0,\alpha)\}, \, M\varphi(y_0) - \varphi(y_0) \right) \geq 0. \qquad (9.3.12)$$

(ii) We say that φ is a viscosity supersolution *of (9.3.10) and (9.3.11) if (9.3.11) holds and for every $h \in C^2(\mathbb{R}^k)$ and every $y_0 \in \mathcal{S}$ such that $h \leq \varphi$ on $\mathcal{S}$ and $h(y_0) = \varphi(y_0)$ we have*

$$\max \left(\sup_{\alpha \in U} \{L^\alpha h(y_0) + f(y_0,\alpha)\}, \, M\varphi(y_0) - \varphi(y_0) \right) \leq 0. \qquad (9.3.13)$$

(iii) We say that φ is a viscosity solution *of (9.3.10) and (9.3.11) if φ is both a viscosity subsolution and a viscosity supersolution of (9.3.10) and (9.3.11).*

Lemma 9.7. *Let Φ be as in (9.3.3). Then $\Phi(y) \geq M\Phi(y)$ for all $y \in \mathcal{S}$.*

Proof. Suppose there exists $y \in \mathcal{S}$ with

$$\Phi(y) < M\Phi(y),$$

i.e.,

$$\Phi(y) < \sup_{\zeta \in \mathcal{Z}} \{\Phi(\Gamma(y, \zeta)) + K(y, \zeta)\}.$$

Then there exist $\varepsilon > 0$ and $\hat{\zeta} \in \mathcal{Z}$ such that, with $\hat{y} = \Gamma(y, \hat{\zeta})$,

$$\Phi(y) < \Phi(\hat{y}) + K(y, \hat{\zeta}) - 2\varepsilon.$$

Let $w = (u, v)$, with $v = (\tau_1, \tau_2, \ldots; \zeta_1, \zeta_2, \ldots)$ be ε-optimal for Φ at $\hat{y}$, in the sense that

$$J^{(w)}(\hat{y}) > \Phi(\hat{y}) - \varepsilon.$$

Define $\hat{w} := (u, \hat{v})$, where $\hat{v} = (0, \tau_1, \tau_2, \ldots; \hat{\zeta}, \zeta_1, \zeta_2, \ldots)$. Then, with $\tau_0 = 0$ and $\zeta_0 = \hat{\zeta}$,

$$\Phi(y) \geq J^{(\hat{w})}(y) = E^y \left[\int_0^{\tau_S} f(Y(t), u(t))dt + g(Y(\tau_S))\chi_{\{\tau_S < \infty\}} \right.$$
$$\left. + \sum_{i=0}^{\infty} K(\check{Y}(\tau_i^-), \zeta_i) \right] = K(y, \hat{\zeta}) + J^{(w)}(\hat{y}).$$

Combining the above we get

$$\Phi(\hat{y}) + K(y, \hat{\zeta}) < J^{(w)}(\hat{y}) + K(y, \hat{\zeta}) + \varepsilon$$
$$\leq \Phi(y) + \varepsilon$$
$$< \Phi(\hat{y}) + K(y, \hat{\zeta}) - \varepsilon,$$

a contradiction. □

Our first main result in this section is the following.

Theorem 9.8. *The value function*

$$\Phi(y) = \sup_{w \in \mathcal{W}} J^{(w)}(y)$$

of the combined stochastic control and impulse control problem (9.3.3) *is a viscosity solution of* (9.3.6) *and* (9.3.7).

Proof. By (9.3.2) and (9.3.5) we see that Φ satisfies (9.3.7).

(a) We first prove that Φ is a viscosity subsolution. To this end, choose $h \in C^2(\mathbb{R}^k)$ and $y_0 \in \mathcal{S}$ such that $h \geq \Phi$ on $\mathcal{S}$ and $h(y_0) = \Phi(y_0)$. We must prove that

$$\max \left(\sup_{\alpha \in U} \{L^\alpha h(y_0) + f(y_0, \alpha)\}, \ \mathcal{M}\Phi(y_0) - \Phi(y_0) \right) \geq 0. \tag{9.3.14}$$

If $\mathcal{M}\Phi(y_0) \geq \Phi(y_0)$ then (9.3.14) holds trivially, so we may assume that

$$\mathcal{M}\Phi(y_0) < \Phi(y_0). \qquad (9.3.15)$$

Choose $\epsilon > 0$ and let $w = (u, v) \in \mathcal{W}$, with $v = (\tau_1, \tau_2, \ldots; \zeta_1, \zeta_2, \ldots) \in \mathcal{V}$, be an ϵ-optimal portfolio, i.e.,

$$\Phi(y_0) < J^{(w)}(y_0) + \epsilon.$$

Since τ_1 is a stopping time we know that $\{\omega; \tau_1(\omega) = 0\}$ is $\mathcal{F}_0$-measurable and hence either

$$\tau_1(\omega) = 0 \text{ a.s.} \quad \text{or} \quad \tau_1(\omega) > 0 \text{ a.s.} \qquad (9.3.16)$$

If $\tau_1 = 0$ a.s. then $Y^{(w)}$ makes an immediate jump from y_0 to the point $y' = \Gamma(y_0, \zeta_1) \in \mathcal{S}$ and hence

$$\Phi(y_0) - \epsilon \leq J^{(w')}(y') + K(y_0, \zeta_1) \leq \Phi(y') + K(y_0, \zeta_1) \leq \mathcal{M}\Phi(y_0),$$

where $w' = (\tau_2, \tau_3, \ldots; \zeta_2, \zeta_3, \ldots)$.

This is a contradiction if $\epsilon < \Phi(y_0) - \mathcal{M}\Phi(y_0)$. This proves that (9.3.15) implies that it is impossible to have $\tau_1 = 0$ a.s.

So by (9.3.16), we can now assume that $\tau_1 > 0$ a.s. Choose $R < \infty$, $\rho > 0$ and define

$$\tau := \tau_1 \wedge R \wedge \inf\{t > 0 \,; \,| Y^{(w)}(t) - y_0| \geq \rho\}.$$

By the dynamic programming principle (see Lemma 7.3) we have: for each $\varepsilon > 0$, there exists a control u such that

$$\Phi(y_0) \leq E^{y_0}\left[\int_0^\tau f(Y(t), u(t))dt + \Phi(\check{Y}(\tau^-))\right] + \varepsilon, \qquad (9.3.17)$$

where, as before, $\check{Y}(\tau^-) = Y(\tau^-) + \Delta_N Y(\tau)$.

Choose $h_m \in C_0^2(\mathbb{R}^k)$ such that $h_m \to h$ and $L^u h_m \to L^u h$ pointwise dominatedly as $m \to \infty$. Then by (9.3.17) and the Dynkin formula we have, using that $\Phi \leq h$,

$$\Phi(y_0) \leq E^{y_0}\left[\int_0^\tau f(Y(t), u(t))dt + h(\check{Y}(\tau^-))\right] + \varepsilon$$

$$\leq \liminf_{m \to \infty} E^{y_0}\left[\int_0^\tau f(Y(t), u(t))dt + h_m(\check{Y}(\tau^-))\right] + \varepsilon$$

$$= h(y_0) + \liminf_{m \to \infty} E^{y_0}\left[\int_0^\tau L^u h_m(Y(t)) + f(Y(t), u(t))dt\right] + \varepsilon$$

$$= h(y_0) + E^{y_0}\left[\int_0^\tau L^u h(Y(t)) + f(Y(t), u(t))dt\right] + \varepsilon. \qquad (9.3.18)$$

Using that $h(y_0) = \Phi(y_0)$, we obtain

$$E^{y_0}\left[\int_0^\tau \{L^u h(Y(t)) + f(Y(t), u(t))\}dt\right] \geq -\varepsilon.$$

Dividing by $E^{y_0}[\tau]$ and letting $\rho \to 0$ we get

$$L^{\alpha_0} h(y_0) + f(y_0, \alpha_0) \geq -\varepsilon,$$

where

$$\alpha_0 = \lim_{s \to 0+} u(s).$$

Since ε is arbitrary, this proves (9.3.14) and hence that Φ is a viscosity subsolution.

(b) Next we prove that Φ is a viscosity supersolution. So we choose $h \in C^2(\mathbb{R}^k)$ and $y_0 \in S$ such that $h \leq \Phi$ on S and $h(y_0) = \Phi(y_0)$. We must prove that

$$\max\left(\sup_{\alpha \in U}\{L^\alpha h(y_0) + f(y_0, \alpha)\}, \mathcal{M}\Phi(y_0) - \Phi(y_0)\right) \leq 0. \qquad (9.3.19)$$

Since $\Phi \geq \mathcal{M}\Phi$ always (Lemma 9.7) it suffices to prove that

$$L^\alpha h(y_0) + f(y_0, \alpha) \leq 0 \text{ for all } \alpha \in U.$$

To this end, fix $\alpha \in U$ and let $w_\alpha = (\alpha, 0)$, i.e., w_α is the combined control $(u_\alpha, v_\alpha) \in \mathcal{W}$ where $u_\alpha = \alpha$ (constant) and $v_\alpha = 0$ (no interventions). Then by the dynamic programming principle and the Dynkin formula we have, with $Y(t) = Y^{(w_\alpha)}(t)$, $\tau = \tau_S \wedge \rho$, and h_m as in (a),

$$\Phi(y_0) \geq E^{y_0}\left[\int_0^\tau f(Y(s), \alpha)ds + \Phi(\check{Y}(\tau^-))\right]$$

$$\geq E^{y_0}\left[\int_0^\tau f(Y(s), \alpha)ds + h(\check{Y}(\tau^-))\right]$$

$$= h(y_0) + \lim_{m \to \infty} E^{y_0}\left[\int_0^\tau \{L^\alpha h_m(Y(t)) + f(Y(t), \alpha)\}dt\right]$$

$$= h(y_0) + E^{y_0}\left[\int_0^\tau \{L^\alpha h(Y(t)) + f(Y(t), \alpha)\}dt\right].$$

Hence

$$E\left[\int_0^\tau \{L^\alpha h(Y(t)) + f(Y(t), \alpha)\}dt\right] \leq 0.$$

Dividing by $E[\tau]$ and letting $\rho \to 0$ we get (9.3.19).

This completes the proof of Theorem 9.8. $\qquad\qquad\qquad\square$

Next we turn to the question of *uniqueness* of viscosity solutions of (9.3.10) and (9.3.11). Many types of uniqueness results can be found in the literature. See the references in the end of this section.

Here we give a proof in the case when the process $Y(t)$ has no jumps, i.e. when $N(\cdot, \cdot) = \nu(\cdot) = 0$. The method we use is a generalization of the method in [ØS, Theorem 3.8].

First we introduce some convenient notation:

Define $\Lambda : \mathbb{R}^{k \times k} \times \mathbb{R}^k \times \mathbb{R}^{\mathcal{S}} \times \mathbb{R}^k \to \mathbb{R}$ by

$$
\Lambda(R, r, \varphi, y) := \sup_{\alpha \in U} \left\{ \sum_{i=1}^{k} b_i(y, \alpha) r_i + \frac{1}{2} \sum_{i,j=1}^{k} (\sigma \sigma^T)_{ij}(y, \alpha) R_{ij} \right.
$$

$$
+ \sum_{j=1}^{\ell} \int_{\mathbb{R}} \Big\{ \varphi(y + \gamma^{(j)}(y, \alpha, z_j)) - \varphi(y)
$$

$$
\left. - r \cdot \gamma^{(j)}(y, \alpha, z_j) \Big\} \nu_j(\mathrm{d}z_j) + f(y, \alpha) \right\} \qquad (9.3.20)
$$

for $R = [R_{ij}] \in \mathbb{R}^{k \times k}$, $r = (r_i, \ldots, r_k) \in \mathbb{R}^k$, $\varphi : \mathcal{S} \to \mathbb{R}$, $y \in \mathbb{R}^k$, and define $F : \mathbb{R}^{k \times k} \times \mathbb{R}^k \times \mathbb{R}^{\mathcal{S}} \times \mathbb{R}^k \to \mathbb{R}$ by

$$
F(R, r, \varphi, y) = \max\{\Lambda(R, r, \varphi, y), \mathcal{M}\varphi(y) - \varphi(y)\}. \qquad (9.3.21)
$$

Note that if $\varphi \in C^2(\mathbb{R}^k)$ then

$$
\Lambda(D^2\varphi, D\varphi, \varphi, y) = \sup_{\alpha \in U} \left\{ L^\alpha \varphi(y) + f(y, \alpha) \right\},
$$

where

$$
D^2\varphi(y) = \left[\frac{\partial^2 \varphi}{\partial y_i \partial y_j} \right](y) \text{ and } D\varphi(y) = \left[\frac{\partial \varphi}{\partial y_i} \right](y).
$$

We recall the concepts of "superjets" $J_{\mathcal{S}}^{2,+}$, $J_{\mathcal{S}}^{2,-}$ and $\bar{J}_{\mathcal{S}}^{2,+}$, $\bar{J}_{\mathcal{S}}^{2,-}$ (see [CIL, Sect. 2]):

$$
J_{\mathcal{S}}^{2,+}\varphi(y) := \left\{ (R, r) \in \mathbb{R}^{k \times k} \times \mathbb{R}^k ; \right.
$$

$$
\left. \limsup_{\substack{\eta \to y \\ \eta \in \mathcal{S}}} \left[\varphi(\eta) - \varphi(y) - (\eta - y)^T r - \frac{1}{2}(\eta - y)^T R(\eta - y) \right] \cdot |\eta - y|^{-2} \le 0 \right\},
$$

$$
\bar{J}_{\mathcal{S}}^{2,+}\varphi(y) := \{ (R, r) \in \mathbb{R}^{k \times k} \times \mathbb{R}^k; \text{ for all } n \text{ there exists}
$$
$$
(R^{(n)}, r^{(n)}, y^{(n)}) \in \mathbb{R}^{k \times k} \times \mathbb{R}^k \times \mathcal{S} \text{ such that } (R^{(n)}, r^{(n)}) \in J_{\mathcal{S}}^{2,+}\varphi(y^{(n)}) \text{ and}
$$
$$
(R^{(n)}, r^{(n)}, \varphi(y^{(n)}), y^{(n)}) \to (R, r, \varphi(y), y) \text{ as } n \to \infty\}
$$

and
$$J_{\mathcal{S}}^{2,-}\varphi = -J_{\mathcal{S}}^{2,+}(-\varphi), \quad \bar{J}_{\mathcal{S}}^{2,-}\varphi = -\bar{J}_{\mathcal{S}}^{2,+}(-\varphi).$$

In terms of these superjets one can give an equivalent definition of viscosity solutions as follows.

Theorem 9.9. *[CIL, Sect. 2]*

(i) A function $\varphi \in C(\mathcal{S})$ is a viscosity subsolution of (9.3.10) and (9.3.11) if and only if (9.3.11) holds and

$$\max(\Lambda(R, r, \varphi, y), \mathcal{M}\varphi(y) - \varphi(y)) \geq 0 \text{ for all } (R, r) \in \bar{J}_{\mathcal{S}}^{2,+}\varphi(y), \ y \in \mathcal{S}.$$

(ii) A function $\varphi \in C(\mathcal{S})$ is a viscosity supersolution of (9.3.10) and (9.3.11) if and only if (9.3.11) holds and

$$\max(\Lambda(R, r, \varphi, y), \mathcal{M}\varphi(y) - \varphi(y)) \leq 0 \text{ for all } (R, r) \in \bar{J}_{\mathcal{S}}^{2,-}\varphi(y), \ y \in \mathcal{S}.$$

We have now ready for the second main theorem of this section.

Theorem 9.10 (Comparison Theorem).
Assume that
$$N(\cdot, \cdot) = 0. \tag{9.3.22}$$

Suppose that there exists a positive function $\beta \in C^2(\bar{\mathcal{S}})$ which satisfies the strict quasivariational inequality

$$\max\left(\sup_{\alpha \in U}\{L^\alpha \beta(y)\}, \sup_{\zeta \in \mathcal{Z}} \beta(\Gamma(y, \zeta)) - \beta(y)\right) \leq -\delta(y) < 0, \quad y \in \mathcal{S}, \tag{9.3.23}$$

where $\delta(y) > 0$ is bounded away from 0 on compact subsets of $\mathcal{S}$.

Let u be a viscosity subsolution and v be a viscosity supersolution of (9.3.10) and (9.3.11) and suppose that

$$\lim_{|y| \to \infty}\left\{\frac{u^+(y)}{\beta(y)} + \frac{v^-(y)}{\beta(y)}\right\} = 0. \tag{9.3.24}$$

Then
$$u(y) \leq v(y) \quad \text{for all } y \in \mathcal{S}.$$

Proof. (Sketch) We argue by contradiction. Suppose that
$$\sup_{y \in \mathcal{S}}\{u(y) - v(y)\} > 0.$$

Then by (9.3.24) there exists $\epsilon > 0$ such that if we put
$$v_\epsilon(y) := v(y) + \epsilon\beta(y), \quad y \in \mathcal{S}$$

then

$$M := \sup_{y \in \mathcal{S}}\{u(y) - v_\epsilon(y)\} > 0.$$

For $n = 1, 2, \ldots$ and $(x, y) \in \mathcal{S} \times \mathcal{S}$ define

$$H_n(x, y) := u(x) - v(y) - \frac{n}{2}|x - y|^2 - \frac{\epsilon}{2}(\beta(x) + \beta(y))$$

and

$$M_n := \sup_{(x,y) \in \mathcal{S} \times \mathcal{S}} H_n(x, y).$$

Then by (9.3.24) we have

$$0 < M_n < \infty \qquad \text{for all } n,$$

and there exists $(x^{(n)}, y^{(n)}) \in \mathcal{S} \times \mathcal{S}$ such that

$$M_n = H_n(x^{(n)}, y^{(n)}).$$

Then by Lemma 3.1 in [CIL] the following holds:

$$\lim_{n \to \infty} n|x^{(n)} - y^{(n)}|^2 = 0$$

and

$$\lim_{n \to \infty} M_n = u(\hat{y}) - v_\epsilon(\hat{y}) = \sup_{y \in \mathcal{S}}\{u(y) - v_\epsilon(y)\} = M,$$

for any limit point $\hat{y}$ of $\{y^{(n)}\}_{n=1}^{\infty}$.

Since v is a supersolution of (9.3.10), (9.3.11), and (9.3.23) holds, we see that v_ϵ is a *strict* supersolution of (9.3.10), in the sense that $\varphi = v_\epsilon$ satisfies (9.3.13) in the following strict form:

$$\max\left(\sup_{\alpha \in U}\{L^\alpha h(y_0) + f(y_0, \alpha)\}, \ \mathcal{M}v_\epsilon(y_0) - v_\epsilon(y_0)\right) \leq -\delta(y_0),$$

with $\delta(\cdot)$ as in (9.3.23).

By [CIL, Theorem 3.2], there exist $k \times k$ matrices $P^{(n)}, Q^{(n)}$ such that, if we put

$$p^{(n)} = q^{(n)} = n(x^{(n)} - y^{(n)})$$

then

$$(P^{(n)}, p^{(n)}) \in \bar{J}^{2,+}u(x^{(n)}) \quad \text{and} \quad (Q^{(n)}, q^{(n)}) \in \bar{J}^{2,-}v_\epsilon(y^{(n)})$$

and

$$\begin{bmatrix} P^{(n)} & 0 \\ 0 & -Q^{(n)} \end{bmatrix} \leq 3n \begin{bmatrix} I & -I \\ -I & I \end{bmatrix},$$

in the sense that

$$\xi^T P^{(n)} \xi - \eta^T Q^{(n)} \eta \leq 3n|\xi - \eta|^2 \quad \text{for all } \xi, \eta \in \mathbb{R}^k. \tag{9.3.25}$$

Since u is a subsolution we have, by Theorem 9.9,

$$\max\left(\Lambda(P^{(n)}, p^{(n)}, u, x^{(n)}), \mathcal{M}u(x^{(n)}) - u(x^{(n)})\right) \geq 0 \tag{9.3.26}$$

and since v_ϵ is a supersolution we have

$$\max\left(\Lambda(Q^{(n)}, q^{(n)}, v_\epsilon, y^{(n)}), \mathcal{M}v_\epsilon(y^{(n)}) - v_\epsilon(y^{(n)})\right) \leq 0. \tag{9.3.27}$$

By (9.3.25) we get

$$\Lambda(P^{(n)}, p^{(n)}, u, x^{(n)}) - \Lambda(Q^{(n)}, q^{(n)}, v_\epsilon, y^{(n)})$$

$$\leq \sup_{\alpha \in U} \left\{ \sum_{i=1}^k (b_i(x^{(n)}, \alpha) - b_i(y^{(n)}, \alpha))(p_i^{(n)} - q_i^{(n)}) \right.$$

$$\left. + \frac{1}{2} \sum_{i,j=1}^k \left[(\sigma\sigma^T)_{ij}(x^{(n)}, \alpha) - (\sigma\sigma^T)_{ij}(y^{(n)}, \alpha) \right] (P_{ij}^{(n)} - Q_{ij}^{(n)}) \right\}$$

$$\leq 0.$$

Therefore, by (9.3.27),

$$\Lambda(P^{(n)}, p^{(n)}, u, x^{(n)}) \leq \Lambda(Q^{(n)}, q^{(n)}, v_\epsilon, y^{(n)}) \leq 0$$

and hence, by (9.3.26),

$$\mathcal{M}u(x^{(n)}) - u(x^{(n)}) \geq 0. \tag{9.3.28}$$

On the other hand, since v_ϵ is a strict supersolution we have

$$\mathcal{M}v_\epsilon(y^{(n)}) - v_\epsilon(y^{(n)}) < -\delta \quad \text{for all } n, \tag{9.3.29}$$

for some constant $\delta > 0$.

Combining the above we get

$$M_n < u(x^{(n)}) - v_\epsilon(y^{(n)}) < \mathcal{M}u(x^{(n)}) - \mathcal{M}v_\epsilon(y^{(n)}) - \delta$$

and hence

$$M = \lim_{n \to \infty} M_n \leq \lim_{n \to \infty} (\mathcal{M}u(x^{(n)}) - \mathcal{M}v_\epsilon(y^{(n)}) - \delta)$$

$$\leq \mathcal{M}u(\hat{y}) - \mathcal{M}v_\epsilon(\hat{y}) - \delta$$

$$= \sup_{\zeta \in \mathcal{Z}}\{u(\Gamma(\hat{y}, \zeta)) + K(\hat{y}, \zeta)\} - \sup_{\zeta \in \mathcal{Z}}\{v_\epsilon(\Gamma(\hat{y}, \zeta)) + K(\hat{y}, \zeta)\} - \delta$$

$$\leq \sup_{\zeta \in \mathcal{Z}}\{u(\Gamma(\hat{y}, \zeta)) - v_\epsilon(\Gamma(\hat{y}, \zeta))\} - \delta \leq M - \delta.$$

This contradiction proves Theorem 9.10. $\square$

Theorem 9.11 (Uniqueness of Viscosity Solutions).
Suppose that the process $Y(t)$ has no jumps, i.e.,

$$N(\cdot, \cdot) = 0$$

and let $\beta \in C^2(\bar{S})$ be as in Theorem 9.10. Then there is at most one viscosity solution φ of (9.3.10) and (9.3.11) with the property that

$$\lim_{|y| \to \infty} \frac{|\varphi(y)|}{\beta(y)} = 0. \tag{9.3.30}$$

Proof. Let φ_1, φ_2 be two viscosity solutions satisfying (9.3.30). If we apply Theorem 9.10 to $u = \varphi_1$ and $v = \varphi_2$ we get

$$\varphi_1 \leq \varphi_2.$$

If we apply Theorem 9.10 to $u = \varphi_2$ and $v = \varphi_1$ we get

$$\varphi_2 \leq \varphi_1.$$

Hence $\varphi_1 = \varphi_2$.   □

Example 9.12 (Optimal Consumption and Portfolio with Both Fixed and Proportional Transaction Costs (2)).
Let us return to Example 8.3. In this case (9.3.10) takes the form (8.2.21) and (8.2.22) in S^0. For simplicity we assume Dirichlet boundary conditions, e.g., $\psi = 0$, on ∂S. Fix $\gamma' \in (\gamma, 1)$ such that (see (3.1.8))

$$\rho > \gamma' \left[r + \frac{(\mu - r)^2}{2\sigma^2(1 - \gamma)} \right]$$

and define

$$\beta(x_1, x_2) = (x_1 + x_2)^{\gamma'}. \tag{9.3.31}$$

Then with $\mathcal{M}$ as in (8.2.22) we have

$$(\mathcal{M}\beta - \beta)(x_1, x_2) \leq (x_1 + x_2)^{\gamma'} \left[\left(1 - \frac{c}{x_1 + x_2} \right)^{\gamma'} - 1 \right] < 0. \tag{9.3.32}$$

Moreover, with

$$L^u \psi(x_1, x_2) := -\rho\psi(x_1, x_2) + (rx_1 - u)\frac{\partial\psi}{\partial x_1}(x_1, x_2) + \mu x_2 \frac{\partial\psi}{\partial x_2}(x_1, x_2)$$

$$+ \frac{1}{2}\sigma^2 x_2^2 \frac{\partial^2\psi}{\partial x_2^2}(x_1, x_2), \quad \psi \in C^2(\mathbb{R}^2) \tag{9.3.33}$$

we get

$$\max_{u \geq 0} L^u \beta(x_1, x_2) < 0, \tag{9.3.34}$$

and in both (9.3.32) and (9.3.34) the strict inequality is uniform on compact subsets of $\mathcal{S}^0$. The proofs of these inequalities are left as an exercise (Exercise 9.3).

We conclude that the function β in (9.3.31) satisfies the conditions (9.3.23) of Theorem 9.10. Thus by Theorem 9.11 we have in this example uniqueness of viscosity solutions φ satisfying the growth condition

$$\lim_{|(x_1,x_2)|\to\infty} (x_1 + x_2)^{-\gamma'} |\varphi(x_1, x_2)| = 0. \qquad (9.3.35)$$

For other results regarding uniqueness of viscosity solutions of equations associated to impulse control, stochastic control and optimal stopping for jump diffusions, we refer to [Am, AKL, BKR2, CIL, Is1, Is2, Isk, MS, AT, Ph, JK1, FS, BCa, BCe] and the references therein.

9.4 Numerical Analysis of HJBQVI

In this section we give some insights in the numerical solution of HJBQVI. We refer, e.g., to [LST] for details on the finite difference approximations and the description of the algorithms to solve dynamic programming equations. Here we focus on the main problem which arises in the case of quasivariational inequalities, i.e., the presence of a nonexpansive operator due to the intervention operator.

9.4.1 Finite Difference Approximation

We want to solve the following HJBQVI numerically

$$\max\left(\sup_{\alpha\in U}\{L^\alpha\Phi(x) + f(x,\alpha)\}, \mathcal{M}\Phi(x) - \Phi(x) \right) = 0, \quad x \in \mathcal{S} \qquad (9.4.1)$$

with boundary values

$$\Phi(x) = g(x), \quad x \in \partial\mathcal{S}, \qquad (9.4.2)$$

where

$$L^\alpha\Phi(x) = -\rho\Phi + \sum_{i=1}^k b_i(x,\alpha)\frac{\partial\Phi}{\partial x_i} + \frac{1}{2}\sum_{i,j=1}^k a_{ij}(x,\alpha)\frac{\partial^2\Phi}{\partial x_i\partial x_j} \qquad (9.4.3)$$

and

$$\mathcal{M}\Phi(x) = \sup\left\{\Phi(\Gamma(x,\zeta)) + K(x,\zeta); \zeta \in \mathcal{Z}, \Gamma(x,\zeta) \in \mathcal{S}\right\}. \qquad (9.4.4)$$

We have denoted here $a_{ij} := (\sigma\sigma^T)_{ij}$. We shall also write $K^\zeta(x)$ for $K(x,\zeta)$.

We assume that $\mathcal{S}$ is bounded, otherwise a change of variable or a localiza-tion procedure has to be performed in order to reduce to a bounded domain. Moreover we assume for simplicity that $\mathcal{S}$ is a box, i.e., a cartesian prod-uct of bounded intervals in $\mathbb{R}^k$. We can also handle Neumann type boundary conditions without additional difficulty.

We discretize (9.4.1) by using a finite difference approximation. Let δ_i denote the finite difference step in each coordinate direction and set $\delta = (\delta_1, \ldots, \delta_k)$. Denote by e_i the unit vector in the ith coordinate direction, and consider the grid $\mathcal{S}_\delta = \mathcal{S} \cap \prod_{i=1}^{k}(\delta_i \mathbb{Z})$. Set $\partial \mathcal{S}_\delta = \partial \mathcal{S} \cap \prod_{i=1}^{k}(\delta_i \mathbb{Z})$. We use the following approximations:

$$\frac{\partial \Phi}{\partial x_i}(x) \sim \frac{\Phi(x + \delta_i e_i) - \Phi(x - \delta_i e_i)}{2\delta_i} \equiv \partial_i^{\delta_i}\Phi(x) \tag{9.4.5}$$

or (see (9.4.16))

$$\frac{\partial \Phi}{\partial x_i}(x) \sim \begin{cases} \dfrac{\Phi(x + \delta_i e_i) - \Phi(x)}{\delta_i} \equiv \partial_i^{\delta_i+}\Phi(x) \text{ if } b_i(x) \geq 0, \\[4mm] \dfrac{\Phi(x) - \Phi(x - \delta_i e_i)}{\delta_i} \equiv \partial_i^{\delta_i-}\Phi(x) \text{ if } b_i(x) \leq 0. \end{cases} \tag{9.4.6}$$

$$\frac{\partial^2 \Phi}{\partial x_i^2}(x) \sim \frac{\Phi(x + \delta_i e_i) - 2\Phi(x) + \Phi(x - \delta_i e_i)}{\delta_i^2} \equiv \partial_{ii}^{\delta_i}\Phi(x). \tag{9.4.7}$$

If $a_{ij}(x) \geq 0$, $i \neq j$, then

$$\frac{\partial^2 \Phi}{\partial x_i \partial x_j}(x) \sim \frac{2\Phi(x) + \Phi(x + \delta_i e_i + \delta_j e_j) + \Phi(x - \delta_i e_i - \delta_j e_j)}{2\delta_i \delta_j}$$
$$- \left[\frac{\Phi(x + \delta_i e_i) + \Phi(x - \delta_i e_i) + \Phi(x + \delta_j e_j) + \Phi(x - \delta_j e_j)}{2\delta_i \delta_j} \right]$$
$$\equiv \partial_{ij}^{\delta_i \delta_j+}\Phi(x). \tag{9.4.8}$$

If $a_{ij}(x) < 0$, $i \neq j$, then

$$\frac{\partial^2 \Phi}{\partial x_i \partial x_j}(x) \sim -\frac{[2\Phi(x) + \Phi(x + \delta_i e_i - \delta_j e_j) + \Phi(x - \delta_i e_i + \delta_j e_j)]}{2\delta_i \delta_j}$$
$$+ \frac{\Phi(x + \delta_i e_i) + \Phi(x - \delta_i e_i) + \Phi(x + \delta_j e_j) + \Phi(x - \delta_j e_j)}{2\delta_i \delta_j}$$
$$\equiv \partial_{ij}^{\delta_i \delta_j-}\Phi(x). \tag{9.4.9}$$

These approximations can be justified when the function Φ is smooth by Taylor expansions. Using approximations (9.4.5), (9.4.7)–(9.4.9), we obtain the following approximation problem:

$$\max \left(\sup_{\alpha \in U} \{ L_\delta^\alpha \Phi_\delta(x) + f(x, \alpha) \}, \mathcal{M}_\delta \Phi_\delta(x) - \Phi_\delta(x) \right) = 0 \quad \text{for all } x \in \mathcal{S}_\delta,$$

$$\Phi_\delta(x) = g(x) \quad \text{for all } x \in \partial \mathcal{S}_\delta,$$

$$(9.4.10)$$

where

$$L_\delta^\alpha \Phi(x) = \Phi(x) \left\{ \sum_{i=1}^{k} \frac{-a_{ii}(x, \alpha)}{\delta_i^2} + \sum_{j \neq i} \frac{|a_{ij}(x, \alpha)|}{2\delta_i \delta_j} - \rho \right\}$$

$$+ \frac{1}{2} \sum_{i, \kappa = \pm 1} \Phi(x + \kappa \delta_i e_i) \left\{ \frac{a_{ii}(x, \alpha)}{\delta_i^2} - \sum_{j, j \neq i} \frac{|a_{ij}(x, \alpha)|}{\delta_i \delta_j} + \kappa \frac{b_i(x, \alpha)}{\delta_i} \right\}$$

$$+ \frac{1}{2} \sum_{i \neq j, \kappa = \pm 1, \lambda = \pm 1} \Phi(x + \kappa e_i \delta_i + \lambda e_j \delta_j) \frac{a_{ij}(x, \alpha)^{[\kappa \lambda]}}{\delta_i \delta_j}$$

$$(9.4.11)$$

and

$$\mathcal{M}_\delta \Phi_\delta(x) = \sup \left\{ \Phi(\Gamma(x, \zeta)) + K(x, \zeta); \zeta \in \mathcal{Z}_\delta(x) \right\} \qquad (9.4.12)$$

with

$$\mathcal{Z}_\delta(x) = \{ \zeta \in \mathcal{Z}, \Gamma(x, \zeta) \in \mathcal{S}_\delta \}. \qquad (9.4.13)$$

We have used here the notation

$$a_{ij}^{[\kappa \lambda]}(x, \alpha) = \begin{cases} , a_{ij}^+(x, \alpha) \equiv \max(0, a_{ij}(x, \alpha)) & \text{if } \kappa \lambda = 1, \\ a_{ij}^-(x, \alpha) \equiv -\min(0, a_{ij}(x, \alpha)) & \text{if } \kappa \lambda = -1. \end{cases}$$

In (9.4.10), Φ_δ denotes an approximation of Φ at the grid points. This approximation is consistent and stable if the following condition holds: (see [LST] for a proof)

$$|b_i(x, \alpha)| \leq \frac{a_{ii}(x, \alpha)}{\delta_i} - \sum_{j \neq i} \frac{|a_{ij}(x, \alpha)|}{\delta_j} \qquad \text{for all } \alpha \text{ in } U, x \text{ in } \mathcal{S}_\delta, i = 1, \ldots, k.$$

$$(9.4.14)$$

In this case ϕ_δ converges to the viscosity solution of (9.4.1) when the step δ goes to 0. This can be proved by using techniques introduced by Barles and Souganidis [BS], provided a comparison theorem holds for viscosity sub- and supersolutions of the continuous-time problem.

If (9.4.14) does not hold but only the following weaker condition

$$0 \leq \frac{a_{ii}(x, \alpha)}{\delta_i} - \sum_{j \neq i} \frac{|a_{ij}(x, \alpha)|}{\delta_j} \qquad \text{for all } \alpha \text{ in } U, x \text{ in } \mathcal{S}_\delta, i = 1 \ldots k. \quad (9.4.15)$$

is satisfied, then it can be shown that we can also obtain a stable approximation (but of lower order) by using the one-sided approximations (9.4.6) for the approximation of the gradient instead of the centered difference (9.4.5). Instead of (9.4.11), the operator L_δ^α is then equal to

$$
L_\delta^\alpha \Phi(x) = \Phi(x) \left\{ \sum_{i=1}^{k} \frac{-a_{ii}(x,\alpha)}{\delta_i^2} + \sum_{j\neq i} \frac{|a_{ij}(x,\alpha)|}{2\delta_i\delta_j} - \frac{|b_i(x,\alpha)|}{\delta_i} - \rho \right\}
$$

$$
+ \frac{1}{2} \sum_{i,\kappa=\pm 1} \Phi(x + \kappa\delta_i e_i) \left\{ \frac{a_{ii}(x,\alpha)}{\delta_i^2} - \sum_{j,j\neq i} \frac{|a_{ij}(x,\alpha)|}{\delta_i\delta_j} + \frac{b_i(x,\alpha)^{[\kappa]}}{\delta_i} \right\}
$$

$$
+ \frac{1}{2} \sum_{i\neq j,\kappa=\pm 1,\lambda=\pm 1} \Phi(x + \kappa e_i\delta_i + \lambda e_j\delta_j) \frac{a_{ij}(x,\alpha)^{[\kappa\lambda]}}{\delta_i\delta_j}. \tag{9.4.16}
$$

By replacing the values of the function Φ_δ by their known values on the boundary, we obtain the following equation in $\mathcal{S}_\delta$:

$$
\max\left(\sup_{\alpha\in U}\{\bar{L}_\delta^\alpha \Phi_\delta(x) + f_\delta(x,\alpha)\}, \mathcal{M}_\delta\Phi_\delta(x) - \Phi_\delta(x)\right) = 0, \quad x \in \mathcal{S}_\delta, \tag{9.4.17}
$$

where $\bar{L}_\delta^\alpha$ is a square $N_\delta \times N_\delta$ matrix, obtained by retrieving the first and last column from L_δ^α, $N_\delta = \mathrm{Card}(\mathcal{S}_\delta)$, i.e., the number of points of the grid, and $f_\delta(x,\alpha)$ (which will also be denoted by $f_\delta^\alpha(x)$) takes into account the boundary values.

9.4.2 A Policy Iteration Algorithm for HJBQVI

When the stability conditions (9.4.14) or (9.4.15) hold, then the matrix $\bar{L}_\delta^\alpha$ is diagonally dominant, i.e.,

$$
(\bar{L}_\delta^\alpha)_{ij} \geq 0 \text{ for } i \neq j \quad \text{and} \quad \sum_{j=1}^{N_\delta}(\bar{L}_\delta^\alpha)_{ij} \leq -\rho < 0 \quad \text{for all } i = 1,\dots,N_\delta.
$$

Now let h be a positive number such that

$$
h \leq \min_i \frac{1}{|(\bar{L}_\delta^\alpha)_{ii} + \rho|} \tag{9.4.18}
$$

and let I_δ denote the $N_\delta \times N_\delta$ identity matrix. It is easy to check that the matrix

$$
P_\delta^\alpha := I_\delta + h(\bar{L}_\delta^\alpha + \rho I_\delta)
$$

is sub-Markovian, i.e., $(P_\delta^\alpha)_{ij} \geq 0$ for all i,j and $\sum_{j=1}^{N_\delta}(P_\delta^\alpha)_{ij} \leq 1$ for all i. Consequently (9.4.17) can be rewritten as

$$
\max\left(\sup_{\alpha\in U}\left\{\frac{1}{h}\left(P_\delta^\alpha\Phi_\delta(x) - (1+\rho h)\Phi_\delta(x)\right) + f_\delta^\alpha(x)\right\}, \mathcal{M}_\delta\Phi_\delta(x) - \Phi_\delta(x)\right) = 0, \tag{9.4.19}
$$

which is equivalent to

$$\Phi_\delta(x) = \max\left(\sup_{\alpha \in U} \mathcal{L}_\delta^\alpha \Phi_\delta(x), \sup_{\zeta \in \mathcal{Z}_\delta(x)} B^\zeta \Phi_\delta(x)\right), \tag{9.4.20}$$

where

$$\mathcal{L}_\delta^\alpha \Phi(x) := \frac{P_\delta^\alpha \Phi(x) + h f_\delta^\alpha(x)}{1 + \rho h}, \tag{9.4.21}$$

$$B^\zeta \Phi(x) := \Phi(\Gamma(x, \zeta)) + K^\zeta(x). \tag{9.4.22}$$

Let $\mathcal{P}(\mathcal{S}_\delta)$ denote the set of all subsets of $\mathcal{S}_\delta$ and for (T, α, ζ) in $\mathcal{P}(\mathcal{S}_\delta) \times U \times \mathcal{Z}_\delta$, denote by $\mathcal{O}_{T,\alpha,\zeta}$ the operator:

$$\mathcal{O}_{T,\alpha,\zeta} v(x) := \begin{cases} \mathcal{L}_\delta^\alpha v(x) & \text{if } x \in \mathcal{S}_\delta \backslash T, \\ B^\zeta v(x) & \text{if } x \in T. \end{cases} \tag{9.4.23}$$

Problem (9.4.20) is equivalent to the fixed point problem

$$\Phi_\delta(x) = \sup_{T \in \mathcal{P}(\mathcal{S}_\delta), \alpha \in U, \zeta \in \mathcal{Z}_\delta} \mathcal{O}_{T,\alpha,\zeta} \Phi_\delta(x).$$

We define $\mathbf{T}_{\mathrm{ad}}$ as

$$\mathbf{T}_{\mathrm{ad}} := \mathcal{P}(\mathcal{S}_\delta) \backslash \mathcal{S}_\delta$$

and restrict ourselves to the following problem

$$\Phi_\delta(x) = \sup_{T \in \mathbf{T}_{\mathrm{ad}}, \alpha \in U, \zeta \in \mathcal{Z}_\delta} \mathcal{O}_{T,w,z} \Phi_\delta(x) =: \mathcal{O} \Phi_\delta(x). \tag{9.4.24}$$

In other words, it is not admissible to make interventions at all points of $\mathcal{S}_\delta$ (i.e., the continuation region is never the empty set). We can always assume that we order the points of the grid in such a way that it is not admissible to intervene at $x_1 \in \mathcal{S}_\delta$.

The operator $\mathcal{L}_\delta^\alpha$ is contractive (because $\|P\|_\infty \le 1$ and $rh > 0$) and satisfies the discrete maximum principle, i.e.,

$$\mathcal{L}_\delta^\alpha v_1 - \mathcal{L}_\delta^\alpha v_2 \le v_1 - v_2 \Rightarrow v_1 - v_2 \ge 0. \tag{9.4.25}$$

(If v is a function from $\mathcal{S}_\delta$ into $\mathbb{R}$, $v \ge 0$ means $v(x) \ge 0$ for all $x \in \mathcal{S}_\delta$.)

The operator B^ζ is nonexpansive and we need some additional hypothesis in order to be able to use a policy iteration algorithm for computing a solution of (9.4.21). We assume

There exists an integer function $\sigma : \{1, 2, \ldots, N_\delta\} \times \mathcal{Z}_\delta \to \{1, 2, \ldots, N_\delta\}$ such that for all $\zeta \in \mathcal{Z}_\delta$ and all $i = 1, \ldots, N_\delta$

$$\Gamma(x_i, \zeta) = x_{\sigma(i,\zeta)} \quad \text{with } \sigma(i, \zeta) < i. \tag{9.4.26}$$

The operator B_ζ defined in (9.4.22) can be rewritten as

$$B^\zeta v = \mathbf{B}^\zeta v + K^\zeta,$$

where $(\mathbf{B}^\zeta, \zeta \in \mathcal{Z}_\delta)$ is a family of $N_\delta \times N_\delta$ Markovian matrices (except for the first row) defined by: $\mathbf{B}^z_{i,j} = 1$ if $j = \sigma(i, z)$ and $i \neq 1$, and 0 elsewhere.

Let $\zeta(\cdot)$ be a feedback Markovian control from $\mathcal{S}_\delta$ into $\mathcal{Z}_\delta$, and define the function $\bar{\sigma}$ on $\mathcal{S}_\delta$ by $\bar{\sigma}(x) := \sigma(x, \zeta(x))$. Condition (9.4.26) implies that the pth composition of $\bar{\sigma}$ starting in $T \in \mathbf{T}_{\mathrm{ad}}$ will end up in $\mathcal{S}_\delta \backslash T$ after a finite number of iterations.

We can now consider the following *Howard* or *policy iteration* algorithm to solve problem (9.4.20) in the finite set $\mathcal{S}_\delta$. It consists of constructing two sequences of feedback Markovian policies $\{(T_k, \alpha_k, \zeta_k), k \in \mathbb{N}\}$ and functions $\{v_k, k \in \mathbb{N}\}$ as follows. Let v_0 be a given initial function in $\mathcal{S}_\delta$. For $k \geq 0$ we do the following iterations:

- (step $2k$) Given v_k, compute a feedback Markovian admissible policy $(T_{k+1}, \alpha_{k+1}, \zeta_{k+1})$ such that

$$(T_{k+1}, \alpha_{k+1}, \zeta_{k+1}) \in \underset{T, \alpha, \zeta}{\mathrm{Argmax}}\{\mathcal{O}_{T, \alpha, \zeta} v_k\}. \tag{9.4.27}$$

In other words

$$\alpha_{k+1}(x) \in \underset{\alpha \in U}{\mathrm{Argmax}}\, \mathcal{L}^\alpha_\delta v_k(x); \quad \text{for all } x \text{ in } \mathcal{S}_\delta,$$

$$\zeta_{k+1}(x) \in \underset{\beta \in \mathcal{Z}_\delta}{\mathrm{Argmax}}\, B^\zeta_\delta v_k(x); \quad \text{for all } x \text{ in } \mathcal{S}_\delta,$$

$$T_{k+1} = \{x \in \mathcal{S}_\delta, \mathcal{L}^{\alpha_{k+1}(x)}_\delta v_k(x) > B^{\zeta_{k+1}(x)}_\delta v_k(x)\}.$$

- (step $2k + 1$) Compute v_{k+1} as the solution of

$$v_{k+1} = \mathcal{O}_{T_{k+1}, \alpha_{k+1}, \zeta_{k+1}} v_{k+1}. \tag{9.4.28}$$

Set $k \leftarrow k + 1$ and go to step $2k$.

It can be proved that if (9.4.15), (9.4.18), and (9.4.26) hold, then the sequence $\{v_k\}$ converges to the solution Φ_δ of (9.4.20) and the sequence $\{(T_k, \alpha_k, \zeta_k)\}$ converges to the optimal feedback Markovian strategy. See [CMS] for a proof and [BT] for similar problems. For more information on the Howard algorithm, we refer to [Pu, LST]. For complements on numerical methods for HJB equations we refer, e.g., to [KD, LST].

Example 9.13 (Optimal Consumption and Portfolio with Both Fixed and Proportional Transaction Costs (3)). We go back to Example 9.12. We want to solve (8.2.21) numerically. We assume now that $\mathcal{S} = (0, l) \times (0, l)$ with $l > 0$, and that the following boundary conditions hold:

$$\psi(0, x_2) = \psi(x_1, 0) = 0,$$

$$\frac{\partial \psi}{\partial x_1}(l, x_2) = \frac{\partial \psi}{\partial x_2}(x_1, l) = 0 \quad \text{for all } (x_1, x_2) \text{ in } (0, l) \times (0, l).$$

Moreover we assume that the consumption is bounded by $u_{max} > 0$ so that $U = [0, u_{max}]$. Let $\delta > 0$ be a positive step and let $\mathcal{S}_\delta = \{(i\delta, j\delta), i, j \in \{1, \ldots, N\}\}$ be the finite difference grid (we suppose that $N = l/\delta$ is an integer). We denote by ψ_δ the approximation of ψ on the grid. We approximate the operator L^u defined in (9.3.33) by the following finite difference operator on $\mathcal{S}_\delta$:

$$L_\delta^u \psi := -\rho \psi + r x_1 \partial_1^{\delta+} \psi + \mu x_2 \partial_2^{\delta+} \psi - u \partial_1^{\delta-} \psi + \frac{1}{2} \sigma^2 x_2^2 \partial_{22}^{\delta+} \psi$$

and set the following boundary values:

$$\psi_\delta(0, x_2) = \psi_\delta(x_1, 0) = 0,$$

$$\psi_\delta(l - \delta, x_2) = \psi_\delta(l, x_2),$$

$$\psi_\delta(x_1, l - \delta) = \psi_\delta(x_1, l).$$

We then obtain a stable approximation. Take now

$$h \leq \frac{r x_1}{\delta} + \frac{\mu x_2}{\delta} + \left(\frac{\sigma x_2}{\delta}\right)^2 + \frac{u_{max}}{\delta}.$$

We obtain a problem of the form (9.4.20). In order to be able to apply the Howard algorithm described above, it remains to check that (9.4.26) holds. This is indeed the case since a finite number of transactions brings the state to the continuation region. The details are left as an exercise.

This problem is solved in [CØS] by using another numerical method based on the iterative methods of Chap. 7.

9.5 Exercises

Exercise* 9.1. Let $k > 0$ be a constant and define

$$G(x) = \begin{cases} K|x| & \text{for } -\frac{1}{K} \leq x \leq \frac{1}{K}, \\ 1 & \text{for } |x| > \frac{1}{K}. \end{cases}$$

Solve the optimal stopping problem

$$\Phi(s, x) = \sup_{\tau \geq 0} E^x \left[e^{-\rho(s+\tau)} G(B(\tau)) \right],$$

where $B(t)$ is a one-dimensional Brownian motion starting at $x \in \mathbb{R}$. Distinguish between the two cases

(a) $K \leq \sqrt{2\rho}/z$, where $z > 0$ is the unique positive solution of the equation

$$\text{tgh}(z) = \frac{1}{z},$$

and

$$\text{tgh}(z) = \frac{e^z - e^{-z}}{e^z + e^{-z}}.$$

(b) $K > \sqrt{2\rho}/z$.

Exercise* 9.2. Assume that the state $X(t) = X^{(w)}(t)$ at time t obtained by using a combined control $w = (u, v)$, where $u = u(t, \omega) \in \mathbb{R}$ and $v = (\tau_1, \tau_2, \ldots ; \zeta_1, \zeta_2, \ldots)$ with $\zeta_i \in \mathbb{R}$ given by

$$dX(t) = u(t)dt + dB(t) + \int_{\mathbb{R}} z\tilde{N}(dt, dz), \quad \tau_i \leq t < \tau_{i+1},$$

$$X(\tau_{i+1}) = X(\tau_{i+1}^-) + \Delta_N X(\tau_{i+1}) + \zeta_{i+1}, \quad X(0^-) = x \in \mathbb{R}.$$

Assume that the cost of applying such a control is

$$J^{(w)}(s, x) = E^x \left[\int_0^\infty e^{-\rho(s+t)}(X^{(w)}(t)^2 + \theta u(t)^2)dt + c \sum_i e^{-\rho(s+\tau_i)} \right],$$

where ρ, θ, and c are positive constants. Consider the problem to find $\Phi(s, x)$ and $w^* = (u^*, v^*)$ such that

$$\Phi(s, x) = \inf_w J^{(w)}(s, x) = J^{(w^*)}(s, x). \tag{9.5.1}$$

Let

$$\Phi_1(s, x) = \inf_u J^{(u,0)}(s, x)$$

be the value function if we de not allow any impulse control (i.e., $v = 0$) and let

$$\Phi_2(s, x) = \inf_v J^{(0,v)}(s, x)$$

be the value function if u is fixed equal to 0, and only impulse controls are allowed. (See Exercises 3.4 and 6.1, respectively.)

Prove that for $i = 1, 2$, there exists $(s, x) \in \mathbb{R} \times \mathbb{R}$ such that

$$\Phi(s, x) < \Phi_i(s, x).$$

In other words, no matter how the positive parameter values ρ, θ, and c are chosen it is never optimal for the problem (9.5.1) to choose $u = 0$ or $v = 0$ (compare with Exercise 8.2).

[Hint: Use Theorem 9.8].

Exercise 9.3. Prove the inequalities (9.3.32) and (9.3.34) and verify that the inequalities hold uniformly on compact subsets of $\mathcal{S}^0$.

10

Optimal Control of Random Jump Fields and Partial Information Control

10.1 A Motivating Example

Example 10.1. Suppose the density $Y(t, x)$ of a fish population at time $t \in [0, T]$ and at the point $x \in D \subset \mathbb{R}^n$ (where D is a given open set) is modeled by a stochastic *partial* differential equation (SPDE for short) of the form

$$
dY(t, x) = \left[\frac{1}{2} \Delta Y(t, x) + \alpha Y(t, x) - u(t, x) \right] dt
$$

$$
+ \beta Y(t, x) dB(t) + Y(t^-, x) \int_{\mathbb{R}} z \tilde{N}(dt, dz); \quad (t, x) \in (0, T) \times D,
$$

$$
(10.1.1)
$$

where we assume that $z \geq -1 + \varepsilon$ a.s. $\nu(dz)$ for some constant $\varepsilon > 0$. The boundary conditions are:

$$
Y(0, x) = \xi(x); \quad x \in D \tag{10.1.2}
$$

$$
Y(t, x) = \eta(t, x); \quad (t, x) \in [0, T) \times \partial D. \tag{10.1.3}
$$

(See Fig. 10.1.)

Here $dY(t, x) = d_t Y(t, x)$ is the differential with respect to t,

$$
\Delta = \Delta_x = \frac{1}{2} \sum_{i=1}^{n} \frac{\partial^2}{\partial x_i^2}
$$

is the Laplacian operator acting on the variable x. We assume that α and β are constants and $\xi(x)$ and $\eta(t, x)$ are given deterministic functions. The process $u(t, x) \geq 0$ is our *control*, representing the harvesting rate at (t, x). Equation (10.1.1) is an example of a *reaction-diffusion* equation. With $u = 0$ and without the Δ-term, the equation reduces to a geometric Lévy equation describing the growth with respect to t. The Δ-term models the *diffusion in space* of the population.

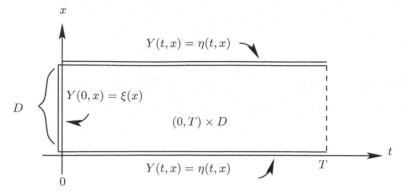

Fig. 10.1. The boundary values of $Y(t, x)$

Let $\mathcal{A}$ be a family of *admissible controls*, contained in the set of all $\mathcal{F}_t$-adapted processes $u(t, \omega)$ such that (10.1.1)–(10.1.3) has a unique solution $Y(t, x)$. Suppose the total expected utility from the harvesting rate $u(\cdot)$ and the corresponding terminal density $Y(T, x)$ is given by

$$J(u) = E\left[\int_0^T \left(\int_D \frac{u^\gamma(t, x)}{\gamma} dx\right) dt + \rho \int_D Y(T, x)dx\right], \qquad (10.1.4)$$

where $\gamma \in (0, 1)$ and $\rho > 0$ are constants.

We want to find $u^* \in \mathcal{A}$ such that

$$\sup_{u \in \mathcal{A}} J(u) = J(u^*). \qquad (10.1.5)$$

Such a control u^* is called an *optimal control*.

This is an example of a stochastic control problem for *random jump fields*, i.e., random fields which are solutions of stochastic partial differential equations driven by Brownian motions and Poisson random measures.

How do we solve problem (10.1.5)? It is possible to use a dynamic programming approach and formulate an infinite-dimensional HJB equation for the value function (see [Mort]) but this HJB equation is difficult to use. Therefore, we will instead formulate a maximum principle for such problems (Theorem 10.2). This principle can be used to solve such stochastic control problems in some cases. We will illustrate this by solving (10.1.5) by this method.

10.2 The Maximum Principle

We first give a general formulation of the stochastic control problem we consider. Suppose that the state $Y(t, x) = Y^{(u)}(t, x)$ at (t, x) is described by a stochastic partial differential equation of the form

$$dY(t,x) = [LY(t,x) + b(t,x,Y(t,x),u(t,x))]\,dt$$
$$+ \sigma(t,x,Y(t,x),u(t,x))dB(t)$$
$$+ \int_{\mathbb{R}} \theta(t,x,Y(t,x),u(t,x),z)\tilde{N}(dt,dz); \quad (t,x) \in (0,T) \times D$$

$$(10.2.1)$$

with boundary conditions

$$Y(0,x) = \xi(x); \quad x \in D \qquad\qquad (10.2.2)$$

$$Y(t,x) = \eta(t,x); \quad (t,x) \in [0,T] \times \partial D \qquad\qquad (10.2.3)$$

(see Fig. 10.1.)

Here L is a linear integro-differential operator acting on x and $b : [0,T] \times D \times \mathbb{R} \times U \to \mathbb{R}$ and $\sigma : [0,T] \times D \times \mathbb{R} \times U \to \mathbb{R}$ and $\theta : [0,T] \times D \times \mathbb{R} \times U \times \mathbb{R} \to \mathbb{R}$ are given functions and $U \subset \mathbb{R}^k$ is a given closed set of admissible control values.

Let $f : [0,T] \times D \times \mathbb{R} \times U \to \mathbb{R}$ and $g : D \times \mathbb{R} \to \mathbb{R}$ be a given *profit rate* function and *bequest rate* function, respectively. Let $\mathcal{A}$ be a given family of *admissible* controls , contained in the set of $\mathcal{F}_t$-adapted right-continuous stochastic processes $u(t,x) \in U$ such that (10.2.1)–(10.2.3) has a unique solution $Y(t,x)$ and such that

$$E\left[\int_0^T \left(\int_D |f(t,x,Y(t,x),u(t,x))|\,dx\right) dt + \int_D |g(x,Y(T,x))|\,dx\right] < \infty.$$

$$(10.2.4)$$

If $U \in \mathcal{A}$ we define its *performance functional* $J(u)$ by

$$J(u) = E\left[\int_0^T \left(\int_D f(t,x,Y(t,x),u(t,x))dx\right) dt + \int_D g(x,Y(T,x))dx\right].$$

$$(10.2.5)$$

The problem is to find $u^* \in \mathcal{A}$ such that

$$\sup_{u \in \mathcal{A}} J(u) = J(u^*). \qquad\qquad (10.2.6)$$

Such a process u^* is called an it optimal control (if it exists). The number

$$J^* = \sup_{u \in \mathcal{A}} J(u) \qquad\qquad (10.2.7)$$

is called the *value* of this problem.

We now state the *maximum principle* for this problem. Let $\mathcal{R}$ be the set of functions $r : \mathbb{R} \to \mathbb{R}$. Define the *Hamiltonian* $H : [0,T] \times D \times \mathbb{R} \times U \times \mathbb{R} \times \mathbb{R} \times \mathcal{R} \to \mathbb{R}$ by

$$H(t,x,y,u,p,q,r) = f(t,x,y,u) + b(t,x,y,u)p$$

$$+ \sigma(t,x,y,u)q + \int_{\mathbb{R}} \theta(t,x,y,u,z)r(z)\nu(dz). \quad (10.2.8)$$

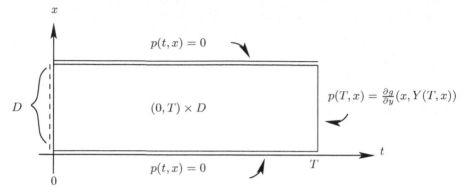

Fig. 10.2. The boundary values of $p(t, x)$

For $u \in \mathcal{A}$ we consider the following *backward* stochastic partial differential equation (the *adjoint* equation) in the three unknown adapted, right-continuous processes $p(t, x) \in \mathbb{R}$, $q(t, x) \in \mathbb{R}$, $r(t, x, z) \in \mathbb{R}$; called the *adjoint processes*:

$$dp(t, x) = - \left\{ L^* p(t, x) + \frac{\partial H}{\partial y}(t, x, Y(t, x), u(t, x), p(t, x), q(t, x), r(t, x, \cdot)) \right\} dt$$

$$+ q(t, x) dB(t) + \int_{\mathbb{R}} r(t, x, z) \tilde{N}(dt, dz); \ (t, x) \in (0, T) \times D,$$

$$(10.2.9)$$

$$p(T, x) = \frac{\partial g}{\partial y}(x, Y(T, x)); \ x \in D \qquad (10.2.10)$$

$$p(t, x) = 0; \ (t, x) \in (0, T) \times \partial D. \qquad (10.2.11)$$

(See Fig. 10.2)

Here L^* is the *adjoint* of the operator L, in the sense that

$$(L^* \varphi, \psi) = (\varphi, L\psi) \text{ for all } \varphi, \psi \in C_0^\infty(\mathbb{R}^n), \qquad (10.2.12)$$

where

$$(\varphi_1, \varphi_2) = \int_{\mathbb{R}^n} \varphi_1(x) \varphi_2(x) dx \text{ is the inner product in } L^2(\mathbb{R}^n).$$

The following result is taken from [ØPZ]. For earlier, related results see [Ø5, FØS3].

Theorem 10.2 (A Maximum Principle for Random Jump Fields [ØPZ]).

Let $\hat{u} \in \mathcal{A}$ with corresponding solution $\hat{Y}(t, x)$ of (10.2.1)–(10.2.3) and suppose that $\hat{p}(t, x)$, $\hat{q}(t, x)$ and $\hat{r}(t, x, z)$ is a solution of the adjoint backward SPDE (10.2.9)–(10.2.11). Suppose that the following, (i)–(iv), hold:

(i) *The functions* $y \to g(x, y)$ *and*

$$(y, u) \to H(y, u) := H(t, x, y, u, \hat{p}(t, x), \hat{q}(t, x), \hat{r}(t, x, \cdot)) \qquad (10.2.13)$$

are concave functions of y *and* (y, u), *respectively, for all* $(t, x) \in [0, T] \times \bar{D}.$

(ii) *(The maximum condition)*

$$H(t, x, \hat{Y}(t, x), \hat{u}(t, x), \hat{p}(t, x), \hat{q}(t, x), \hat{r}(t, x, \cdot))$$

$$= \sup_{v \in U} \left\{ H(t, x, \hat{Y}(t, x), v, \hat{p}(t, x), \hat{q}(t, x), \hat{r}(t, x, \cdot)) \right\} \qquad (10.2.14)$$

for all $(t, x) \in [0, T] \times \bar{D}.$

(iii) *For all* $u \in \mathcal{A}$ *we have*

$$E\left[\int_D \int_0^T (Y(x, t) - \hat{Y}(t, x))^2 \left\{ \hat{q}(t, x)^2 + \int_{\mathbb{R}} \hat{r}(t, x, z)^2 \nu(\mathrm{d}z) \right\} \mathrm{d}t \, \mathrm{d}x \right] < \infty$$

$$(10.2.15)$$

and

(iv)

$$E\left[\int_D \int_0^T \hat{p}(t, x)^2 \left\{ \sigma(t, x, Y(t, x), u(t, x))^2 \right. \right.$$

$$\left. \left. + \int_{\mathbb{R}} \theta(t, x, Y(t, x), u(t, x), z)^2 \nu(\mathrm{d}z) \right\} \mathrm{d}t \mathrm{d}x \right] < \infty. \, (10.2.16)$$

Then $\hat{u}(t)$ *is an optimal control for the random jump field control problem* (10.2.6).

Proof. Choose $u \in \mathcal{A}$ and let $Y(t, x) = Y^{(u)}(t, x)$ be the corresponding solution of (10.2.1)–(10.2.3). Write

$$\hat{f} = f(t, x, \hat{Y}(t, x), \hat{u}(t, x)), \quad f = f(t, x, Y(t, x), u(t, x)),$$

$$\hat{g} = g(x, \hat{Y}(T, x)), \quad g = g(x, Y(T, x)),$$

and similarly

$$\hat{b} = b(t, x, \hat{Y}(t, x), \hat{u}(t, x)), \quad b = b(t, x, Y(t, x), u(t, x)),$$

$$\hat{\sigma} = \sigma(t, x, \hat{Y}(t, x), \hat{u}(t, x)), \quad \sigma = \sigma(t, x, Y(t, x), u(t, x)),$$

and

$$\hat{\theta} = \theta(t, x, \hat{Y}(t, x), \hat{u}(t, x), z), \quad \theta = \theta(t, x, Y(t, x), u(t, x), z).$$

Moreover, put

$$\hat{H} = H(t, x, \hat{Y}(t, x), \hat{u}(t, x), \hat{p}(t, x), \hat{q}(t, x), \hat{r}(t, x, \cdot))$$

and
$$H = H(t, x, Y(t, x), u(t, x), \hat{p}(t, x), \hat{q}(t, x), \hat{r}(t, x, \cdot)).$$

Note that since $f(t, x, y, u)$ does not depend on p, q or r we have

$$\hat{f} = \hat{H} - \hat{b} \cdot \hat{p}(t, x) - \hat{\sigma} \cdot \hat{q}(t, x) - \int_{\mathbb{R}} \hat{\theta} \cdot \hat{r}(t, x, z)\nu(\mathrm{d}z)$$

and

$$f = H - b \cdot \hat{p}(t, x) - \sigma \cdot \hat{q}(t, x) - \int_{\mathbb{R}} \theta \cdot \hat{r}(t, x, z)\nu(\mathrm{d}z).$$

Therefore
$$J(\hat{u}) - J(u) = I_1 + I_2, \tag{10.2.17}$$

where

$$I_1 = E\left[\int_0^T \left(\int_D \{\hat{f} - f\}\mathrm{d}x\right)\mathrm{d}t\right]$$

$$= E\left[\int_0^T \int_D \left\{\hat{H} - H - (\hat{b} - b) \cdot \hat{p} - (\hat{\sigma} - \sigma) \cdot \hat{q} - \int_D (\hat{\theta} - \theta)r\nu(\mathrm{d}z)\right\} \mathrm{d}x\,\mathrm{d}t\right]$$
$$\tag{10.2.18}$$

and

$$I_2 = E\left[\int_D \{\hat{g} - g\}\mathrm{d}x\right]. \tag{10.2.19}$$

Since $y \to g(x, y)$ is concave, we have

$$g - \hat{g} \le \frac{\partial g}{\partial y}(x, \hat{Y}(T, x))\tilde{Y}(T, x), \tag{10.2.20}$$

where
$$\tilde{Y}(t, x) = Y(t, x) - \hat{Y}(t, x); \ 0 \le t \le T. \tag{10.2.21}$$

Therefore, by (10.2.10) and integration by parts for jump diffusions (Lemma 3.6) we get

$$I_2 \ge -E\left[\int_D \frac{\partial g}{\partial y}(x, \hat{Y}(T, x))\tilde{Y}(T, x)\mathrm{d}x\right]$$

$$= -E\left[\int_D \hat{p}(T, x) \cdot \tilde{Y}(T, x)\mathrm{d}x\right]$$

$$= -E\left[\int_D \left(\hat{p}(0, x) \cdot \tilde{Y}(0, x) + \int_0^T \{\tilde{Y}(t, x)d\hat{p}(t, x) + \hat{p}(t, x)d\tilde{Y}(t, x)\right.\right.$$

$$\left.\left. + (\sigma - \hat{\sigma})\hat{q}(t, x)\}\mathrm{d}t + \int_0^T \int_D (\theta - \hat{\theta})\hat{r}(t, x, z)N(\mathrm{d}t, \mathrm{d}z)\right)\mathrm{d}x\right]$$

$$
= -E\left[\int_D \int_0^T \left\{ \tilde{Y}(t,x)\left[-L^*\hat{p}(t,x) - \left(\frac{\partial H}{\partial y}\right)^\wedge\right] + \hat{p}(t,x)[L\tilde{Y}(t,x) - (b - \hat{b})] \right.\right.
$$

$$
\left.\left. + (\sigma - \hat{\sigma})\hat{q}(t,x) + \int_{\mathbb{R}} (\theta - \hat{\theta})\hat{r}(t,x,z)\nu(dz) \right\} dt dx \right], \qquad (10.2.22)
$$

where

$$
\left(\frac{\partial H}{\partial y}\right)^\wedge = \frac{\partial H}{\partial y}(t,x,\hat{Y}(t,x),\hat{u}(t,x),\hat{p}(t,x),\hat{q}(t,x),\hat{r}(t,x,\cdot)).
$$

Combining (10.2.18) and (10.2.22) we obtain

$$
J(\hat{u}) - J(u) \geq E\left[\int_0^T \left(\int_D \{\tilde{Y}(t,x)L^*\hat{p}(t,x) - \hat{p}(t,x)L\tilde{Y}(t,x)\} dx\right) dt \right.
$$

$$
\left. + \int_D \left(\int_0^T \left\{\hat{H} - H + \left(\frac{\partial H}{\partial y}\right)^\wedge \cdot \tilde{Y}(t,x)\right\} dt\right) dx \right].
$$

$$(10.2.23)$$

Since $\tilde{Y}(t,x) = \hat{p}(t,x) = 0$ for all $(t,x) \in (0,T) \times \partial D$, we get by an easy extension of (10.2.12) that

$$
\int_D \tilde{Y}(t,x)L^*\hat{p}(t,x)dx = \int_D \hat{p}(t,x)L\tilde{Y}(t,x)dx
$$

for all $t \in (0,T)$.

Hence

$$
J(\hat{u}) - J(u) \geq E\left[\int_D \left(\int_0^T \left\{\hat{H} - H + \left(\frac{\partial H}{\partial y}\right)^\wedge \cdot \tilde{Y}(t,x)\right\} dt\right) dx\right].
$$

$$(10.2.24)$$

By the concavity assumption (10.2.13) we have

$$
H - \hat{H} \leq \frac{\partial H}{\partial y}(\hat{Y},\hat{u}) \cdot (Y - \hat{Y}) + \frac{\partial H}{\partial u}(\hat{Y},\hat{u}) \cdot (u - \hat{u})
$$

and the maximum condition (10.2.14) implies that

$$
\frac{\partial H}{\partial u}(\hat{Y},\hat{u}) \cdot (u - \hat{u}) \leq 0.
$$

Hence

$$
\hat{H} - H + \left(\frac{\partial H}{\partial y}\right)^\wedge \cdot \tilde{Y} \geq 0,
$$

which gives

$$
J(\hat{u}) \geq J(u).
$$

Since $\hat{u} \in \mathcal{A}$ was arbitrary, this shows that $\hat{u}$ is optimal.   □

10.3 The Arrow Condition

In many cases the Hamiltonian

$$h(y, u) := H(t, x, y, u, \hat{p}(t, x), \hat{q}(t, x), \hat{r}(t, x, \cdot))$$

is not concave in both variables y, u.

In such cases it might be useful to replace the concavity condition in (y, u) (see (10.2.13)) by a weaker condition, called the *Arrow condition*:

For each fixed t, x the function

$$\hat{h}(y) := \max_{v \in V} H(t, x, y, v, \hat{p}(t, x), \hat{q}(t, x), \hat{r}(t, x, \cdot)) \qquad (10.3.1)$$

exists and is a concave function of y.

We now get the following extension of Theorem 10.2:

Corollary 10.3 (Strengthened Maximum Principle). *Let* $\hat{u}(t, x)$, $\hat{Y}(t, x)$, $\hat{p}(t, x)$, $\hat{q}(t, x)$ *and* $\hat{r}(t, x, \cdot)$ *be as in Theorem 10.2. Assume that* $g(x, y)$ *is concave in y for each x and that the Arrow condition (10.3.1) and the maximum condition (10.2.14) hold, in addition to (10.2.15) and (10.2.16). Then* $\hat{u}(t, x)$ *is an optimal control for the stochastic control problem (10.2.6).*

Proof. We proceed as in the proof of Theorem 10.2 up to and including (10.2.22). Then to obtain

$$H - \hat{H} - \left(\frac{\partial H}{\partial y}\right)^{\wedge} \cdot (Y - \hat{Y}) \leq 0 \qquad (10.3.2)$$

we note that

$$H - \hat{H} - \frac{\partial H}{\partial y}(\hat{Y}, \hat{u}) \cdot (Y - \hat{Y}) = h(Y(t, x), u(t, x)) - h(\hat{Y}(t, x), \hat{u}(t, x))$$

$$- \frac{\partial h}{\partial y}(\hat{Y}(t, x), \hat{u}(t, x))(Y(t, x) - \hat{Y}(t, x)). \qquad (10.3.3)$$

This is ≤ 0 by the same argument as in the deterministic case. See [SS], Theorem 5, p. 107–108. For completeness we give the details:

Note that by the maximum condition (10.2.14) we have

$$h(\hat{Y}(t, x), \hat{u}(t, x)) = \hat{h}(\hat{Y}(t, x)). \qquad (10.3.4)$$

Moreover, by definition of $\hat{h}$,

$$h(y, u) \leq \hat{h}(y) \text{ for all } y, u. \qquad (10.3.5)$$

Therefore, subtracting (10.3.4) from (10.3.5) gives

$$h(y, u) - h(\hat{Y}(t, x), \hat{u}(t, x)) \leq \hat{h}(y) - \hat{h}(\hat{Y}(t, x)) \quad \text{for all } y, u. \quad (10.3.6)$$

Accordingly, to prove (10.3.2) it suffices to prove that (see (10.3.3))

$$\hat{h}(Y(t, x)) - \hat{h}(\hat{Y}(t, x)) - \frac{\partial h}{\partial y}(\hat{Y}(t, x), \hat{u}(t, x)) \cdot (Y(t, x) - \hat{Y}(t, x)) \leq 0. \quad (10.3.7)$$

To this end, note that since the function $\hat{h}(y)$ is concave it follows by a standard separating hyperplane argument (see e.g., [R], Chap. 5, Sect. 23) that there exists a *supergradient* $a \in \mathbb{R}$ for $\hat{h}(y)$ at $y = \hat{Y}(t, x)$, i.e.,

$$\hat{h}(y) - \hat{h}(\hat{Y}(t, x)) \leq a \cdot (y - \hat{Y}(t, x)) \quad \text{for all } y. \quad (10.3.8)$$

Define

$$\varphi(y) = h(y, \hat{u}(t, x)) - h(\hat{Y}(t, x), \hat{u}(t, x)) - a \cdot (y - \hat{Y}(t, x)) \; ; \; y \in \mathbb{R}.$$

Then by (10.3.6) and (10.3.8) we have

$$\varphi(y) \leq 0 \quad \text{for all } y \in \mathbb{R}.$$

Moreover, by definition of φ we have

$$\varphi(\hat{Y}(t, x)) = 0.$$

Therefore

$$\varphi'(\hat{Y}(t, x)) = \frac{\partial h}{\partial y}(\hat{Y}(t, x), \hat{u}(t, x)) = a.$$

Combining this with (10.3.8), we obtain (10.3.7) and the proof is complete. □

10.3.1 Return to Example 10.1

As an illustration of the maximum principle let us apply it to solve the problem in Example 10.1.

In this case the Hamiltonian is

$$H(t, x, y, u, p, q, r) = \frac{u^{\gamma}}{\gamma} + (\alpha y - u)p + \beta yq + y \int_{\mathbb{R}} r(z) z \nu(\mathrm{d}z) \quad (10.3.9)$$

which is clearly concave in (y, u).

The adjoint equation is

$$\mathrm{d}p(t, x) = - \left[\frac{1}{2}\Delta p(t, x) + \alpha p(t, x) + \beta q(t, x) + \int_{\mathbb{R}} r(t, x, z) z \nu(\mathrm{d}z) \right] \mathrm{d}t$$

$$+ q(t, x)\mathrm{d}B(t) + \int_{\mathbb{R}} r(t, x, z)\tilde{N}(\mathrm{d}t, \mathrm{d}z); \; (t, x) \in (0, T) \times D$$

$$(10.3.10)$$

$$p(T, x) = \rho; \ x \in D, \tag{10.3.11}$$

$$p(t, x) = 0 \ ; \ (t, x) \in (0, T) \times \partial D. \tag{10.3.12}$$

Since the coefficients α, β and the boundary value ρ are all deterministic, we see that we can choose

$$q(t, x) = r(t, x, z) = 0$$

and solve the resulting deterministic equation

$$\frac{\partial}{\partial t} p(t, x) = -\frac{1}{2} \Delta p(t, x) - \alpha p(t, x) \ ; \ (t, x) \in (0, T) \times D \tag{10.3.13}$$

(together with (10.3.11)–(10.3.12)) for a deterministic solution $p(t, x)$.

This is a classical boundary value problem, and it is well known that the solution can be expressed as follows (Fig. 10.3):

$$p(t, x) = \rho e^{\alpha(T-t)} P[W^x(s) \in D \quad \text{for all } s \in [t, T]], \tag{10.3.14}$$

where $W^x(\cdot)$ denotes an auxiliary n-dimensional Brownian motion starting at $x \in \mathbb{R}^n$ with probability law P. (See e.g., [KS, Chap. 4], or [Ø1, Chap. 9]).

The function

$$u \to H(t, x, y, u, p, q, r) = \frac{u^\gamma}{\gamma} + (\alpha y - u)p + \beta y q + y \int_{\mathbb{R}} r(z) z \nu(dz)$$

is maximal when

$$u = \hat{u}(t, x) = (p(t, x))^{1/(\gamma-1)}, \tag{10.3.15}$$

where $p(t, x)$ is given by (10.3.14).

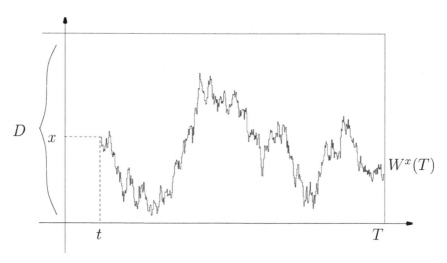

Fig. 10.3. Interpretation of the function $p(t, x)$

   With this choice of $\hat{u}(t,x)$ we see that all the conditions of Theorem 10.2 are satisfied and we conclude that $\hat{u}(t,x)$ is an optimal harvesting rate for Example 10.1.

Example 10.4 ([ØPZ]). The solution $\hat{u}(t,x)$ of Example 10.1 is a bit degenerate, in the sense that it is deterministic and hence independent of the history of the population density $Y(t,x)$. The mathematical reason for this is the deterministic parameters of the adjoint equation, including the terminal condition $p(T,x) = \rho$. Therefore, let us consider a more general situation, where the performance functional $J(u)$ of (10.1.4) is replaced by

$$J(u) = E\left[\int_0^T \int_{\mathbb{R}} \frac{u^\gamma(t,x)}{\gamma}\,dxdt + \int_{\mathbb{R}} g(x, Y(T,x))dx\right], \qquad (10.3.16)$$

where $g : \mathbb{R}^2 \to \mathbb{R}$ is a given C^1-function . The Hamiltonian remains the same as in Example 10.1, and hence the candidate $\hat{u}(t,x)$ for the optimal control has the same form as in (10.3.15), i.e.,

$$\hat{u}(t,x) = (p(t,x))^{\frac{1}{\gamma-1}}. \qquad (10.3.17)$$

   The difference is that now we have to work harder to find $p(t,x)$. The backward SPDE is now

$$dp(t,x) = -\left[\alpha p(t,x) + \beta q(t,x) + \int_{\mathbb{R}} r(t,x,z)z\nu(dz) + \frac{1}{2}\Delta p(t,x)\right]dt$$

$$+ q(t,x)dB(t) + \int_{\mathbb{R}} r(t,x,z)\tilde{N}(dt,dz); \ (t,x) \in [0,T] \times \mathbb{R},$$
$$(10.3.18)$$

$$p(T,x) = F(x); \ x \in \mathbb{R}. \qquad (10.3.19)$$

$$\lim_{|x|\to\infty} p(t,x) = 0; \ t \in [0,T], \qquad (10.3.20)$$

where
$$F(x) = F(x,\omega) = \frac{\partial g}{\partial y}(x, Y(T,x)); \ x \in \mathbb{R}. \qquad (10.3.21)$$

   To solve this equation we proceed as follows:
Put
$$\tilde{p}(t,x) = e^{\alpha t}p(t,x). \qquad (10.3.22)$$
This transforms (10.3.18)–(10.3.20) into

$$d\tilde{p}(t,x) = -\beta e^{\alpha t}q(t,x)dt - \frac{1}{2}\Delta\tilde{p}(t,x)dt$$

$$+ e^{\alpha t}\int_{\mathbb{R}} r(t,x,z)z\nu(dz)dt + e^{\alpha t}q(t,x)dB(t)$$

$$+ e^{\alpha t}\int_{\mathbb{R}} r(t,x,z)\tilde{N}(dt,dz); \ t < T, \qquad (10.3.23)$$

$$\tilde{p}(T, x) = e^{\alpha T} F(x), \tag{10.3.24}$$

$$\lim_{|x| \to \infty} \tilde{p}(t, x) = 0. \tag{10.3.25}$$

Define the probability measure P_0 on $\mathcal{F}_T$ by

$$dP_0(\omega) = Z(T)dP(\omega),$$

where

$$Z(t) \quad = \exp\left[\beta B(t) - \frac{1}{2}\beta^2 t + \int_0^t \int_{\mathbb{R}} \ln(1 + z)\tilde{N}(ds, dz)\right.$$

$$\left. + \int_0^t \int_{\mathbb{R}} \{\ln(1 + z) - z\}\nu(dz)ds\right]; \ 0 \leq t \leq T. \tag{10.3.26}$$

Then by the Girsanov theorem (use Theorem 1.35 with $u = -\beta$ and $\theta(t, z) = -z$) the process $B_0(t) := B(t) - \beta t$ is a Brownian motion with respect to P_0 and the random measure $\tilde{N}_0(dt, dz) := \tilde{N}(dt, dz) - z\nu(dz)dt$ is a compensated Poisson random measure with respect to P_0.

In terms of $dB_0(t)$ and $\tilde{N}_0(dt, dz)$ (10.3.23) gets the form

$$d\tilde{p}(t, x) = -\frac{1}{2}\Delta\tilde{p}(t, x)dt + e^{\alpha t}q(t, x)dB_0(t)$$

$$+ e^{\alpha t} \int_{\mathbb{R}} r(t, x, z)\tilde{N}_0(dt, dz). \tag{10.3.27}$$

Suppose

$$E_0\left[\int_{\mathbb{R}} F^2(x)dx\right] < \infty, \tag{10.3.28}$$

where E_0 denotes the expectation with respect to P_0. Then by the Itô representation theorem (see e.g., [I]), there exists for a.a., $x \in \mathbb{R}$ a unique pair of adapted processes $(\varphi(t, x), \psi(t, x, z))$ such that

$$E_0\left[\int_0^T \varphi^2(t, x)dt + \int_0^T \int_{\mathbb{R}} \psi^2(t, x, z)\nu(dz)dt\right] < \infty$$

and

$$e^{\alpha T} F(x) = h(x) + \int_0^T \varphi(t, x)dB_0(t) + \int_0^T \int_{\mathbb{R}} \psi(t, x, z)\tilde{N}_0(dt, dz), \tag{10.3.29}$$

where

$$h(x) = E_0\left[e^{\alpha T} F(x)\right].$$

Let Q_t be the *heat operator*, defined by

$$(Q_t f)(x) = (2\pi t)^{-1/2} \int_{\mathbb{R}} f(y) \exp\left(-\frac{(x-y)^2}{2t}\right) \, dy; \ f \in \mathcal{D}$$

where $\mathcal{D}$ is the set of functions $f : \mathbb{R} \to \mathbb{R}$ such that the integral exists. Define

$$\hat{p}(t,x) := Q_{T-t}\left(\int_0^t \varphi(s,\cdot) r m d B_0(s) + \int_0^t \int_{\mathbb{R}} \psi(s,\cdot,z) \tilde{N}_0(ds,dz) + h(\cdot)\right)(x)$$

$$= \int_0^t Q_{T-t}\varphi(s,\cdot)(x) dB_0(s) + \int_0^t \int_{\mathbb{R}} Q_{T-t}\psi(s,\cdot,z)(x)\tilde{N}_0(ds,dz)$$

$$+ Q_{T-t}h(x). \tag{10.3.30}$$

Then, since

$$\frac{d}{dt}Q_t f = \frac{1}{2}\Delta(Q_t f),$$

we see that

$$d\hat{p}(t,x)x = Q_{T-t}\varphi(t,\cdot)(x)dB_0(t) + \left[\int_0^t \left(-\frac{1}{2}\Delta(Q_{T-t}\varphi(s,\cdot))(x)\right)dB_0(s)\right]dt$$

$$+ \int_{\mathbb{R}} Q_{T-t}\psi(t,\cdot,z)(x)\tilde{N}_0(ds,dz)$$

$$+ \left[\int_0^t \int_{\mathbb{R}} -\frac{1}{2}\Delta(Q_{T-t}\psi(s,\cdot,z))(x)\tilde{N}_0(ds,dz)\right]dt$$

$$- \frac{1}{2}\Delta(Q_{T-t}h(x))dt$$

$$= -\frac{1}{2}\Delta\left(\int_0^t Q_{T-t}\varphi(s,\cdot)(x)dB_0(s)\right.$$

$$\left. + \int_0^t \int_{\mathbb{R}} Q_{T-t}\psi(s,\cdot z)(x)\tilde{N}_0(ds,dz) + Q_{T-t}h(x)\right)dt$$

$$+ Q_{T-t}\varphi(t,\cdot)(x)dB_0(t) + \int_{\mathbb{R}} Q_{T-t}\psi(t,\cdot,z)(x)\tilde{N}_0(dt,dz)$$

$$= -\frac{1}{2}\Delta\hat{p}(t,x)dt + Q_{T-t}\varphi(t,\cdot)(x)dB_0(t)$$

$$+ \int_{\mathbb{R}} Q_{T-t}\psi(t,\cdot,z)(x)\tilde{N}_0(dt,dz). \tag{10.3.31}$$

Comparing with (10.3.27) we see that the triple $(\tilde{p}, q, r)$ given by

$$\tilde{p}(t,x) := \hat{p}(t,x), \tag{10.3.32}$$

$$q(t,x) := e^{-\alpha t}Q_{T-t}\varphi(t,\cdot)(x), \tag{10.3.33}$$

$$r(t,x,z) := e^{-\alpha t}Q_{T-t}\psi(t,\cdot,z)(x) \tag{10.3.34}$$

solves the backward SPDE (10.3.27), and hence it solves (10.3.23), together with the terminal values (10.3.24) and (10.3.25).

We have proved:

Theorem 10.5. *Assume that* (10.3.28) *holds Then the optimal control of the problem* (10.1.1)–(10.1.3), (10.1.5), *with performance functional $J(u)$ as in* (10.3.16), *satisfies*

$$\hat{u}(t,x) = (p(t,x))^{1/(\gamma-1)},$$

where

$$p(t,x) = \mathrm{e}^{-\alpha t}\hat{p}(t,x)$$

with $\hat{p}(t,x)$ defined by (10.3.30) *and* (10.3.29).

10.4 Controls Which do not Depend on x

In some cases, for example in the application to partial observation control (see Sect. 10.5), it is of interest to consider only controls $u(t) = u(t,x)$ which do not depend on the space variable x. Thus we let the set $\mathcal{A}_0$ of admissible controls be defined by

$$\mathcal{A}_0 = \{u \in \mathcal{A};\ u(t,x) = u(t) \text{ does not depend on } x\}. \tag{10.4.1}$$

The performance functional $J(u)$ is as before,

$$J(u) = E\left[\int_0^T \left(\int_D f(t,x,Y(t,x),u(t))\mathrm{d}x\right)\mathrm{d}t + \int_D g(x,,Y(T,x))\mathrm{d}x\right]. \tag{10.4.2}$$

We want to find $J_0^* \in \mathbb{R}$ and $u_0^* \in \mathcal{A}_0$ such that

$$J_0^* = \sup_{u \in \mathcal{A}_0} J(u) = J(u_0^*). \tag{10.4.3}$$

It turns out that one can formulate an analog of Theorem 10.2 for this case:

Theorem 10.6 (Controls Which do not Depend on x).
Suppose $\hat{u} \in \mathcal{A}_0$ with corresponding solutions $\hat{Y}(t,x)$ of (10.2.1)–(10.2.3) *and $\hat{p}(t,x)$, $\hat{q}(t,x)$, $\hat{r}(t,x,z)$ of* (10.2.9)–(10.2.11), *respectively. Assume that* (10.2.15)–(10.2.16) *hold, together with the following,* (10.4.4)–(10.4.5):

-   *The functions $y \rightarrow g(x,y)$ and $(y,u) \rightarrow$*

    $$H(y,u) := H(t,x,y,u,\hat{p}(t,x),\hat{q}(t,x),\hat{r}(t,x,z))\ ;\ (y,u) \in \mathbb{R} \times U \tag{10.4.4}$$

    are concave, for all $(t,x) \in (0,T) \times D$

– *(the average maximum condition)*

$$\int_D H(t, x, \hat{Y}(t, x), \hat{u}(t), \hat{p}(t, x), \hat{r}(t, x, \cdot)) dx$$

$$= \sup_{v \in U} \int_D H(t, x, \hat{Y}(t, x), v, \hat{p}(t, x), \hat{q}(t, x), \hat{r}(t, x, \cdot)) dx. \quad (10.4.5)$$

Then $\hat{u}(t)$ is an optimal control for the problem (10.4.3).

Proof. We proceed as in the proof of Theorem 10.2. Let $u \in \mathcal{A}_0$ with corresponding solution $Y(t, x)$ of (10.2.1)–(10.2.3). With $\hat{u} \in \mathcal{A}_0$ as in (10.4.5), consider

$$J(\hat{u}) - J(u) = E\left[\int_0^T \int_D \{\hat{f} - f\} dx dt + \int_D \{\hat{g} - g\} dx\right], \quad (10.4.6)$$

where

$$\hat{f} = f(t, x, \hat{Y}(t, x), \hat{u}(t)), \quad f = f(t, x, Y(t, x), u(t))$$
$$\hat{g} = g(x, \hat{Y}(T, x)) \quad \text{and} \quad g = g(x, Y(T, x)).$$

Using a similar shorthand notation for $\hat{b}, b, \hat{\sigma}, \sigma$ and $\hat{\theta}, \theta$ and setting

$$\hat{H} = H(t, x, \hat{Y}(t, x), \hat{u}(t), \hat{p}(t, x), \hat{q}(t, x), \hat{r}(t, x, \cdot))$$

and

$$H = H(t, x, Y(t, x), u(t), \hat{p}(t, x), \hat{q}(t, x), \hat{r}(t, x, \cdot))$$

we see that (10.4.6) can be written

$$J(\hat{u}) - J(u) = I_1 + I_2, \quad (10.4.7)$$

where

$$I_1 = E\left[\int_0^T \left(\int_D \left\{\hat{H} - H - (\hat{b} - b)\hat{p} - (\hat{\sigma} - \sigma)\hat{q} - \int_{\mathbb{R}} (\hat{\theta} - \theta)\hat{r}\nu(dz)\right\} dx\right) dt\right] \quad (10.4.8)$$

and

$$I_2 = E\left[\int_D \{\hat{g} - g\} dx\right]. \quad (10.4.9)$$

By concavity of the function $y \to g(x, y)$ we have

$$\int_D \left\{g(x, Y(T, x)) - g(x, \hat{Y}(T, x))\right\} dx \leq \int_D \frac{\partial g}{\partial y}(x, \hat{Y}(T, x)) \cdot \tilde{Y}(T, x) dx, \quad (10.4.10)$$

where
$$\tilde{Y}(t, x) = Y(t, x) - \hat{Y}(t, x).$$

Therefore, as in the proof of Theorem 10.2,

$$
I_2 \geq -E\left[\int_0^T \left(\int_D \left\{\tilde{Y}(t, x)\left[-\left(\frac{\partial H}{\partial y}\right)^\wedge - L^*\hat{p}(t, x)\right]\right.\right.\right.
$$
$$
+ \hat{p}(t, x)[L\tilde{Y}(t, x) + (b - \hat{b})] + (\sigma - \hat{\sigma})\hat{q}(t, x) \qquad (10.4.11)
$$
$$
\left.\left.\left. + \int_{\mathbb{R}} (\theta - \hat{\theta})\hat{r}(t, x, z)\nu(\mathrm{d}z)\right\}\mathrm{d}x\right)\mathrm{d}t\right],
$$

where

$$
\left(\frac{\partial H}{\partial y}\right)^\wedge = \frac{\partial H}{\partial y}(t, x, \hat{Y}(t, x), \hat{u}(t), \hat{p}(t, x), \hat{q}(t, x), \hat{r}(t, x, \cdot)).
$$

Summing (10.4.8) and (10.4.11) we get, as in (10.2.24),

$$
J(\hat{u}) - J(u) \geq E\left[\int_0^T \left(\int_D \left\{\hat{H} - H + \tilde{Y} \cdot \left(\frac{\partial H}{\partial y}\right)^\wedge\right\}\mathrm{d}x\right)\mathrm{d}t\right]. \qquad (10.4.12)
$$

Since $H(y, u)$ is concave, we have

$$
H - \hat{H} \leq \frac{\partial H}{\partial y}(\hat{Y}, \hat{u}) \cdot (Y - \hat{Y}) + \frac{\partial H}{\partial u}(\hat{Y}, \hat{u}) \cdot (u - \hat{u}). \qquad (10.4.13)
$$

Combining (10.4.12) and (10.4.13) lead to

$$
J(\hat{u}) - J(u) \geq E\left[\int_0^T \left(\int_D -\frac{\partial H}{\partial u}(\hat{Y}, \hat{u}) \cdot (u - \hat{u})\mathrm{d}x\right)\mathrm{d}t\right]
$$
$$
= -E\left[\int_0^T \frac{\partial}{\partial v}\left(\int_D H(t, x, \hat{Y}, v, \hat{p}, \hat{q}, \hat{r})\mathrm{d}x\right)_{v=\hat{u}(t)} \cdot ((u(t) - \hat{u}(t))\mathrm{d}t\right]
$$
$$
\geq 0, \text{ since } v = \hat{u}(t) \text{ maximizes}
$$
$$
v \to \int_G H(t, x, \hat{Y}(t, x), v, \hat{p}(t, x), \hat{q}(t, x), \hat{r}(t, x, \cdot))\mathrm{d}x. \qquad \square
$$

10.5 Connection with Partial Observation Control

In this section we present the well-known connection between *partial observation* control of *ordinary* stochastic differential equations and *full observation* control of stochastic *partial* differential equations, discussed above. See e.g., [Mort, Ben1, Ben2, Ben3], and the references therein.

To this end we need to recall briefly some key concepts and results from nonlinear filtering theory. We refer to [D2, DM, Kalli, Pa1, Pa2] for more details. See also [MP].

Suppose the *state process* $x(t) = x^{(u)}(t) \in \mathbb{R}^n$ and its corresponding *observation process* $\zeta(t) \in \mathbb{R}^n$ are given by the following stochastic differential equations, (10.5.1)–(10.5.4):

– (*State process*)

$$dx(t) = \alpha(x(t), \zeta(t), u(t))dt + \beta(x(t), \zeta(t), u(t))dv(t)$$
$$+ \int_{\mathbb{R}} \xi(x(t^-), \zeta(t^-), u(t^-), z)\tilde{N}(dt, dz); \ t \in [0, T], \quad (10.5.1)$$

where $T > 0$ is a fixed constant

$$x(0) \text{ has density } F(x), \text{ i.e.,}$$
$$E[\varphi(x(0))] = \int_{\mathbb{R}} \varphi(x)F(x)dx; \ \varphi \in C_0(\mathbb{R}^n). \quad (10.5.2)$$

– (*Observation process*)

$$d\zeta(t) = h(x(t))dt + dw(t); \ t \in [0, T], \quad (10.5.3)$$
$$\zeta(0) = 0. \quad (10.5.4)$$

Here $\alpha : \mathbb{R}^n \times \mathbb{R}^m \times U \to \mathbb{R}^n$, $\beta : \mathbb{R}^n \times \mathbb{R}^m \times U \to \mathbb{R}^{n \times n}$, $\xi : \mathbb{R}^n \times \mathbb{R}^m \times U \times \mathbb{R}^n \to \mathbb{R}^{n \times n}$ and $h : \mathbb{R}^n \to \mathbb{R}^m$ are given deterministic functions, $v(t) = v(t, \omega) \in \mathbb{R}^n$ and $w(t) = w(t, \omega) \in \mathbb{R}^m$ are independent Brownian motions on $\mathbb{R}^n$ and $\mathbb{R}^m$, respectively, and $\tilde{N}(dt, dz)$ is a compensated Poisson random measure of dimension n.

For simplicity we assume that h is a bounded function. The process $u(t) = u(t, \omega)$ is our *control* process, assumed to have values in a given closed set $U \subset \mathbb{R}^k$. We require that $u(t)$ be adapted to the filtration $\mathcal{Z}_t$ generated by the observations $\zeta(s); \ s \leq t$. We call $u(t)$ *admissible* if, in addition, (10.5.1)–(10.5.4) has a unique strong solution $(x(t), \zeta(t))$ such that

$$E\left[\int_0^T |\ell(x(t), u(t))|dt + |k(x(T))|\right] < \infty, \quad (10.5.5)$$

where $\ell : \mathbb{R}^n \times U \to \mathbb{R}$ and $k : \mathbb{R}^n \to \mathbb{R}$ are given functions, called the *profit rate* and the *bequest* function, respectively. The set of all admissible controls is denoted by $\mathcal{A}_{\mathcal{Z}}$.

For $u \in \mathcal{A}_{\mathcal{Z}}$ we define the performance functional

$$J(u) = E\left[\int_0^T \ell(x(t), u(t))dt + k(x(T))\right]. \quad (10.5.6)$$

The stochastic control problem with partial observation is to find $J^* \in \mathbb{R}$ and $u^* \in \mathcal{A}_{\mathcal{Z}}$ such that

$$J^* = \sup_{u \in \mathcal{A}_{\mathcal{Z}}} J(u) = J(u^*). \qquad (10.5.7)$$

We now show that (under some conditions) this problem can be transformed into a full information control problem of an associated SPDE, with controls $u(t)$ not depending on x, as described in Sect. 10.4.

Define

$$M_t(\omega) = \exp\left(-\int_0^t h(x(s))dw(s) - \frac{1}{2}\int_0^t h^2(x(s))ds\right) \qquad (10.5.8)$$

and define the probability measure Q on $\mathcal{F}_T$ by

$$dQ(\omega) = M_t(\omega)dP(\omega) \text{ on } \mathcal{F}_T. \qquad (10.5.9)$$

Then M_t is a martingale with respect to P and we have

$$dQ(\omega) = M_t(\omega)dP(\omega) \text{ on } \mathcal{F}_t; \; 0 \le t \le T. \qquad (10.5.10)$$

Then by the Girsanov theorem the observation process

$$\zeta(t) = \int_0^t h(x(s))ds + w(t)$$

is a Brownian motion with respect to Q.

Put

$$K_t(\omega) := M_t^{-1}(\omega) = \exp\left(\int_0^t h(x(s))dw(s) + \frac{1}{2}\int_0^t h^2(x(s))ds\right)$$
$$= \exp\left(\int_0^t h(x(s))d\zeta(s) - \frac{1}{2}\int_0^t h^2(x(s))ds\right). \qquad (10.5.11)$$

Then K_t is a martingale with respect to Q and

$$dP(\omega) = K_t(\omega)dQ(\omega) \text{ on } \mathcal{F}_t; \; 0 \le t \le T. \qquad (10.5.12)$$

For fixed $\zeta \in \mathbb{R}$, $u \in U$ define the integro-differential operator $A = A_{\zeta,u}$ by

$$A_{\zeta,u}\varphi(x) = \sum_{i=1}^n \alpha_i(x, \zeta, u)\frac{\partial\varphi}{\partial x_i}(x) + \frac{1}{2}\sum_{i,j=1}^n (\beta\beta^T)_{i,j}(x, \zeta, u)\frac{\partial^2\varphi}{\partial x_i \partial x_j}$$
$$+ \sum_{j=1}^n \int_{\mathbb{R}} \left\{\varphi(x + \xi^{(j)}(x, \zeta, u, z_j)) - \varphi(x)\right.$$
$$\left. - \nabla\varphi(x) \cdot \xi^{(j)}(x, \zeta, u, z_j)\right\}\nu_j(dz_j) \qquad (10.5.13)$$

for $\varphi \in C_0^2(\mathbb{R}^n)$, i.e., $A_{\zeta,u}$ is the generator of $x(t)$ if we regard ζ and u as known constants. (Here $\xi^{(j)}$ denotes the jth column of the $n \times n$ matrix ξ).

Let A^* be the adjoint of A, in the sense that

$$(A\varphi, \psi)_{L^2(\mathbb{R}^n)} = (\varphi, A^*\psi)_{L^2(\mathbb{R}^n)}; \quad \varphi, \psi \in C_0^2(\mathbb{R}^n). \tag{10.5.14}$$

Suppose there exists a stochastic process $y(t, x) = y(t, x, \omega)$; $(t, x, \omega) \in [0, T] \times \mathbb{R}^n \times \Omega$ such that

$$E_Q[\varphi(x(t))K_t \mid \mathcal{Z}_t] = \int_{\mathbb{R}^n} \varphi(x)y(t, x)dx ; \quad \varphi \in C_0(\mathbb{R}^n). \tag{10.5.15}$$

Then $y(t, x)$ is called the *unnormalized conditional density* of $x(t)$ *given* $\mathcal{Z}_t$. Conditions which imply the existence of $y(t, x)$ can be found in [GK].

Under certain conditions the process $y(t, x)$ satisfies the *Duncan–Mortensen–Zakai* equation

$$dy(t, x) = A^*_{\zeta(t),u(t)}y(t, x)dt + h(x)y(t, x)d\zeta(t); \quad t > 0, \tag{10.5.16}$$

$$y(0, x) = F(x), \tag{10.5.17}$$

where $F(x)$ is the density of $x(0)$ (see (10.5.2)).

In terms of $y(t, x)$ we can now rewrite the performance functional (10.5.6) as follows:

$$
\begin{aligned}
J(u) &= E\left[\int_0^T \ell(x(t), u(t))dt + k(x(T))\right] \\
&= \int_0^T E_Q[\ell(x(t), u(t))K_t]dt + E_Q[k(x(T))K_T] \\
&= E_Q\left[\int_0^T E_Q[\ell(x(t), u(t))K_t \mid \mathcal{Z}_t]dt + k(x(T))K_T\right] \\
&= E_Q\left[\int_0^T E_Q[\ell(x(t), v)K_t \mid \mathcal{Z}_t]_{v=u(t)}dt + \int_{\mathbb{R}} k(x)y(T, x)dx\right] \\
&= E_Q\left[\int_0^T \left(\int_{\mathbb{R}} \ell(x, v)y(t, x)dx\right)_{v=u(t)} dt + \int_{\mathbb{R}} k(x)y(T, x)dx\right] \\
&= E_Q\left[\int_0^T \left(\int_{\mathbb{R}} \ell(x, u(t))y(t, x)dx\right) dt + \int_{\mathbb{R}} k(x)y(T, x)dx\right].
\end{aligned}
$$
$$\tag{10.5.18}$$

This is a performance criterion of the type discussed in Sects. 10.1–10.4, with the control $u(t)$ not depending on x.

We summarize this as follows:

Theorem 10.7. *The partial observation SDE control problem* (10.5.1)–(10.5.7) *can be rewritten as the full observation SPDE control problem*

$$J^* = \sup_{u \in \mathcal{A}_{\mathcal{Z}}} J(u) = J(u^*)$$

where

$$J(u) = E_Q \left[\int_0^T \left(\int_{\mathbb{R}} \ell(x, u(t)) y(t, x) \mathrm{d}x \right) \mathrm{d}t + \int_{\mathbb{R}} k(x) y(T, x) \mathrm{d}x \right],$$

with $y(t, x)$ *(the unnormalized conditional density of* $x(t)$ *given* $\mathcal{Z}_t$*) given by the Duncan–Mortensen–Zakai equation* (10.5.16), *which is driven by the observation process* $\zeta(t)$. *This process is a Brownian motion with respect to the measure* Q *defined by* (10.5.9).

10.6 Exercises

Exercise* 10.1. Transform the following *partial observation SDE* control problems into *complete observation SPDE* control problems:

(a) (The partially observed linear-quadratic control problem)

(state): $\begin{cases} \mathrm{d}x(t) = (\alpha x(t) + u(t))\mathrm{d}t + \sigma \mathrm{d}v(t); \ t > 0, \\ x(0) \text{ has density } F(x), \end{cases}$

(observations): $\begin{cases} \mathrm{d}\zeta(t) = h(x(t))\mathrm{d}t + \mathrm{d}w(t); \ t > 0, \\ \zeta(0) = 0, \end{cases}$

performance: $J(u) = E \left[\int_0^T \{x^2(t) + \theta u^2(t)\} \, \mathrm{d}t \right], \quad J^* = \inf_{u \in \mathcal{A}_{\mathcal{Z}}} J(u).$

Here $\alpha, \sigma \neq 0$, $\theta > 0$ are constants and F and h are given functions.

(b) (The partially observed optimal portfolio problem)

(state): $\begin{cases} \mathrm{d}x(t) = x(t)[\alpha u(t)\mathrm{d}t + \beta u(t)\mathrm{d}v(t)]; \ t > 0 \\ x(0) \text{ has density } F(x), \end{cases}$

(observations): $\begin{cases} \mathrm{d}\zeta(t) = x(t)\mathrm{d}t + \mathrm{d}w(t); \ t > 0, \\ \zeta(0) = 0. \end{cases}$

performance: $J(u) = E[X^\gamma(T)], \quad J^* = \sup_{u \in \mathcal{A}_{\mathcal{Z}}} J(u).$

Here $\alpha > 0$, $\beta > 0$ and $\gamma \in (0, 1)$ are constants.

We may interpret $u(t)$ as the *portfolio* representing the fraction of the total wealth $x(t)$ invested in the risky asset with price dynamics

$$dS_1(t) = S_1(t)[\alpha dt + \beta dv(t)]; \ t > 0,$$
$$S_1(0) > 0.$$

The remaining fraction $1 - \pi(t)$ is then invested in the other investment alternative, being a risk free asset with price $S_0(t) = 1$ for all t.

Exercise 10.2 (Terminal Conditions). Let $Y(t,x) = Y^{(u)}(t,x)$ be as in (10.1.1)–(10.1.3) and define

$$J(u) = E\left[\int_0^T \left(\int_D \ln u(t,x)dx\right) dt\right]; \ u \in \tilde{A}, \tag{10.6.1}$$

where $\tilde{A}$ is the set of controls in A (see (10.1.3)) such that the *terminal constraint*

$$E\left[\int_D Y^{(u)}(T,x)dx\right] \ge 0 \tag{10.6.2}$$

holds. We consider the constrained stochastic control problem to find $\tilde{J} \in \mathbb{R}$ and $\tilde{u} \in \tilde{A}$ such that

$$\tilde{J} = \sup_{u\in\tilde{A}} J(u) = J(\tilde{u}). \tag{10.6.3}$$

To solve this problem we use the *Lagrange multiplier method* as follows:

(a) Fix $\lambda > 0$ and solve the *unconstrained* control problem to find $J_\lambda^* \in \mathbb{R}$ and $u_\lambda^* \in A$ such that

$$J_\lambda^* = \sup_{u\in A} J_\lambda(u) = J_\lambda(u_\lambda^*), \tag{10.6.4}$$

where

$$J_\lambda(u) = E\left[\int_0^T \left(\int_D \ln u(t,x)dx\right) dt + \int_D \lambda Y(T,x)dx\right]. \tag{10.6.5}$$

[Hint: Use the method in Example 10.1.]
(b) Suppose there exists $\hat{\lambda} > 0$ such that

$$E\left[\int_D Y^{(u_{\hat\lambda}^*)}(T,x)dx\right] = 0, \tag{10.6.6}$$

where u_λ^* is the corresponding solution of the *unconstrained* problem (10.6.4)–(10.6.5) with $\lambda = \hat{\lambda}$. Show that then in fact $\tilde{u} := u_{\hat\lambda}^* \in \tilde{A}$ solves the *constrained* problem (10.6.1)–(10.6.3) and hence $\tilde{J} = J_{\hat\lambda}^*$.
(c) Use the above to solve the constrained stochastic control problem (10.6.3).

Exercise 10.3 (Controls Which do not Depend on x). Consider Example 10.1 again, but this time we only allow controls $u(t,x) = u(t)$ which do not depend on x. Use Theorem 10.6 to find the optimal control $u^*(t)$ in this case.

11

Solutions of Selected Exercises

11.1 Exercises of Chapter 1

Exercise 1.1

Choose $f \in C^2(\mathbb{R})$ and put $Y(t) = f(X(t))$. Then by the Itô formula

$$dY(t) = f'(X(t))[\alpha\, dt + \sigma\, dB(t)] + \tfrac{1}{2}\sigma^2 f''(X(t))dt$$

$$+ \int_{|z|<R} \{f(X(t^-) + \gamma(z)) - f(X(t^-)) - \gamma(z)f'(X(t^-))\}\, \nu(dz)dt$$

$$+ \int_{\mathbb{R}} \{f(X(t^-) + \gamma(z)) - f(X(t^-))\}\, \bar{N}(dt, dz). \qquad (11.1.1)$$

(i) In particular, if $f(x) = \exp(x)$ this gives

$$dY(t) = Y(t)[\alpha\, dt + \sigma\, dB(t)] + \tfrac{1}{2}\sigma^2 Y(t)dt$$

$$+ \int_{|z|<R} \{\exp(X(t^-) + \gamma(z)) - \exp(X(t^-)) - \gamma(z)\exp(X(t^-))\}\nu(dz)dt$$

$$+ \int_{\mathbb{R}} \{\exp(X(t^-) + \gamma(z)) - \exp(X(t^-))\}\, \bar{N}(dt, dz)$$

$$= Y(t^-)\left[\left(\alpha + \tfrac{1}{2}\sigma^2 + \int_{|z|<R} \{e^{\gamma(z)} - 1 - \gamma(z)\}\nu(dz)\right)dt\right.$$

$$\left. + \sigma\, dB(t) + \int_{\mathbb{R}} \{e^{\gamma(z)} - 1\}\, \bar{N}(dt, dz)\right]. \qquad (11.1.2)$$

(ii) By (i) we see that $Y(t)$ solves the equation

$$dY(t) = Y(t^-)\left[\beta\, dt + \theta\, dB(t) + \lambda \int_{\mathbb{R}} z\bar{N}(dt, dz)\right]$$

if and only if

$$\alpha + \tfrac{1}{2}\sigma^2 + \int_{|z|<R} \left\{ e^{\gamma(z)} - 1 - \gamma(z) \right\} \nu(dz) = \beta, \quad \sigma = \theta$$

and $e^{\gamma(z)} - 1 = \lambda z$ (i.e., $\gamma(z) = \ln(1 + \lambda z)$) a.e. ν.

Exercise 1.2

We first make some general remarks:

Suppose $dX_i(t) = \alpha_i(t, \omega)dt + \sigma_i(t, \omega)dB(t) + \int_{\mathbb{R}} \gamma_i(t, z, \omega)\bar{N}(dt, dz)$ for $i = 1, 2$. Define $Y(t) = X_1(t) \cdot X_2(t)$. Then, by the Itô formula with $f(x_1, x_2) = x_1 \cdot x_2$,

$$
\begin{aligned}
dY(t) = & \; X_2(t)[\alpha_1 dt + \sigma_1 dB(t)] + X_1(t)[\alpha_2 dt + \sigma_2 dB(t)] + \tfrac{1}{2} \cdot 2\sigma_1\sigma_2 dt \\
& + \int_{|z|<R} \left\{ (X_1(t^-) + \gamma_1(t, z))(X_2(t^-) + \gamma_2(t, z)) - X_1(t^-)X_2(t^-) \right. \\
& \qquad \left. - X_2(t^-)\gamma_1(t, z) - X_1(t^-)\gamma_2(t, z) \right\} \nu(dz)dt \\
& + \int_{\mathbb{R}} \left\{ (X_1(t^-) + \gamma_1(t, z))(X_2(t^-) + \gamma_2(t, z)) - X_1(t^-)X_2(t^-) \right\} \bar{N}(dt, dz) \\
= & \; X_2(t)[\alpha_1 dt + \sigma_1 dB(t)] + X_1(t)[\alpha_2 dt + \sigma_2 dB(t)] + \sigma_1\sigma_2 dt \\
& + \int_{|z|<R} \gamma_1(t, z)\gamma_2(t, z)\nu(dz)dt \\
& + \int_{\mathbb{R}} \left\{ \gamma_1(t, z)\gamma_2(t, z) + X_1(t^-)\gamma_2(t, z) + X_2(t^-)\gamma_1(t, z) \right\} \bar{N}(dt, dz).
\end{aligned}
$$

$$(11.1.3)$$

In particular, if $dX(t) = \alpha \, dt + \sigma \, dB(t) + \int_{\mathbb{R}} \gamma(t, z)\bar{N}(dt, dz)$, we get

$$
\begin{aligned}
d(e^{\lambda t} X(t)) &= X(t)\lambda e^{\lambda t} dt + e^{\lambda t}[\alpha \, dt + \sigma \, dB(t)] + \int_{\mathbb{R}} e^{\lambda t}\gamma(t, z)\bar{N}(dt, dz) \\
&= e^{\lambda t} dX(t) + \lambda X(t)e^{\lambda t} dt.
\end{aligned}
$$

(i) Now consider the equation

$$dX(t) = (m - X(t))dt + \sigma \, dB(t) + \gamma \int_{\mathbb{R}} z\tilde{N}(dt, dz),$$

where we assume that $\gamma z > -1$ for a.a. $z \, (\nu)$. It can be written

$$d(e^t X(t)) = m \, e^t dt + \sigma \, e^t dB(t) + \gamma \, e^t \int_{\mathbb{R}} z \widetilde{N}(dt, dz).$$

This gives the solution

$$X(t) = X(0)e^{-t} + m \int_0^t e^{(s-t)} ds + \sigma \int_0^t e^{(s-t)} dB(s) + \gamma \int_0^t \int_{\mathbb{R}} z e^{(s-t)} \widetilde{N}(dt, dz)$$

or

$$X(t) = m + (X_0 - m)e^{-t} + \sigma \int_0^t e^{s-t} dB(s) + \gamma \int_0^t \int_{\mathbb{R}} z e^{s-t} \widetilde{N}(dt, dz). \quad (11.1.4)$$

(ii) Next consider the equation

$$dX(t) = \alpha \, dt + \gamma \, X(t^-) \int_{\mathbb{R}} z \bar{N}(dt, dz); \qquad X(0) = x \in \mathbb{R}. \qquad (11.1.5)$$

Define, for a given function $\theta(z)$,

$$G(t) = \exp \left(\int_0^t \int_{\mathbb{R}} \theta(z) \bar{N}(dt, dz) - \int_{|z| < R} \left\{ e^{\theta(z)} - 1 - \theta(z) \right\} \nu(dz) \cdot t \right).$$

Then by Itô's formula (see Exercise 1.1)

$$dG(t) = G(t^-) \int_{\mathbb{R}} \left\{ e^{\theta(z)} - 1 \right\} \bar{N}(dt, dz).$$

Hence, if we put

$$\widetilde{X}(t) = X(0)G(t) + \alpha \, G(t) \int_0^t G^{-1}(s) ds, \qquad (11.1.6)$$

we have

$$d\widetilde{X}(t) = X(0)dG(t) + \alpha \, G(t)G^{-1}(t)dt + \alpha \int_0^t G^{-1}(s) ds \cdot dG(t)$$

$$= \alpha \, dt + X(0)G(t^-) \int_{\mathbb{R}} \left\{ e^{\theta(z)} - 1 \right\} \bar{N}(dt, dz)$$

$$+ \alpha \cdot \int_0^t G^{-1}(s) ds \cdot \left[G(t^-) \int_{\mathbb{R}} \left\{ e^{\theta(z)} - 1 \right\} \bar{N}(dt, dz) \right]$$

$$= \alpha \, dt + \left[X(0)G(t^-) + \alpha \, G(t^-) \int_0^t G^{-1}(s)ds \right] \int_{\mathbb{R}} \left\{ e^{\theta(z)} - 1 \right\} \bar{N}(dt, dz)$$

$$= \alpha \, dt + \tilde{X}(t^-) \int_{\mathbb{R}} \left\{ e^{\theta(z)} - 1 \right\} \bar{N}(dt, dz).$$

So $X(t) := \tilde{X}(t)$ solves (11.1.5) if we choose $\theta(z)$ such that

$$e^{\theta(z)} - 1 = \gamma \, z \quad \text{a.s. } \nu$$

i.e.,

$$\theta(z) = \ln(1 + \gamma \, z) \quad \text{a.s. } \nu. \tag{11.1.7}$$

Exercise 1.6

By (1.2.5) (Example 1.15) we know that the equation

$$dX(t) = X(t^-) \int_{\mathbb{R}} (e^{\gamma(t,z)} - 1)\tilde{N}(dt, dz); \quad X(0) = 1 \tag{11.1.8}$$

has the solution

$$X(t) = \exp\left\{ \int_0^t \int_{\mathbb{R}} \gamma(s, z)N(ds, dz) - \int_0^t \int_{\mathbb{R}} (e^{\gamma(s,z)} - 1)\nu(dz)ds \right\}$$

$$= \exp\left\{ \int_0^t \int_{\mathbb{R}} \gamma(s, z)\tilde{N}(ds, dz) - \int_0^t \int_{\mathbb{R}} (e^{\gamma(s,z)} - 1 - \gamma(s,t))\nu(dz)ds \right\}. \tag{11.1.9}$$

If we assume that

$$\int_0^t \int_{\mathbb{R}} (e^{\gamma(s,z)} - 1)^2 \nu(dz)ds < \infty \tag{11.1.10}$$

then by (11.1.8) we see that $E[X(t)] = 1$ and hence by (11.1.9) we get

$$E\left[\exp\left\{ \int_0^t \int_{\mathbb{R}} \gamma(s, z)\tilde{N}(ds, dz) \right\} \right] = \exp\left\{ \int_0^t \int_{\mathbb{R}} (e^{\gamma(s,z)} - 1 - \gamma(s,z))\nu(dz)ds \right\}. \tag{11.1.11}$$

Exercise 1.7

By (11.1.3) in the solution of Exercise 1.2 we have (with $R = \infty$)

$$d(X_1(t)X_2(t)) = \int_{\mathbb{R}} \gamma_1(t,z)\gamma_2(t,z)\nu(dz)dt$$

$$+ \int_{\mathbb{R}} \{\gamma_1(t,z) + X_1(t^-)\gamma_2(t,z) + X_2(t^-)\gamma_1(t,z)\} \tilde{N}(dt,dz)$$

$$= X_1(t^-) \int_{\mathbb{R}} \gamma_2(t,z)\tilde{N}(dt,dz)$$

$$+ X_2(t^-) \int_{\mathbb{R}} \gamma_1(t,z)\tilde{N}(dt,dz) + \int_{\mathbb{R}} \gamma_1(t,z)\gamma_2(t,z)\nu(dz)$$

$$= X_1(t^-)dX_2(t) + X_2(t^-)dX_1(t) + \int_{\mathbb{R}} \gamma_1(t,z)\gamma_2(t,z)N(dt,dz),$$

$$(11.1.12)$$

which is (1.6.1).

Exercise 1.8

To find Q we apply Theorem 1.34. So we must find a solution $(\theta_1(z), \theta_2(z))$ of the two equations:

(i) $\gamma_{11} \int_{\mathbb{R}} z\theta_1(z)\nu_1(dz) + \gamma_{12} \int_{\mathbb{R}} z\theta_2(z)\nu_2(dz) = \mu_1$

(ii) $\gamma_{21} \int_{\mathbb{R}} z\theta_1(z)\nu_1(dz) + \gamma_{22} \int_{\mathbb{R}} z\theta_2(z)\nu_2(dz) = \mu_2$

and such that $\theta_j(z) < 1$ for $j = 1, 2$.

This system is equivalent to

(iii) $\int_{\mathbb{R}} z\theta_1(z)\nu_1(dz) = \lambda_{11}\mu_1 + \lambda_{12}\mu_2$

(iv) $\int_{\mathbb{R}} z\theta_2(z)\nu_2(dz) = \lambda_{21}\mu_1 + \lambda_{22}\mu_2.$

By our assumption (1.6.3) we see that we can choose $A_i \in \mathbf{B}_0$ with

$$|\lambda_{i1}\mu_1 + \lambda_{i2}\mu_2| < \int_{A_i} |z|\nu_i(dz) = K_i < \infty$$

and then the functions

$$\theta_i(z) = \text{sign} z \frac{\lambda_{i1}\mu_1 + \lambda_{i2}\mu_2}{K_i} \mathcal{X}_{A_i}(z) ; \quad i = 1, 2$$

solve (i), (ii). With this choice of $\theta_i(z)$; $i = 1, 2$ we define

$$Z(t) = \exp\left\{ \sum_{i=1}^{2} \left[\int_0^t \int_{\mathbb{R}} \ln(1 - \theta_i(z_i)) \tilde{N}_i(ds, dz_i) \right.\right.$$

$$\left.\left. + \int_0^t \int_{\mathbb{R}} \{\ln(1 - \theta_i(z)) + \theta_i(z)\} \nu_i(dz) \right] \right\}; \quad 0 \leq t \leq T$$

and we put

$$dQ = Z(T)dP \quad \text{on } \mathcal{F}_T.$$

Then Q is an equivalent local martingale measure for $(S_1(t), S_2(t))$. Just as in Sect. 1.5 we can now deduce that the market has no arbitrage.

Exercise 1.9

$$Z(t) = \exp\left(\int_0^t \int_{\mathbb{R}} \ln(1 - \theta(s, z)) \tilde{N}(ds, dz) + \int_0^t \int_{\mathbb{R}} \{\ln(1 - \theta(s, z)) + \theta(s, z)\} \nu(dz)ds \right).$$

Define

$$dX(t) = \alpha(t)dt + \int_{\mathbb{R}} \gamma(t, z) \tilde{N}(dt, dz)$$

where

$$\alpha(t) = \int_{\mathbb{R}} \{\ln(1 - \theta(t, z)) + \theta(t, z)\} \nu(dz)$$

and

$$\gamma(t, z) = \ln(1 - \theta(t, z)).$$

Then by the Itô formula, with $Z(t) = f(X(t))$, $f(x) = \exp(x)$

$$dZ(t) = f'(X(t))\alpha(t)dt$$

$$+ \int_{\mathbb{R}} \{f(X(t^-) + \gamma(t, z)) - f(X(t^-)) - f'(X(t^-))\gamma(t, z)\} \nu(dz)dt$$

$$+ \int_{\mathbb{R}} \{f(X(t^-) + \gamma(t, z)) - f(X(t^-))\} \tilde{N}(dt, dz)$$

$$= Z(t)\alpha(t)dt$$

$$+ \int_{\mathbb{R}} \{\exp(X(t^-) + \gamma(t, z)) - \exp(X(t^-)) - \exp(X(t^-))\gamma(t, z)\} \nu(dz)dt$$

$$+ \int_{\mathbb{R}} \{\exp(X(t^-) + \gamma(t, z)) - \exp(X(t^-))\} \tilde{N}(dt, dz)$$

$$= Z(t)\alpha(t)dt + \int_{\mathbb{R}} Z(t^-)\left\{\exp\gamma(t,z) - 1 - \gamma(t,z)\right\}\nu(dz)dt$$

$$+ \int_{\mathbb{R}} Z(t^-)\left\{\exp\gamma(t,z) - 1\right\}\tilde{N}(dt,dz)$$

$$= Z(t)\int_{\mathbb{R}}\left\{\ln(1 - \theta(t,z)) + \theta(t,z)\right\}\nu(dz)dt$$

$$+ Z(t^-)\int_{\mathbb{R}}\left\{1 - \theta(t,z) - 1 - \ln(1 - \theta(t,z))\right\}\nu(dz)dt$$

$$+ Z(t^-)\int_{\mathbb{R}}\left\{1 - \theta(t,z) - 1\right\}\tilde{N}(dt,dz)$$

$$= -Z(t)\int_{\mathbb{R}}\theta(t,z)\tilde{N}(dt,dz).$$

11.2 Exercises of Chapter 2

Exercise 2.1

We seek

$$\Phi(s,x) = \sup_{\tau \geq 0} E^{(s,x)}\left[e^{-\rho(s+\tau)}(X(\tau) - a)\right],$$

where

$$dX(t) = dB(t) + \gamma\int_{\mathbb{R}} z\tilde{N}(dt,dz); \quad X(0) = x \in \mathbb{R}.$$

We intend to apply Theorem 2.2 and start by putting

$$Y(t) = \begin{bmatrix} s+t \\ X(t) \end{bmatrix}; \quad Y(0) = \begin{bmatrix} s \\ x \end{bmatrix} = y \in \mathbb{R}^2 = \mathcal{S}.$$

The generator of Y is

$$A\phi(s,z) = \frac{\partial\phi}{\partial s} + \frac{1}{2}\frac{\partial^2\phi}{\partial x^2} + \int_{\mathbb{R}}\left\{\phi(s, x+\gamma z) - \phi(s,x) - \frac{\partial\phi}{\partial x}(s,x)\gamma z\right\}\nu(dz).$$

According to Theorem 2.2 (ix) we should look for a function ϕ such that $A\phi(s,x) = 0$ in D. We try

$$\phi(s,x) = e^{-\rho s}\psi(x) \quad \text{for some function } \psi.$$

Then
$$A\phi(s,x) = e^{-\rho s} A_0 \psi(x),$$

where
$$A_0\psi(x) = -\rho\psi(x) + \tfrac{1}{2}\psi''(x) + \int_{\mathbb{R}} \{\psi(x+\gamma z) - \psi(x) - \psi'(x)\gamma z\}\, \nu(dz).$$

Choose
$$\psi(x) = e^{\lambda x} \quad \text{for some constant } \lambda > 0.$$

Then
$$A_0\psi(x) = -\rho e^{\lambda x} + \tfrac{1}{2}\lambda^2 e^{\lambda x} + \int_{\mathbb{R}} \left\{ e^{\lambda(x+\gamma z)} - e^{\lambda x} - \lambda e^{\lambda x} \cdot \gamma z \right\} \nu(dz)$$

$$= e^{\lambda x}\left[-\rho + \tfrac{1}{2}\lambda^2 + \int_{\mathbb{R}} \left\{ e^{\lambda \gamma z} - 1 - \lambda\gamma z \right\} \nu(dz) \right].$$

Put
$$h(\lambda) := -\rho + \tfrac{1}{2}\lambda^2 + \int_{\mathbb{R}} \left\{ e^{\lambda\gamma z} - 1 - \lambda\gamma z \right\} \nu(dz).$$

Note that $h(0) = -\rho < 0$. Therefore, since
$$e^{\lambda\gamma z} - 1 - \lambda\gamma z \geq 0 \quad \text{for all } x \in \mathbb{R},$$

we see that $\lim_{\lambda\to\infty} h(\lambda) = \infty$.

So the equation $h(\lambda) = 0$ has at least one solution $\lambda_1 > 0$. Define
$$\psi(x) = \begin{cases} x - a & \text{for } x \geq x^*, \\ C\, e^{\lambda_1 x} & \text{for } x < x^*, \end{cases} \tag{11.2.1}$$

where $C > 0$, $x^* > 0$ are two constants to be determined.

If we require ψ to be continuous at $x = x^*$ we get the equation
$$C\, e^{\lambda_1 x^*} = x^* - a. \tag{11.2.2}$$

If we require ψ to be differentiable at $x = x^*$ we get the additional equation
$$\lambda_1 C\, e^{\lambda_1 x^*} = 1. \tag{11.2.3}$$

Dividing (11.2.1) by (11.2.2) we get
$$x^* = a + \frac{1}{\lambda_1}, \quad C = \frac{1}{\lambda_1}\, e^{-(\lambda_1 a + 1)}. \tag{11.2.4}$$

We now propose that the function

$$\phi(s,x) := e^{-\rho s}\psi(x)$$

with $\psi(x)$ given by (11.2.1), (11.2.2), and (11.2.3) satisfies all the requirements of Theorem 2.2 (possibly under some assumptions) and hence that

$$\phi(s,x) = \Phi(s,x)$$

and that

$$\tau^* := \inf\{t > 0; X(t) \geq x^*\}$$

is an optimal stopping time.

We proceed to check if the conditions (i)–(xi) of Theorem 2.2 hold. Many of these conditions are satisfied trivially or by construction of ϕ. We only discuss the remaining ones:

(ii) We know that $\phi = g$ for $x > x^*$, by construction. For $x < x^*$ we must check that

$$Ce^{\lambda_1 x} \geq x - a .$$

To this end, put

$$k(x) = Ce^{\lambda_1 x} - x + a ; \quad x \leq x^*.$$

Then

$$k(x^*) = k'(x^*) = 0 \quad \text{and}$$
$$k''(x^*) = \lambda_1^2 Ce^{\lambda_1 x} > 0 \quad \text{for } x \leq x^*.$$

Therefore $k'(x) < 0$ for $x < x^*$ and hence $k(x) > 0$ for $x < x^*$. Hence (ii) holds.

(vi) We know that $A\phi + f = A\phi = 0$ for $x < x^*$, by construction. For $x > x^*$ we have

$$A\phi(s,x) = e^{-\rho s}A_0(x - a)$$

$$= e^{-\rho s}\left(-\rho(x - a) + \int_{x+\gamma z < x^*}\left\{Ce^{\lambda_1(x+\gamma z)} - (x + \gamma z - a)\right\}\nu(dz)\right)$$

$$< e^{-\rho s}\left(-\rho(x^* - a) + \int_{\mathbb{R}}\left\{Ce^{\lambda_1(x^*+\gamma z)} - (x^* + \gamma z - a)\right\}\nu(dz)\right).$$

Therefore, using (11.2.2) and (11.2.4) we see that (vi) holds if

$$\int_{\mathbb{R}}\left\{e^{\lambda_1\gamma z} - 1 - \lambda_1\gamma z\right\}\nu(dz) \leq \rho. \tag{11.2.5}$$

(viii) In our case this condition gets the form

$$E\left[\int_0^{\infty}\left\{\sigma^2 e^{-2\rho t}X^2(t) + \int_{\mathbb{R}}e^{-2\rho t}\left|(X(t) + \gamma z)^2 - X^2(t)\right|^2\nu(dz)\right\}dt\right] < \infty$$

i.e.,

$$E\left[\int_0^\infty e^{-2\rho t}\left\{\sigma^2 X^2(t) + \int_{\mathbb{R}} |2X(t)\gamma z + \gamma^2 z^2| \nu(\mathrm{d}z)\right\}\mathrm{d}t\right] < \infty.$$

This will hold if

$$\gamma z \le 0 \quad \text{a.s. } \nu \tag{11.2.6}$$

or if

$$\sup_{\tau \in \mathcal{T}} E^x\left[e^{-2\rho\tau}\left(\int_0^\tau \int_{\mathbb{R}} zN(\mathrm{d}s, \mathrm{d}z)\right)^2\right] < \infty. \tag{11.2.7}$$

We will not discuss this condition further here.

(x) With our proposed solution ϕ we have

$$D = \{(s, x) \in \mathbb{R}^2; x < x^*\}.$$

So condition (x) states that

$$\tau_D := \inf\{t > 0; X(t) > x^*\} < \infty \text{ a.s.} \tag{11.2.8}$$

Some conditions are needed on σ, γ, and ν for (11.2.8) to hold. For example, it suffices that

$$\varlimsup_{t\to\infty} X(t) = \varlimsup_{t\to\infty}\left\{\sigma B(t) + \int_0^t \int_{\mathbb{R}} \gamma zN(\mathrm{d}s, \mathrm{d}z)\right\} = \infty \quad \text{a.s.} \tag{11.2.9}$$

(xi) For (xi) to hold it suffices that

$$\sup_{\tau \in \mathcal{T}} E^x[e^{-2\rho\tau} X^2(\tau)] < \infty. \tag{11.2.10}$$

Again it suffices to assume that (11.2.7) holds.

Conclusion

Assume that (11.2.5), (11.2.6), and (11.2.8) hold. Then the value function is

$$\Phi(s, x) = e^{-\rho s}\psi(x),$$

where $\psi(x)$ is given by (11.2.1) and (11.2.4). An optimal stopping time is

$$\tau^* = \inf\{t > 0; X(t) \ge x^*\}.$$

Exercise 2.2

Define

$$\mathrm{d}Y(t) = \begin{bmatrix} \mathrm{d}t \\ \mathrm{d}P(t) \\ \mathrm{d}Q(t) \end{bmatrix} = \begin{bmatrix} 1 \\ \alpha P(t) \\ -\lambda Q(t) \end{bmatrix}\mathrm{d}t + \begin{bmatrix} 0 \\ \beta P(t) \\ 0 \end{bmatrix}\mathrm{d}B(t) + \begin{bmatrix} 0 \\ \gamma\int_{\mathbb{R}} P(t^-)z\tilde{N}(\mathrm{d}t, \mathrm{d}z) \\ 0 \end{bmatrix}.$$

Then the generator A of $Y(t)$ is

$$A\phi(y) = A\phi(s, p, q) = \frac{\partial \phi}{\partial s} + \alpha p \frac{\partial \phi}{\partial p} - \lambda q \frac{\partial \phi}{\partial q} + \tfrac{1}{2}\beta^2 p^2 \frac{\partial^2 \phi}{\partial p^2}$$

$$+ \int_{\mathbb{R}} \left\{ \phi(s, p + \gamma pz, q) - \phi(s, p, q) - \frac{\partial \phi}{\partial p}(s, p, q)\gamma \, zp \right\} \nu(dz) \ .$$

If we try

$$\phi(s, p, q) = e^{-\rho s}\psi(w) \qquad \text{with } w = p \cdot q \ ,$$

then

$$A\phi(s, p, q) = e^{-\rho s} A_0 \psi(w),$$

where

$$A_0\psi(w) = -\rho\psi(w) + (\alpha - \lambda)w\psi'(w) + \tfrac{1}{2}\beta^2 w^2 \psi''(w)$$

$$+ \int_{\mathbb{R}} \left\{ \psi((1 + \gamma \, z)w) - \psi(w) - \gamma \, wz\psi'(w) \right\} \nu(dz).$$

Consider the set U defined in Proposition 2.3:

$$U = \{y; Ag(y) + f(y) > 0\} = \{(s, p, q); A_0(\theta \, w) + \lambda \, w - K > 0\}$$

$$= \{(s, p, q); [\theta(\alpha - \rho - \lambda) + \lambda]w - K > 0\}$$

$$= \begin{cases} \left\{ (s, p, q) : w > \frac{K}{\theta(\alpha - \rho - \lambda) + \lambda} \right\} & \text{if } \theta(\alpha - \rho - \lambda) + \lambda > 0 \\ \emptyset & \text{if } \theta(\alpha - \rho - \lambda) + \lambda \leq 0 \end{cases}$$

By Proposition 2.4 we therefore get

Case 1:

Assume $\lambda \leq \theta(\lambda + \rho - \alpha)$.
Then $\tau^* = 0$ is optimal and $\Phi(y) = g(y) = e^{-\rho s}p \cdot q$ for all y.

Case 2:

Assume $\theta(\lambda + \rho - \alpha) < \lambda$.
Then $U = \left\{ (s, w); w > \frac{K}{\lambda - \theta(\lambda + \rho - \alpha)} \right\} \subset D$.
   In view of this it is natural to guess that the continuation region D has
the form

$$D = \{(s, w); 0 < w^* < w\}$$

for some constant w^*; $0 < w^* < \frac{K}{\lambda - \theta(\lambda + \rho - \alpha)}$. In D we try to solve the equation

$$A_0\psi(w) + f(w) = 0.$$

The homogeneous equation $A_0\psi_0(w) = 0$ has a solution $\psi_0(w) = w^r$ if and only if

$$h(r) := -\rho + (\alpha - \lambda)r + \tfrac{1}{2}\beta^2 r(r-1) + \int_{\mathbb{R}} \{(1 + \gamma z)^r - 1 - r\gamma z\}\,\nu(dz) = 0.$$

Since $h(0) = -\rho < 0$ and $\lim_{|r| \to \infty} h(r) = \infty$, we see that the equation $h(r) = 0$ has two solutions r_1, r_2 such that

$$r_2 < 0 < r_1.$$

Let r be a solution of this equation. To find a particular solution $\psi_1(w)$ of the non-homogeneous equation

$$A_0\psi_1(w) + \lambda w - K = 0$$

we try

$$\psi_1(w) = aw + b$$

and find

$$a = \frac{\lambda}{\lambda + \rho - \alpha}, \qquad b = -\frac{K}{\rho}.$$

This gives that for all constants C the function

$$\psi(w) = C\,w^r + \frac{\lambda}{\lambda + \rho - \alpha}w - \frac{K}{\rho}$$

is a solution of

$$A_0\psi(w) + \lambda w - K = 0.$$

Therefore we try to put

$$\psi(w) = \begin{cases} \theta\,w; & 0 < w \le w^* \\ C\,w^r + \frac{\lambda}{\lambda + \rho - \alpha}w - \frac{K}{\rho}; & w \ge w^*, \end{cases} \qquad (11.2.11)$$

where $w^* > 0$ and C remain to be determined.

Continuity and differentiability at $w = w^*$ give

$$\theta\,w^* = C(w^*)^r + \frac{\lambda}{\lambda + \rho - \alpha}w^* - \frac{K}{\rho} \qquad (11.2.12)$$

$$\theta = C\,r(w^*)^{r-1} + \frac{\lambda}{\lambda + \rho - \alpha}. \qquad (11.2.13)$$

Combining (11.2.12) and (11.2.13) we get

$$w^* = \frac{(-r)K(\lambda + \rho - \alpha)}{(1-r)\rho(\lambda - \theta(\lambda + \rho - \alpha))} \qquad (11.2.14)$$

and

$$C = \frac{\lambda - \theta(\lambda + \rho - \alpha)}{-r} \cdot (w^*)^{1-r}. \qquad (11.2.15)$$

Since we need to have $w^* > 0$ we are led to the following condition:

Case 2a)

$\theta(\lambda + \rho - \alpha) < \lambda$ and $\lambda + \rho - \alpha > 0$.

Then we choose $r = r_2 < 0$, and with the corresponding values (11.2.14), (11.2.15) of w^* and C the function $\phi(s, p, q) = e^{-\rho s}\psi(p \cdot q)$, with ψ given by (11.2.11), is the value function of the problem. The optimal stopping time τ^* is

$$\tau^* = \inf \{t > 0; P(t) \cdot Q(t) \le w^*\}, \qquad (11.2.16)$$

provided that all the other conditions of Theorem 2.2 are satisfied. For condition (vi) to hold it suffices that

$$w^*(\lambda - \theta(\lambda + \rho - \alpha)) - K$$
$$+ \int_{\gamma zw > w^*} \left\{ C(w + \gamma zw)^r + \frac{\lambda}{\lambda + \rho - \alpha}(w + \gamma zw) - \frac{K}{\rho} - \theta(w + \gamma zw) \right\} \nu(dz) \le 0.$$
$$(11.2.17)$$

See also Remark 11.1.

Case 2b)

$\theta(\lambda + \rho - \alpha) < \lambda$ and $\lambda + \rho - \alpha \le 0$, i.e.,

$$\alpha \ge \lambda + \rho .$$

In this case we have $\Phi^*(y) = \infty$.

To see this note that since

$$P(t) = p + \int_0^t \alpha P(s)ds + \int_0^t \beta P(s)dB(s) + \int_0^t \int_{\mathbb{R}} \gamma P(s^-)z\tilde{N}(ds, dz),$$

we have

$$E[P(t)] = p + \int_0^t \alpha E[P(s)]ds$$

which gives

$$E[P(t)] = p e^{\alpha t}.$$

Therefore

$$E[e^{-\rho t}P(t)Q(t)] = E[pq\, e^{-\rho t}e^{-\lambda t}P(t)] = pq \exp \{(\alpha - \lambda - \rho)t\} .$$

Hence

$$\lim_{T \to \infty} E\left[\int_0^T e^{-\rho t}P(t)Q(t)dt \right] = \lim_{T \to \infty} pq \int_0^T \exp\{(\alpha - \lambda - \rho)t\}\, dt = \infty$$

if and only if $\alpha \ge \lambda + \rho$.

Remark 11.1 (On condition (viii) of Theorem 2.2). Consider

$$\phi(Y(t)) = e^{-\rho t}\psi(P(t)Q(t)),$$

where

$$P(t) = p\exp\left\{\left(\alpha - \tfrac{1}{2}\beta^2 - \gamma\int_{\mathbb{R}} z\,\nu(\mathrm{d}z)\right)t + \int_0^t\int_{\mathbb{R}} \ln(1+\gamma z)N(\mathrm{d}t,\mathrm{d}z) + \beta\,B(t)\right\}$$

and

$$Q(t) = q\exp(-\lambda t).$$

We have

$$P(t)Q(t) = pq\exp\left\{\left(\alpha - \lambda - \tfrac{1}{2}\beta^2 - \gamma\int_{\mathbb{R}} z\,\nu(\mathrm{d}z)\right)t\right.$$
$$\left. + \int_0^t\int_{\mathbb{R}} \ln(1+\gamma z)N(\mathrm{d}t,\mathrm{d}z) + \beta\,B(t)\right\}$$

and

$$e^{-\rho t}P(t)Q(t) = pq\exp\left\{\left(\alpha - \lambda - \rho - \tfrac{1}{2}\beta^2 - \gamma\int_{\mathbb{R}} z\,\nu(\mathrm{d}z)\right)t\right.$$
$$\left. + \int_0^t\int_{\mathbb{R}} \ln(1+\gamma z)N(\mathrm{d}s,\mathrm{d}z) + \beta\,B(t)\right\}.$$

Hence

$$E[(e^{-\rho t}P(t)Q(t))^2] = (pq)^2 E\left[\exp\left\{\left(2\alpha - 2\lambda - 2\rho - \beta^2 - 2\gamma\int_{\mathbb{R}} z\,\nu(\mathrm{d}z)\right)t\right.\right.$$
$$\left.\left. + 2\int_0^t\int_{\mathbb{R}} \ln(1+\gamma z)N(\mathrm{d}s,\mathrm{d}z) + 2\beta\,B(t)\right\}\right]$$
$$= (pq)^2 \exp\left\{\left(2\alpha - 2\lambda - 2\rho - \beta^2 - 2\gamma\int_{\mathbb{R}} z\,\nu(\mathrm{d}z)\right)t + 2\beta^2 t\right\}$$
$$\cdot E\left[\exp\left(2\int_0^t\int_{\mathbb{R}} \ln(1+\gamma z)N(\mathrm{d}t,\mathrm{d}z)\right)\right]$$

Using Exercise 1.6 we get

$$E[(\mathrm{e}^{-\rho t}P(t)Q(t))^2] = p^2q^2 \exp\left\{\left(2\alpha - 2\lambda - 2\rho + \beta^2 - 2\gamma\int_{\mathbb{R}} z\,\nu(\mathrm{d}z)\right.\right.$$

$$\left.\left.+ \int_{\mathbb{R}}\{(1+\gamma z)^2 - 1 - 2\ln(1+\gamma z)\}\,\nu(\mathrm{d}z)\right)t\right\}.$$

So condition (viii) of Theorem 2.2 holds if

$$2\alpha - 2\lambda - 2\rho + \beta^2 + \int_{\mathbb{R}}\{\gamma^2 z^2 - 2\ln(1+\gamma z)\}\,\nu(\mathrm{d}z) < 0.$$

Exercise 2.3

In this case we have

$$g(s, x) = \mathrm{e}^{-\rho s}|x|$$

and

$$\mathrm{d}X(t) = \mathrm{d}B(t) + \int_{\mathbb{R}} z\tilde{N}(\mathrm{d}t, \mathrm{d}z).$$

We look for a solution of the form

$$\phi(s, x) = \mathrm{e}^{-\rho s}\psi(x).$$

The continuation region is given by

$$D = \{(s, x) \in \mathbb{R} \times \mathbb{R} : \phi(s, x) > g(s, x)\} = \{(s, x) \in \mathbb{R} \times \mathbb{R} : \psi(x) \geq |x|\}$$

Because of the symmetry we assume that D is of the form

$$D = \{(s, x) \in \mathbb{R} \times \mathbb{R}; \ -x^* < x < x^*\}$$

where $x^* > 0$. It is trivial that D is a Lipschitz surface and $X(t)$ spends 0 time on ∂D. We must have

$$A\phi \equiv 0 \quad \text{on } D \tag{11.2.18}$$

where the generator A is given by

$$A\phi = \frac{\partial\phi}{\partial s} + \frac{1}{2}\frac{\partial^2\phi}{\partial x^2} + \int_{\mathbb{R}}\left\{\phi(s, x+z) - \phi(s, x) - \frac{\partial\phi}{\partial x}(s, x)z\right\}\nu(\mathrm{d}z).$$

Hence (11.2.18) becomes

$$-\rho\psi(x) + \tfrac{1}{2}\psi''(x) + \int_{\mathbb{R}} \{\psi(x+z) - \psi(x) - z\psi'(x)\}\,\nu(dz) = 0. \qquad (11.2.19)$$

For $|x| < \xi$ this equation becomes, by (2.4.1)

$$-\rho\psi(x) + \frac{1}{2}\psi''(x) = 0,$$

which has the general solution

$$\psi(x) = C_1 e^{\sqrt{2\rho}\,x} + C_2 e^{-\sqrt{2\rho}\,x},$$

where C_1, C_2 are arbitrary constants. Let $\lambda = \sqrt{2\rho}$ and $-\lambda$ be the two roots of the equation

$$F(\lambda) := -\rho + \frac{1}{2}\lambda^2 = 0.$$

Because of the symmetry we guess that

$$\psi(x) = \frac{C}{2}\left(e^{\lambda x} + e^{-\lambda x}\right) = C\cosh(\lambda x); \quad x \in D = \{(s,x); |x| < x^*\}$$

for some constants $C > 0$ and $x^* \in (0,\xi)$. Therefore

$$\psi(x) = \begin{cases} C\cosh(\lambda x) & \text{for } |x| < x^* \\ |x| & \text{for } |x| \ge x^*. \end{cases}$$

In order to find x^* and C, we impose the continuity and C^1-conditions on $\psi(x)$ at $x = x^*$:

- Continuity: $1 = |x^*| = C\cosh\left(\lambda x^*\right)$
- C^1: $1 = C\lambda\sinh(\lambda x^*)$

It follows that:

$$C = \frac{x^*}{\cosh(\lambda x^*)} \qquad (11.2.20)$$

and x^* is the solution of

$$\operatorname{tgh}\left(\lambda x^*\right) = \frac{1}{\lambda x^*}. \qquad (11.2.21)$$

Figure 11.1 illustrates that there exists a unique solution for (11.2.21).

Finally we have to verify that the conditions of Theorem 2.2 hold. We check some:

(ii) $\psi(x) \ge |x|$ for $(s,x) \in D$.

Define

$$h(x) = C\cosh(\lambda x) - x \; ; \; x > 0.$$

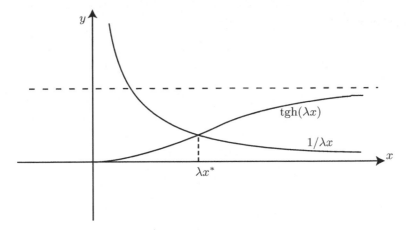

Fig. 11.1. The value of x^*

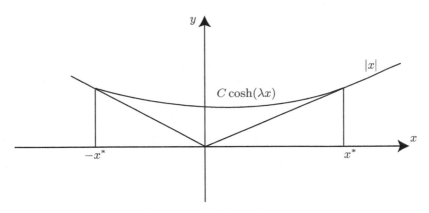

Fig. 11.2. The function ψ

Then $h(x^*) = h'(x^*) = 0$ and $h''(x) = C\lambda^2 \cosh(\lambda x) > 0$ for all x. Hence $h(x) > 0$ for $0 < x < x^*$, so (ii) holds. See Fig. 11.2.
(vi) $A\psi \leq 0$ outside $\bar{D}$.
This holds since, by (2.4.2) and (2.4.3),

$$A\psi(x) = -\rho|x| + \int_{\mathbb{R}} \{|x + z| - x - z\}\nu(dz) \leq 0 \text{ for all } x > x^*.$$

Since all the conditions of Theorem 2.2 are satisfied, we conclude that

$$\phi(s, y) = e^{-\rho s}\psi(y)$$

is the optimal value function and $\tau^* = \inf\{t > 0; |X(t)| = x^*\}$

Exercise 2.4

From Example 2.5 we know that in the *no delay case* ($\delta = 0$) the solution of
the problem (2.4.8) is the following (under some additional assumptions on
the Lévy measure ν):

$$\Phi_0(s, x) = e^{-\rho s} \Psi_0(x), \qquad (11.2.22)$$

where

$$\Psi_0(x) = \begin{cases} x - q & ; & x \geq x_0^* \\ C_0 x^\lambda & ; & 0 < x < x_0^* \end{cases}. \qquad (11.2.23)$$

Here $\lambda > 1$ is uniquely determined by the equation

$$-\rho + \mu\lambda + \tfrac{1}{2}\sigma^2\lambda(\lambda - 1) + \int_{\mathbb{R}} \{(1 + z)^\lambda - 1 - \lambda z\} \nu(dz) = 0, \qquad (11.2.24)$$

and x_0^* and C_0 are given by

$$x_0^* = \frac{\lambda q}{\lambda - 1}, \qquad (11.2.25)$$

$$C_0 = \frac{1}{\lambda}(x_0^*)^{1-\lambda}. \qquad (11.2.26)$$

The corresponding optimal stopping time $\tau^* \in \mathcal{T}_0$ is

$$\tau^* = \inf \{t > 0; X(t) \geq x_0^*\}. \qquad (11.2.27)$$

Thus it is optimal to sell at the first time the price $X(t)$ equals or exceeds the
value x_0^*.

To find the solution in the *delay* case ($\delta > 0$) we note that we have $f = 0$
and

$$g(y) = g(s, x) = e^{-\rho s}(x - q)$$

Hence, by (2.4.5),

$$\tilde{g}_\delta(y) = E^y[g(Y(\delta))] = E^{s,x}[e^{-\rho(s+\delta)}(X(\delta) - q)]$$

$$= e^{-\rho(s+\delta)}(E^x[X(\delta)] - q) = e^{-\rho(s+\delta)}(x e^{\mu\delta} - q)$$

$$= e^{-\rho s + \delta(\mu-\rho)}(x - q e^{-\mu\delta}) = K e^{-\rho s}(x - \tilde{q}), \qquad (11.2.28)$$

where

$$K = e^{\delta(\mu-\rho)} \quad \text{and} \quad \tilde{q} = q e^{-\mu\delta}. \qquad (11.2.29)$$

Thus $\tilde{g}_\delta$ has the same form as g, so we can apply the results (11.2.22)–(11.2.27)
to find $\tilde{\Phi}(y)$ and the corresponding optimal τ^*:

$$\tilde{\Phi}(y) = \tilde{\Phi}(s, x) = e^{-\rho s}\tilde{\Psi}(x) \tag{11.2.30}$$

where

$$\tilde{\Psi}(x) = \begin{cases} K(x - \tilde{q}) & ; \quad x \geq \tilde{x}^* \\ \tilde{C}x^\lambda & ; \quad 0 < x < \tilde{x}^* \end{cases} \tag{11.2.31}$$

with λ as in (11.2.24). Here $\tilde{x}^*$ and $\tilde{C}$ are given by

$$\tilde{x}^* = \frac{\lambda\tilde{q}}{\lambda - 1}, \tag{11.2.32}$$

$$\tilde{C} = \frac{1}{\lambda}(\tilde{x}^*)^{1-\lambda}. \tag{11.2.33}$$

The corresponding optimal stopping time for problem (2.3.7) and (2.3.6), respectively, is

$$\tilde{\tau}^* = \inf\{t > 0; X(t) \geq \tilde{x}^*\} \tag{11.2.34}$$

$$\alpha^* = \tilde{\tau}^* + \delta. \tag{11.2.35}$$

Using Theorem 2.11 we conclude the following:

Conclusion

The value function $\Phi_\delta(y)$ for the delayed optimal stopping problem (2.4.7) is given by

$$\Phi_\delta(y) = \tilde{\Phi}(y),$$

where $\tilde{\Phi}$ is as in (11.2.30)–(11.2.33). The corresponding optimal stopping time $\alpha^* \in \mathcal{T}_\delta$ is

$$\alpha^* = \inf\{t > 0; X(t) \geq \tilde{x}^*\} + \delta.$$

Remark 11.2. Assume for example that

$$\mu > 0.$$

Then comparing (11.2.32) with the non-delayed case (11.2.25) we see that $\tilde{q} > q$ and hence

$$\tilde{x}^* < x_0^*$$

Thus, in terms of the *delayed effect of the stopping time* formulation (see (2.3.4)), it is optimal to stop at the first time $t = \tilde{\tau}^*$ when $X(t) \geq \tilde{x}^*$. This is sooner than in the non-delayed case, because of the anticipation that during the delay time interval $[\tau^*, \tau^* + \delta]$ $X(t)$ is likely to increase (since $\mu > 0$). See Fig. 11.3.

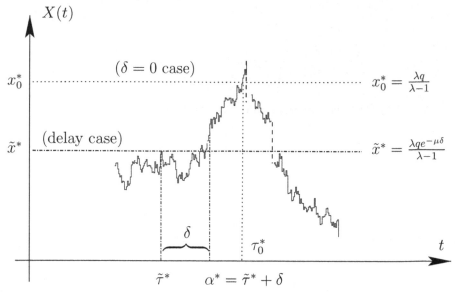

Fig. 11.3. The optimal stopping times for Exercise 2.4 $(\mu > 0)$

11.3 Exercises of Chapter 3

Exercise 3.1

Put

$$Y(t) = \begin{bmatrix} s+t \\ X(t) \end{bmatrix}.$$

Note: If we put

$$\mathcal{S} = \{(s,x); s < T\}$$

then

$$\tau_{\mathcal{S}} = \inf\{t > 0; \ Y^{s,x}(t,x) \notin \mathcal{S}\} = T - s.$$

The generator of $Y(t)$ is

$$A^u \phi(y) = A^u \phi(s,x) = \frac{\partial \phi}{\partial s} + (\mu - \rho x - u)\frac{\partial \phi}{\partial x} + \tfrac{1}{2}\sigma^2 \frac{\partial^2 \phi}{\partial x^2}$$

$$+ \int_{\mathbb{R}} \left\{ \phi(s, x + \theta z) - \phi(s,x) - \frac{\partial \phi}{\partial x} \cdot \theta z \right\} \nu(dz).$$

So the conditions of Theorem 3.1 get the form

(i) $A^u \phi(s,x) + e^{-\delta s} \frac{u^\gamma}{\gamma} \le 0$ for all $u \ge 0$, $s < T$

(ii) $\lim_{s \to T-} \phi(s,x) = \lambda x$

(iv) $\{\phi^-(Y(\tau))\}_{\tau \leq \tau_S}$ is uniformly integrable

(v) $A^{\hat{u}}\phi(s,x) + e^{-\delta s}\frac{u^\gamma}{\gamma} = 0$ for $s < T$

in addition to requirements (iii) and (vi).

We try a function ϕ of the form

$$\phi(s,x) = h(s) + k(s)x$$

for suitable functions $h(s), k(s)$. Then the conditions above get the form

(i)' $h'(s) + k'(s)x + (\mu - \rho\,x - u)k(s) + e^{-\delta s}\frac{u^\gamma}{\gamma}$

$$+ \int_{\mathbb{R}} \{h(s) + k(s)(x + \gamma\,z) - h(s) - k(s)x - k(s)\gamma\,z\}\,\nu(\mathrm{d}z) \leq 0$$

i.e.

$$e^{-\delta s}\frac{u^\gamma}{\gamma} + h'(s) + k'(s)x + (\mu - \rho\,x - u)k(s) \leq 0 \quad \text{for all } s < T,\ u \geq 0$$

(ii)' $h(T) = 0,\ k(T) = \lambda$

(iv)' $\{h(\tau) + k(\tau)X(\tau)\}_{\tau \leq \tau_S}$ is uniformly integrable.

(v)' $h'(s) + k'(s)x + (\mu - \rho\,x - \hat{u})k(s) + e^{-\delta s}\frac{\hat{u}^\gamma}{\gamma} = 0$

From (i)' and (v)' we get

$$-k(s) + e^{-\delta s}\,\hat{u}^{\gamma-1} = 0$$

or

$$\hat{u} = \hat{u}(s) = \left(e^{\delta s}k(s)\right)^{\frac{1}{\gamma-1}}.$$

Combined with (v)' this gives

(1) $k'(s) - \rho\,k(s) = 0$ so $k(s) = \lambda\,e^{\rho(s-T)}$

(2) $h'(s) = (\hat{u}(s) - \mu)k(s) - e^{-\delta s}\frac{\hat{u}^\gamma(s)}{\gamma}$, $h(T) = 0$

Note that

$$h'(s) = \left(e^{\delta s}k(s)\right)^{\frac{1}{\gamma-1}}k(s) - \mu\,k(s) - e^{-\delta s}\frac{\left(e^{\delta s}k(x)\right)^{\frac{\gamma}{\gamma-1}}}{\gamma}$$

$$= e^{\frac{\delta s}{\gamma-1}}k(s)^{\frac{\gamma}{\gamma-1}} - \mu\,k(s) - e^{-\delta s\left(1-\frac{\gamma}{\gamma-1}\right)} \cdot \frac{1}{\gamma} \cdot k(s)^{\frac{\gamma}{\gamma-1}}$$

$$= e^{\frac{\delta s}{\gamma-1}}k(s)^{\frac{\gamma}{\gamma-1}}\left[1 - \frac{1}{\gamma}\right] - \mu\,k(s) < 0.$$

Hence, since $h(T) = 0$, we have $h(s) > 0$ for $s < T$. Therefore

$$\phi(s,x) = h(s) + k(s)x \geq 0.$$

Clearly ϕ satisfies (i), (ii), (iv) and (v). It remains to check (vi), i.e., that

$$\{h(\tau) + k(\tau)X(\tau)\}_{\tau \leq T}$$

is uniformly integrable, and to check (iii).

For these properties to hold some conditions on ν must be imposed. We omit the details.

We conclude that if these conditions hold then

$$\hat{u}(s) = \lambda^{\frac{1}{\gamma-1}} \exp\left\{\frac{(\delta+\rho)s - \rho T}{\gamma-1}\right\}; \quad s \leq T \qquad (11.3.1)$$

is the optimal control.

Exercise 3.2

Define

$$J(u) = E\left[\int_0^{T_0} e^{-\delta t}\frac{u^\gamma(t)}{\gamma}dt + \lambda X(T_0)\right]$$

where

$$dX(t) = (\mu - \rho X(t) - u(t))dt + \sigma B(t) + \gamma \int_{\mathbb{R}} z \tilde{N}(dt, dz); \quad 0 \leq t \leq T_0.$$

The Hamiltonian is

$$H(t, x, u, p, q, r) = e^{-\delta t}\frac{u^\gamma}{\gamma} + (\mu - \rho x - u)p + \sigma q + \int_{\mathbb{R}} \gamma z r(t, z)\nu(dz).$$

The adjoint equation is

$$\begin{cases} d\hat{p}(t) = \rho\hat{p}(t)dt + \sigma\hat{q}(t)dB(t) + \int_{\mathbb{R}} \hat{r}(t, z)\tilde{N}(dt, dz); \quad t < T_0 \\ \hat{p}(T_0) = \lambda \end{cases}$$

Since λ and ρ are deterministic, we guess that $\hat{q} = \hat{r} = 0$ and this gives

$$\hat{p}(t) = \lambda e^{\rho(t-T_0)}.$$

Hence

$$H(t, \hat{X}(t), u, \hat{p}(t), \hat{q}(t), \hat{r}(t)) = e^{-\delta t}\frac{u^\gamma}{\gamma} + (\mu - \rho\hat{X}(t) - u)\hat{p}(t),$$

which is maximal when

$$u = \hat{u}(t) = \left(e^{\delta t}\hat{p}(t)\right)^{\frac{1}{\gamma-1}} = \lambda^{\frac{1}{\gamma-1}}\exp\left\{\frac{(\delta+\rho)t - \rho T_0}{\gamma-1}\right\}. \qquad (11.3.2)$$

Exercise 3.3

In this case we have

$$dX(t) = \begin{bmatrix} \int_{\mathbb{R}} u(t^-,\omega)z \\ \int_{\mathbb{R}} z^2 \end{bmatrix} \tilde{N}(dt, dz) = \begin{bmatrix} \int_{\mathbb{R}} \gamma_1(t, X(t^-), u(t^-), z)\tilde{N}(dt, dz) \\ \int_{\mathbb{R}} \gamma_2(t, X(t^-), u(t^-), z)\tilde{N}(dt, dz) \end{bmatrix}$$

so the Hamiltonian is

$$H(t, x, u, p, q, r) = \int_{\mathbb{R}} \{u\, z\, r_1(t, z) + z^2 r_2(t, z)\}\, \nu(dz)$$

and the adjoint equations are $(g(x_1, x_2) = -(x_1 - x_2)^2)$

$$\begin{cases} dp_1(t) = \int_{\mathbb{R}} r_1(t^-, z)\tilde{N}(dt, dz) \; ; \quad t < T \\ p_1(T) = -2(X_1(T) - X_2(T)) \end{cases}$$

$$\begin{cases} dp_2(t) = \int_{\mathbb{R}} r_2(t^-, z)\tilde{N}(dt, dz) \\ p_2(T) = 2(X_1(T) - X_2(T)). \end{cases}$$

Now $X_1(T) - X_2(T) = \int_0^T \int_{\mathbb{R}} \{u(t^-) - z\}\, \tilde{N}(dt, dz)$. So if $\hat{u}$ is a given candidate for an optimal control we get

$$\hat{r}_1(t, z) = -2(\hat{u}(t) - z)z,$$
$$\hat{u}_2(t, z) = 2(\hat{u}(t) - z)z.$$

This gives

$$H(t, x, u, \hat{p}, \hat{q}, \hat{r}) = \int_{\mathbb{R}} \{u\, z(-2(\hat{u}(t) - z)z) + z^2 2(\hat{u}(t) - z)z\}\, \nu(dz)$$

$$= -2u \int_{\mathbb{R}} \{\hat{u}(t)z^2 - z^3\}\, \nu(dz) + 2 \int_{\mathbb{R}} \{\hat{u}(t)z^3 - z^4\}\, \nu(dz).$$

This is a linear expression in u, so we guess that the coefficient of u is 0, i.e., that

$$\hat{u}(t) = \frac{\int_{\mathbb{R}} z^3 \nu(dz)}{\int_{\mathbb{R}} z^2 \nu(dz)} \quad \text{for all } (t, \omega) \in [0, T] \times \Omega. \tag{11.3.3}$$

With this choice of $\hat{u}(t)$ all the conditions of the stochastic maximum principle are satisfied and we conclude that $\hat{u}$ is optimal.

Note that this implies that

$$\inf_u E\left[\left(F - \int_0^T u(t)\,dS_1(t)\right)^2\right] = E\left[\left(\int_0^T \int_\mathbb{R} \{z^2 - \hat{u}(t)z\}\,\tilde{N}(dt,dz)\right)^2\right]$$

$$= \int_0^T \int_\mathbb{R} E[(z^2 - \hat{u}(t)z)^2]\nu(dz)\,dt$$

$$= T\int_\mathbb{R}\left[z^2 - \frac{\int_\mathbb{R} z^3\nu(dz)}{\int_\mathbb{R} z^2\nu(dz)}z\right]^2 \nu(dz).$$

We see that this is 0 if and only if

$$\int_\mathbb{R} z^3\nu(dz) = z\int_\mathbb{R} z^2\nu(dz) \quad \text{for a.a. } z(\nu) \qquad (11.3.4)$$

i.e. iff ν is supported on one point $\{z_0\}$. *Only then is the market complete!* See [BDLØP] for more information.

Exercise 3.4

We try to find a, b such that the function

$$\varphi(s, x) = e^{-\rho s}\psi(x) := e^{-\rho s}(ax^2 + b)$$

satisfies the conditions of (the minimum version of) Theorem 3.1. In this case the generator is

$$A_\varphi^v(s, x) = e^{-\rho s}A_0^v\psi(x),$$

where

$$A_0^v\psi(x) = -\rho\psi(x) + v\psi'(x) + \frac{1}{2}\sigma^2\psi''(x)$$

$$+ \int_\mathbb{R}\{\psi(x+z) - \psi(x) - z\psi'(x)\}\,\nu(dz).$$

Hence condition (i) of Theorem 3.1 becomes

$$A_0^v\psi(x) + x^2 + \theta v^2 = -\rho(ax^2 + b) + v2ax + \frac{1}{2}\sigma^2 2a + a\int_\mathbb{R} z^2\nu(dz) + x^2 + \theta v^2$$

$$= \theta v^2 + 2axv + x^2(1 - \rho a) + a\left(\sigma^2 + \int_\mathbb{R} z^2\nu(dz)\right) - \rho b =: h(v).$$

The function h is minimal when

$$v = u^*(x) = -\frac{ax}{\theta}. \tag{11.3.5}$$

With this value of v condition (v) becomes

$$x^2\left[1 - \rho a - \frac{a^2}{\theta}\right] + a\left(\sigma^2 + \int_{\mathbb{R}} z^2 \nu(\mathrm{d}z)\right) - \rho b = 0.$$

Hence we choose $a > 0$ and b such that

$$a^2 + \rho\theta a - \theta = 0 \tag{11.3.6}$$

and

$$b = \frac{a}{\rho}\left(\sigma^2 + \int_{\mathbb{R}} z^2 \nu(\mathrm{d}z)\right). \tag{11.3.7}$$

With these values of a and b we can easily check that

$$\varphi(s, x) := e^{-\rho s}(ax^2 + b)$$

satisfies all the conditions of Theorem 3.1. The corresponding optimal control is given by (11.3.5).

Exercise 3.5

(b) The Hamiltonian for this problem is

$$H(t, x, u, p, q, r) = x^2 + \theta u^2 + up + \sigma q + \int_{\mathbb{R}} r(t^-, z)\tilde{N}(\mathrm{d}t, \mathrm{d}z).$$

The adjoint equation is

$$\begin{cases} \mathrm{d}p(t) = -2X(t)\mathrm{d}t + q(t)\mathrm{d}B(t) + \int_{\mathbb{R}} r(t^-, z)\tilde{N}(\mathrm{d}t, \mathrm{d}z) \; ; \; t < T \\ p(T) = 2\lambda X(T). \end{cases} \tag{11.3.8}$$

By imposing the first and second-order conditions, we see that $H(t, x, u, p, q, r)$ is minimal for

$$u = u(t) = \hat{u}(t) = -\frac{1}{2\theta}p(t). \tag{11.3.9}$$

In order to find a solution of (11.3.8), we consider $p(t) = h(t)X(t)$, where $h : \mathbb{R} \to \mathbb{R}$ is a deterministic function such that

$$h(T) = 2\lambda.$$

Note that $u(t) = -\dfrac{h(t)X(t)}{2\theta}$ and

$$dX(t) = -\frac{h(t)X(t)}{2\theta}dt + \sigma dB(t) + \int_{\mathbb{R}} z\tilde{N}(dt, dz); \ X(0) = x.$$

Moreover, (11.3.8) turns into

$$dp(t) = h(t)dX(t) + X(t)h'(t)dt$$

$$= X(t)\left[-\frac{h(t)^2}{2\theta} + h'(t)\right]dt + h(t)\sigma dB(t) + h(t)\int_{\mathbb{R}} z\tilde{N}(dt, dz).$$

Hence $h(t)$ is the solution of

$$\begin{cases} h'(t) = \frac{h(t)^2}{2\theta} - 2; \ t < T \\ h(T) = 2\lambda. \end{cases} \tag{11.3.10}$$

The general solution of (11.3.10) is

$$h(t) = 2\sqrt{\theta}\,\frac{1 + \beta e^{\frac{2t}{\sqrt{\theta}}}}{1 - \beta e^{\frac{2T}{\sqrt{\theta}}}} \tag{11.3.11}$$

with $\beta = \frac{\lambda - \sqrt{\theta}}{\lambda + \sqrt{\theta}}e^{-\frac{2T}{\sqrt{\theta}}}$. By using the stochastic maximum principle, we can conclude that

$$u^*(t) = -\frac{h(t)}{2\theta}X(t)$$

is the optimal control, $p(t) = h(t)X(t)$ and $q(t) = \sigma h(t)$, $r(t^-, z) = h(t)z$, where $h(t)$ is given by (11.3.11).

Exercise 3.6

If we try a function of the form

$$\varphi(s, x) = e^{-\delta s}\psi(x)$$

then equations (i) and (v) for Theorem 3.1 combine to give the equation

$$\sup_{c \geq 0}\left\{\ln c - \delta\psi(x) + (\mu x - c)\psi'(x) + \frac{1}{2}\sigma^2 x^2\psi''(x)\right.$$

$$\left. + \int_{\mathbb{R}}\{\psi(x + x\theta z) - \psi(x) - x\theta z\psi'(x)\}\nu(dz)\right\} = 0.$$

The function

$$h(c) := \ln c - c\psi'(x); \ c > 0$$

is maximal when

$$c = \hat{c}(x) = \frac{1}{\psi'(x)}.$$

If we set
$$\psi(x) = a \, \ln x + b$$
where a, b are constants, $a > 0$, then this gives
$$\hat{c}(x) = \frac{x}{a},$$
and hence the above equation becomes
$$\ln x - \ln a - \delta(a \ln x + b) + \mu x \cdot \frac{a}{x} - 1 + \frac{1}{2}\sigma^2 x^2 \left(-\frac{a}{x^2}\right)$$
$$+ a \int_{\mathbb{R}} \left\{\ln(x + x\theta z) - \ln x - x\theta z \cdot \frac{1}{x}\right\} \nu(dz) = 0$$

or

$$(1 - \delta a) \ln x - \ln a - \delta b + \mu a - 1 - \frac{1}{2}\sigma^2 a$$
$$+ a \int_{\mathbb{R}} \left\{\ln(1 + \theta z) - \theta z\right\} \nu(dz) = 0, \quad \text{for all } x > 0.$$

This is possible if and only if
$$a = \frac{1}{\delta}$$
and
$$b = \delta^{-2}\left[\delta \ln \delta - \delta + \mu - \frac{1}{2}\sigma^2 + \int_{\mathbb{R}} \left\{\ln(1 + \theta z) - \theta z\right\} \nu(dz)\right].$$

One can now verify that if $\delta > \mu$ then with these values of a and b the function
$$\varphi(s, x) = e^{-\delta t}(a \ln x + b)$$
satisfies all the conditions of Theorem 3.1. We conclude that
$$\Phi(s, x) = e^{-\delta t}(a \ln x + b)$$
and that
$$c^*(x) = \hat{c} = \frac{x}{a}$$
(in feedback form) is an optimal consumption rate.

11.4 Exercises of Chapter 4

Exercise 4.1

(a) The HJB equation, i.e., (vi) and (ix) of Theorem 4.2, for this problem gets the form

$$0 = \sup_{u \geq 0} \left\{ e^{-\delta s} \frac{u^\gamma}{\gamma} + \frac{\partial \phi}{\partial s} + (\mu x - u) \frac{\partial \phi}{\partial x} + \frac{\sigma^2 x^2}{2} \frac{\partial^2 \phi}{\partial x^2} \right.$$

$$\left. + \int_{\mathbb{R}} \left\{ \phi(s, x + \theta x z) - \phi(s, x) - \theta x z \frac{\partial \phi}{\partial x}(s, x) \right\} d\nu(z) \right\} \qquad (11.4.1)$$

for $x > 0$.

We impose the first-order conditions to find the supremum, which is obtained for

$$u = u^*(s, x) = \left(e^{\delta s} \frac{\partial \phi}{\partial x} \right)^{1/(\gamma - 1)}. \qquad (11.4.2)$$

We guess that $\phi(s, x) = K e^{-\delta s} x^\gamma$ with $K > 0$ to be determined. Then

$$u^*(s, x) = (K\gamma)^{1/(\gamma - 1)} x \qquad (11.4.3)$$

and (11.4.1) turns into

$$\frac{1}{\gamma}(K\gamma)^{\gamma/(\gamma - 1)} - K\delta + \left(\mu - (K\gamma)^{1/(\gamma - 1)} \right) K\gamma + \frac{1}{2}\sigma^2 K\gamma(\gamma - 1)$$

$$+ K \int_{\mathbb{R}} \{ (1 + \theta z)^\gamma - 1 - \gamma \theta z \} \nu(dz) = 0$$

or

$$\gamma^{\gamma/(\gamma - 1)} K^{1/(\gamma - 1)} - \delta + \mu\gamma - \gamma^{\gamma/(\gamma - 1)} K^{1/(\gamma - 1)} + \frac{1}{2}\sigma^2 \gamma(\gamma - 1)$$

$$+ \int_{\mathbb{R}} \{ (1 + \theta z)^\gamma - 1 - \gamma \theta z \} \nu(dz) = 0.$$

Hence

$$K = \frac{1}{\gamma} \left[\frac{1}{1 - \gamma} \left(\delta - \mu\gamma + \frac{\sigma^2}{2}\gamma(1 - \gamma) \right. \right.$$

$$\left. \left. - \int_{\mathbb{R}} \{ (1 + \theta z)^\gamma - 1 - \gamma \theta z \} \nu(dz) \right) \right]^{\gamma - 1} \qquad (11.4.4)$$

provided that

$$\delta - \mu\gamma + \frac{\sigma^2}{2}\gamma(1 - \gamma) - \int_{\mathbb{R}} \{ (1 + \theta z)^\gamma - 1 - \gamma \theta z \} \nu(dz) > 0.$$

With this choice of K the conditions of Theorem 4.2 are satisfied and we can conclude that $\phi = \Phi$ is the value function.

(b)

(i) First assume $\lambda \geq K$. Choose $\phi(s, x) = \lambda e^{-\delta s} x^\gamma$. By the same computations as in a), condition (vi) of Theorem 4.2 gets the form

$$\lambda \geq \frac{1}{\gamma} \left[\frac{1}{\gamma - 1} \left(\delta - \mu\gamma + \frac{1}{2}\sigma^2\gamma(1 - \gamma) \right. \right.$$

$$\left. \left. - \int_{\mathbb{R}} \{(1 + \theta z)^\gamma - 1 - \gamma\theta z\} \nu(\mathrm{d}z) \right) \right]^{\gamma - 1}. \tag{11.4.5}$$

Since $\lambda \geq K$, the inequality (11.4.5) holds by (11.4.4).

By Theorem 4.2a), it follows that:

$$\phi(s, x) = \lambda e^{-\delta s} x^\gamma \geq \Phi(s, x)$$

where Φ is the value function for our problem. On the other hand, $\phi(s, x)$ is obtained by the (admissible) control of stopping immediately ($\tau = 0$). Hence we also have

$$\phi(s, x) \leq \Phi(s, x).$$

We conclude that

$$\Phi(s, x) = \lambda e^{-\delta s} x^\gamma$$

in this case and $\tau^* = 0$ is optimal. Note that $D = \emptyset$.

(ii) Assume now $\lambda < K$. Choose $\phi(s, x) = K e^{-\delta s} x^\gamma$. Then for all $(s, x) \in \mathbb{R} \times (0, \infty)$ we have

$$\phi(s, x) > \lambda e^{-\delta s} x^\gamma.$$

Hence we have $D = \mathbb{R} \times (0, \infty)$ and by Theorem 4.2a) we conclude that

$$\Phi(s, x) \leq K e^{-\delta s} x^\gamma.$$

On the other hand, we have seen in (a) above that if we apply the control

$$u^*(s, x) = (K\gamma)^{1/(\gamma - 1)} x$$

and never stop, then we achieve the performance $J^{(u^*)}(s, x) = K e^{-\delta s} x^\gamma$. Hence

$$\Phi(s, x) = K e^{-\delta s} x^\gamma$$

and it is optimal never to stop ($\tau^* = \infty$).

11.5 Exercises of Chapter 5

Exercise 5.1

In this case we put

$$\mathrm{d}Y(t) = \begin{bmatrix} \mathrm{d}t \\ \mathrm{d}X(t) \end{bmatrix} = \begin{bmatrix} 1 \\ \alpha \end{bmatrix} \mathrm{d}t + \begin{bmatrix} 0 \\ \sigma \end{bmatrix} \mathrm{d}B(t) + \begin{bmatrix} 0 \\ \beta \int_{\mathbb{R}} z \, \tilde{N}(\mathrm{d}t, \mathrm{d}z) \end{bmatrix} + \begin{bmatrix} 0 \\ -(1 + \lambda) \end{bmatrix} \mathrm{d}\xi(t).$$

The generator if $\xi = 0$ is

$$A\phi = \frac{\partial \phi}{\partial s} + \alpha \frac{\partial \phi}{\partial x} + \frac{1}{2}\sigma^2 \frac{\partial^2 \phi}{\partial x^2} + \int_{\mathbb{R}} \left\{ \phi(s, x + \beta z) - \phi(s, x) - \beta z \frac{\partial \phi}{\partial x}(s, x) \right\} \nu(dz).$$

The non-intervention region D is described by (see (5.2.5)

$$D = \left\{ (s, x); \sum_{i=1}^{k} \kappa_{ij} \frac{\partial \phi}{\partial y_i}(y) + \theta_j < 0 \quad \text{for all} \ \ j = 1, \ldots, p \right\}$$

$$= \left\{ (s, x); -(1 + \lambda) \frac{\partial \phi}{\partial x}(s, x) + e^{-\rho s} < 0 \right\}.$$

If we guess that D has the form

$$D = \{(s, x); 0 < x < x^*\} \quad \text{for some} \ \ x^* > 0$$

then by Theorem 5.2 we should have

$$A\phi(s, x) = 0 \quad \text{for} \ 0 < x < x^*.$$

We try a solution ϕ of the form

$$\phi(s, x) = e^{-\rho s} \psi(x)$$

and get

$$A_0 \psi(x) := -\rho \psi(x) + \alpha \psi'(x) + \frac{1}{2}\sigma^2 \psi''(x) + \int_{\mathbb{R}} \{\psi(x + \beta z) - \psi(x)$$

$$-\beta z \psi'(x)\} \nu(dz) = 0.$$

We now choose

$$\psi(x) = e^{r x} \quad \text{for some constant} \ r \in \mathbb{R}$$

and get the equation

$$h(r) := -\rho + \alpha r + \frac{1}{2}\sigma^2 r^2 + \int_{\mathbb{R}} \{e^{r \beta z} - 1 - r \beta z\} \nu(dz) = 0.$$

Since $h(0) < 0$ and $\lim_{r \to \infty} h(r) = \lim_{r \to -\infty} h(r) = \infty$, we see that the equation $h(r) = 0$ has two solutions r_1, r_2 such that

$$r_2 < 0 < r_1.$$

Outside D we require that

$$-(1+\lambda)\psi'(x) + 1 = 0$$

or

$$\psi(x) = \frac{x}{1+\lambda} + C_3, \quad C_3 \text{ constant.}$$

Hence we put

$$\psi(x) = \begin{cases} C_1 e^{r_1 x} + C_2 e^{r_2 x}; & 0 < x < x^* \\ \frac{x}{1+\lambda} + C_3; & x^* \le x \end{cases} \tag{11.5.1}$$

where C_1, C_2 are constants.

To determine C_1, C_2, C_3 and x^* we have the four equations:

$$\psi(0) = 0 \Rightarrow C_1 + C_2 = 0. \tag{11.5.2}$$

Put $C_2 = -C_1$

$$\psi \text{ continuous at } x = x^* \Rightarrow C_1(e^{r_1 x^*} - e^{r_2 x^*}) = \frac{x^*}{1+\lambda} + C_3 \tag{11.5.3}$$

$$\psi \in C^1 \text{ at } x = x^* \Rightarrow C_1(r_1 e^{r_1 x^*} - r_2 e^{r_2 x^*}) = \frac{1}{1+\lambda} \tag{11.5.4}$$

$$\psi \in C^2 \text{ at } x = x^* \Rightarrow C_1(r_1^2 e^{r_1 x^*} - r_2^2 e^{r_2 x^*}) = 0. \tag{11.5.5}$$

From (11.5.4) and (11.5.5) we deduce that

$$x^* = \frac{2(\ln|r_2| - \ln r_1)}{r_1 - r_2}. \tag{11.5.6}$$

Then by (11.5.4) we get the value for C_1, and hence the value of C_3 by (11.5.3).

With these values of C_1, C_2, C_3 and x^* we must verify that $\phi(s,x) = e^{-\rho s}\psi(x)$ satisfies all the requirements of Theorem 5.2:

(i) We have constructed ϕ such that $A\phi + f = 0$ in D. Outside D, i.e., for $x \ge x^*$, we have

$$e^{\rho s}(A\phi(s,x) + f(s,x)) = A_0\psi(x)$$

$$= -\rho\left(\frac{x}{1+\lambda} + C_3\right)$$

$$+ \alpha \cdot \frac{1}{1+\lambda} + \int_{x+\beta z < x^*} \left\{ C_1(e^{r_1(x+\beta z)} - e^{r_2(x+\beta z)}) - (\frac{x+\beta z}{1+\lambda} + C_3) \right\} \nu(dz)$$

$$\le -\frac{\rho}{1+\lambda}x + \frac{\alpha}{1+\lambda} - \rho C_3 , \text{ which is decreasing in } x.$$

So we need only to check that this holds for $x = x^*$, i.e., that

$$A_0 \psi(x^*) \le 0.$$

But this follows from the fact that $A_0 \psi(x) = 0$ for all $x < x^*$ and $\psi \in C^2$.
(ii) By construction we have

$$-(1 + \lambda)\psi'(x) + 1 = 0 \quad \text{for } x \ge x^*.$$

For $x < x^*$ the condition (ii) gets the form

$$F(x) := -(1 + \lambda)C_1(r_1 e^{r_1 x} - r_2 e^{r_2 x}) + 1 \le 0.$$

We know that $F(x^*) = 0$ by (11.5.4) and

$$F'(x) = -(1 + \lambda)C_1(r_1^2 e^{r_1 x} - r_2^2 e^{r_2 x}).$$

So $F'(x^*) = 0$ by (11.5.5) and hence (since $C_1 > 0$)

$$F'(x) > F'(x^*) = 0 \qquad \text{for } x < x^*.$$

Hence

$$F(x) < 0 \quad \text{for } 0 < x < x^*.$$

The conditions (iii), (iv), and (v) are left to the reader to verify.
(vi) This holds by construction of ϕ.
(vii)–(x) These conditions claim the existence of an increasing process $\hat{\xi}$ such that $Y^{\hat{\xi}}(t)$ stays in $\bar{D}$ for all times t, $\hat{\xi}(t)$ is strictly increasing only when $Y(t) \notin D$, and if $Y(t) \notin \bar{D}$ then $\hat{\xi}(t)$ brings $Y(t)$ down to a point on ∂D. Such a singular control is called a *local time* at ∂D of the process $Y(t)$ reflected downwards at ∂D. The existence and uniqueness of such a local time is proved in [CEM].
(xi) This is left to the reader.

We conclude that the optimal dividend policy $\xi^*(t)$ is to take out exactly the amount of money needed to keep $X(t)$ on or below the value x^*. If $X(t) < x^*$ we take out nothing. If $X(t) > x^*$ we take out $X(t) - x^*$.

Exercise 5.2

It suffices to prove that the function

$$\Phi_0(s, x_1, x_2) := K e^{-\delta s}(x_1 + x_2)^\gamma$$

satisfies conditions (i)–(iv) of Theorem 5.2. In this case we have (see Sect. 5.3)

$$A^{(v)}\Phi_0(y) = A^{(c)}\Phi_0(y) = \frac{\partial \Phi_0}{\partial s} + (rx_1 - c)\frac{\partial \Phi_0}{\partial x_1} + \alpha x_2 \frac{\partial \Phi_0}{\partial x_2} + \frac{1}{2}\beta^2 x_2^2 \frac{\partial^2 \Phi_0}{\partial x_2^2}$$

$$+ \int_{\mathbb{R}} \Bigg\{ \Phi_0(s, x_1, x_2 + x_2 z) - \Phi_0(s, x_1, x_2)$$

$$- x_2 z \frac{\partial \Phi_0}{\partial x_2}(s, x_1, x_2) \Bigg\} \nu(dz)$$

and $f(s, x_1, x_2, c) = e^{-\delta s}\frac{c^\gamma}{\gamma}$, so condition (i) becomes

(i)' $A^c \Phi_0(s, x_1, x_2) + e^{-\delta s}\frac{c^\gamma}{\gamma} \le 0$ for all $c \ge 0$.

This holds because we know by Example 3.2 that (see (3.1.21))

$$\sup_{c \ge 0} \Bigg\{ A^{(c)}\Phi(s, x_1, x_2) + e^{-\delta s}\frac{c^\gamma}{\gamma} \Bigg\} = 0.$$

Since in this case $\theta = 0$ and

$$\kappa = \begin{bmatrix} -(1+\lambda) & 1-\mu \\ 1 & -1 \end{bmatrix}$$

we see that condition (ii) of Theorem 5.2 becomes

(ii)' $-(1+\lambda)\frac{\partial \Phi_0}{\partial x_1} + \frac{\partial \Phi_0}{\partial x_2} \le 0$

(ii)" $(1-\mu)\frac{\partial \Phi_0}{\partial x_1} - \frac{\partial \Phi_0}{\partial x_2} \le 0.$

Since

$$\frac{\partial \Phi_0}{\partial x_1} = \frac{\partial \Phi_0}{\partial x_2} = Ke^{-\delta s}\gamma(x_1 + x_2)^{\gamma-1}$$

we see that (ii)' and (ii)" hold trivially.

We leave the verification of conditions (iii)–(v) to the reader.

11.6 Exercises of Chapter 6

Exercise 6.1

By using the same notation as in Chap. 6, we have here

$$Y^{(v)}(t) = \begin{bmatrix} s+t \\ X^{(v)}(t) \end{bmatrix}; \ t \geq 0; \ Y^{(v)}(0^-) = \begin{bmatrix} s \\ x \end{bmatrix} = y \in \mathbb{R}^2$$

$$\Gamma(y,\zeta) = \Gamma(s,x,\zeta) = \begin{bmatrix} s \\ x+\zeta \end{bmatrix}; \ (s,x,\zeta) \in \mathbb{R}^3$$

$$K(y,\zeta) = K(s,x,\zeta) = e^{-\rho s}(x + \lambda|\zeta|)$$

$$f(y) = f(s,x) = e^{-\rho s}x^2, \ g(y) = 0.$$

By symmetry we expect the continuation region to be of the form

$$D = \{(s,x) : -\bar{x} < x < \bar{x}\}$$

for some $\bar{x} > 0$, to be determined.

As soon as $X(t)$ reaches the unknown value $\bar{x}$ or $-\bar{x}$, there is an intervention and $X(t)$ is brought down (or up) to a certain value $\hat{x}$ (or $-\hat{x}$) where $-\bar{x} < -\hat{x} < 0 < \hat{x} < \bar{x}$. We determine $\bar{x}$ and $\hat{x}$ in the following computations (Fig. 11.4).

We guess that the value function is of the form

$$\phi(s,x) = e^{-\rho s}\psi(x).$$

In the continuation region D, we have by Theorem 6.2 (x)

$$A\phi + f = 0, \tag{11.6.1}$$

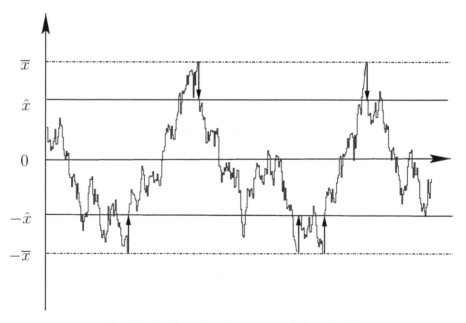

Fig. 11.4. The optimal strategy of Exercise 6.1

where A is the generator of Y, i.e.,

$$A\phi(s,x) = \frac{\partial\phi}{\partial s} + \frac{1}{2}\frac{\partial^2\phi}{\partial x^2} + \int_{\mathbb{R}}\left\{\phi(s,x+\theta(x,z)) - \phi(s,x) - \theta(x,z)\frac{\partial\phi}{\partial x}(s,x)\right\}\nu(dz).$$

In this case, if $|x| < \xi$, (11.6.1) becomes, by (6.3.1),

$$A_0\psi(x) + f(x) := -\rho\psi(x) + \frac{1}{2}\psi''(x) + x^2 = 0.$$

We try a solution of the form

$$\psi(x) = C\cosh(\gamma x) + \frac{1}{\rho}x^2 + \frac{1}{\rho^2}$$

where C is a constant (to be determined) and $\gamma = \sqrt{2\rho}$ is the positive solution of the equation

$$F(\gamma) := -\rho + \frac{1}{2}\gamma^2 = 0.$$

We try to set $C = -a$ where $a > 0$. Define

$$\psi_0(x) := \frac{1}{\rho}x^2 + \frac{1}{\rho^2} - a\cosh(\gamma x)$$

and put

$$\psi(x) = \psi_0(x); \ x \in D.$$

We recall that

$$D = \{(s,x) : \phi(s,x) < \mathcal{M}\phi(s,x)\}$$
$$= \{x : \psi(x) < \mathcal{M}\psi(x)\}$$

and the intervention operator is in this case

$$\mathcal{M}\psi(x) = \inf\{\psi(x+\zeta) + c + \lambda|\zeta|; \ \zeta \in \mathbb{R}\}.$$

The first-order condition for a minimum $\hat{\zeta} = \hat{\zeta}(x)$ of the function

$$G(\zeta) = \begin{cases} \psi(x+\zeta) + c + \lambda\zeta & \zeta > 0 \\ \psi(x+\zeta) + c - \lambda\zeta & \zeta < 0 \end{cases}$$

is the following:

(i) $\zeta > 0$: $\psi'(x+\zeta) + \lambda = 0 \Rightarrow \psi'(x+\zeta) = -\lambda$
(ii) $\zeta < 0$: $\psi'(x+\zeta) - \lambda = 0 \Rightarrow \psi'(x+\zeta) = \lambda$.

Hence we look for points $\hat{x}, \bar{x}$ such that

$$-\bar{x} < -\hat{x} < 0 < \hat{x} < \bar{x}$$

and

$$\psi'(\hat{x}) = -\lambda$$
$$\psi'(-\hat{x}) = \lambda. \tag{11.6.2}$$

Note that since $\hat{x} < \bar{x}$, $\psi'(\hat{x}) = \psi_0'(\hat{x})$.

Arguing as in Example 6.5, we put

$$\psi(x) = \begin{cases} \psi_0(x) & ; -\bar{x} \leq x \leq \bar{x} \\ \psi_0(\hat{x}) + c + \lambda(x - \hat{x}) & ; x > \bar{x} \\ \psi_0(-\hat{x}) + c - \lambda(x + \hat{x}) & ; x < -\bar{x}. \end{cases} \tag{11.6.3}$$

We have to show that there exist $0 < \hat{x} < \bar{x}$ and a value of a such that $\phi(s, x) := e^{-\rho s} \psi(x)$ satisfies all the requirements of (the minimum version of) Theorem 6.2. By symmetry we may assume $x > 0$ and $\zeta > 0$ in the following.

Continuity at $x = \bar{x}$ gives the equation

$$\psi_0(\hat{x}) + c + \lambda(\bar{x} - \hat{x}) = \psi_0(\bar{x}).$$

Differentiability at $x = \bar{x}$ gives the equation

$$\lambda = \psi_0'(\bar{x}).$$

Substituting for ψ_0 these equations give

$$\frac{\hat{x}^2}{\rho} - a \cosh(\gamma \hat{x}) - \lambda \hat{x} + c = \frac{\bar{x}^2}{\rho} - a \cosh(\gamma \bar{x}) - \gamma \bar{x} \tag{11.6.4}$$

and

$$\lambda = \frac{2\bar{x}}{\rho} - a\gamma \sinh(\gamma \bar{x}). \tag{11.6.5}$$

In addition we have required

$$\lambda = \psi_0'(\hat{x}) = \frac{2\hat{x}}{\rho} - a\gamma \sinh(\gamma \hat{x}). \tag{11.6.6}$$

As in Example 6.5 one can prove that for each $c > 0$ there exist $a = a^*(c) > 0$, $\hat{x} = \hat{x}(x) > 0$ and $\bar{x} = \bar{x}(c) > \hat{x}$ such that (11.6.3)–(11.6.5) hold. With these values of $a, \hat{x}$ and $\bar{x}$ it remains to verify that the conditions of Theorem 6.2 hold. We check some of them:

(ii) $\psi \leq \mathcal{M}\psi = \inf \{\psi(x - \zeta) + c + \lambda\zeta; \ \zeta > 0\}$.
First suppose $x \geq \bar{x}$.
If $x - \zeta \geq \bar{x}$ then

$$\psi(x - \zeta) + c + \lambda\zeta = \psi_0(\hat{x}) + c + \lambda(x - \zeta - \hat{x}) + c + \lambda\zeta = c + \psi(x) > \psi(x).$$

If $0 < x - \zeta < \bar{x}$ then

$$\psi(x - \zeta) + c + \lambda\zeta = \psi_0(x - \zeta) + c + \lambda\zeta,$$

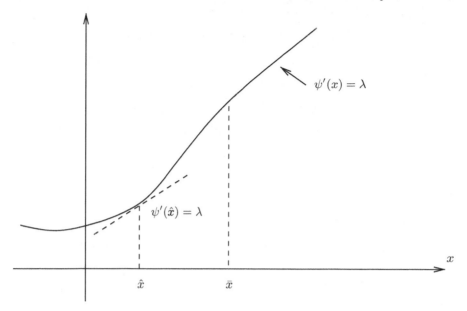

Fig. 11.5. The function $\psi(x)$ for $x > 0$

which is minimal when

$$-\psi_0'(x - \zeta) + \lambda = 0$$

i.e., when

$$\zeta = \hat{\zeta} = x - \hat{x}.$$

This is the minimum point because

$$\psi_0''(\hat{x}) > 0.$$

See Fig. 11.5.
This shows that

$$\mathcal{M}\psi(x) = \psi(x - \hat{\zeta}) + c + \lambda\hat{\zeta} = \psi(\hat{x}) + c + \lambda(x - \hat{x}) = \psi(x)$$

for $x > \hat{x}$.
Next suppose $0 < x < \bar{x}$.
Then

$$\mathcal{M}\psi(x) = \psi_0(\hat{x}) + c + \lambda(x - \hat{x}) > \psi(x)$$

if and only if

$$\psi(x) - \lambda x < \psi(\hat{x}) - \lambda\hat{x} + c.$$

Now the minimum of

$$H(x) := \psi(x) - \lambda x \quad \text{for } 0 < x < \bar{x}$$

is attained when

$$\psi'(x) = \lambda$$

i.e. when $x = \hat{x}$.
Therefore

$$\psi(x) - \lambda x \le \psi(\hat{x}) - \lambda\hat{x} < \psi(\hat{x}) - \lambda\hat{x} + c.$$

This shows that $\mathcal{M}\psi(x) > \psi(x)$ for all $0 < x < \bar{x}$.
Combined with the above we can conclude that

$$\mathcal{M}\psi(x) \ge \psi(x) \quad \text{for all } x > 0,$$

which proves (ii). Moreover,

$$\mathcal{M}\psi(x) > \psi(x) \quad \text{if and only if } 0 < x < \bar{x}.$$

Hence

$$D \cap (0, \infty) = (0, \bar{x}).$$

Finally we verify
(vi) $A\phi + f \ge 0$ for $x > \bar{x}$.
    For $x > \bar{x}$, we have, if $\bar{x} \le \xi$ (using (6.3.2)),

$$A_0\psi(x) + f(x) = -\rho(\psi_0(\hat{x}) + c + \lambda(x - \hat{x})) + x^2.$$

This is nonnegative for all $x > \bar{x}$ iff it is nonnegative for $x = \bar{x}$, i.e., iff

$$-\rho\psi_0(\bar{x}) + \bar{x}^2 \ge 0. \tag{11.6.7}$$

By construction of ψ_0 we know that, for $x < \bar{x}$

$$-\rho\psi_0(x) + \frac{1}{2}\psi_0''(x) + \int_{\mathbb{R}} \{\psi_0(x + z) - \psi_0(x) - z\psi_0'(x)\}\nu(\mathrm{d}z) + x^2 = 0.$$

Therefore (11.6.7) holds iff

$$\frac{1}{2}\psi_0''(\bar{x}) + \int_{\mathbb{R}} \{\psi_0(\bar{x} + z) - \psi_0(\bar{x}) - z\psi_0'(\bar{x})\}\nu(\mathrm{d}z) \le 0.$$

For this it suffices that

$$\int_{\mathbb{R}} z^2 \nu(\mathrm{d}z) \le -\frac{\rho}{2}\psi_0''(\bar{x}). \tag{11.6.8}$$

Conclusion

Suppose $\bar{x} \le \xi$ and that (11.6.8) holds. Then $\Phi(s, x) = \mathrm{e}^{-\rho s}\psi(x)$, with $\psi(x)$ given by (11.6.3) and $a, \hat{x}, \bar{x}$ given by (11.6.4)-(11.6.6). The optimal impulse control is to do nothing while $|X(t)| < \bar{x}$, then move $X(t)$ down to $\hat{x}$ (respectively, up to $-\hat{x}$) as soon as $X(t)$ reaches a value $\ge \bar{x}$ (respectively, a value $\le -\bar{x}$).

Exercise 6.2

Here we put

$$Y^{(v)}(t) = \begin{bmatrix} s+t \\ X^{(v)}(t) \end{bmatrix}$$

$$Y^{(v)}(0^-) = \begin{bmatrix} s \\ x \end{bmatrix} = y$$

$$\Gamma(y,\zeta) = x - c - (1+\lambda)\zeta$$

$$K(y,\zeta) = e^{-\rho s}\zeta$$

$$f \equiv g \equiv 0$$

$$S = \{(s,x) : x > 0\}.$$

We guess that the value function ϕ is of the form

$$\phi(s,x) = e^{-\rho s}\psi(x)$$

and consider the intervention operator

$$\mathcal{M}\psi(x) = \sup\left\{ \psi(x - c - (1+\lambda)\zeta) + \zeta; 0 \le \zeta \le \frac{x-c}{1+\lambda} \right\}. \qquad (11.6.9)$$

Note that the condition on ζ is due to the fact that the impulse must be positive and $x - c - (1+\lambda)\zeta$ must belong to S. We distinguish between two cases:

(1) $\mu > \rho$.

    In this case, suppose we wait until time t_1 and then take out

$$\zeta_1 = \frac{X(t_1) - c}{1 + \lambda}.$$

The corresponding value is

$$J^{(v_1)}(s,x) = E^x\left[\frac{e^{-\rho(t_1+s)}}{1+\lambda}(X(t_1) - c) \right]$$

$$= E^x\left[\frac{1}{1+\lambda}(xe^{-\rho s}e^{(\mu-\rho)t_1} - ce^{-\rho(s+t_1)}) \right]$$

$$\to \infty \text{ as } t_1 \to \infty.$$

Therefore we obtain $\Phi(s,x) = +\infty$ in this case.

(2) $\mu < \rho$.

We look for a solution by using the results of Theorem 6.2. In this case condition (x) becomes

$$A_0\psi(x) := -\rho\psi(x) + \mu x\psi'(x) + \tfrac{1}{2}\sigma^2\psi''(x)$$

$$+ \int_{\mathbb{R}} \{\psi(x + \gamma xz) - \psi(x) - \gamma z\psi'(x)\}\, \nu(dz) = 0 \text{ in } D. \quad (11.6.10)$$

We try a solution of the form

$$\psi(x) = C_1\, x^{\gamma_1} + C_2\, x^{\gamma_2},$$

where $\gamma_1 > 1$, $\gamma_2 < 0$ are the solutions of the equation

$$F(\gamma) := -\rho + \mu\gamma + \frac{1}{2}\sigma^2\gamma(\gamma - 1) + \int_{\mathbb{R}} \{(1 + \theta z)^\gamma - 1 - \theta z\gamma\}\, \nu'(dz) = 0.$$

We guess that the continuation region is of the form

$$D = \{(s, x) : 0 < x < \bar{x}\}$$

for some $\bar{x} > 0$ (to be determined).

We see that $C_2 = 0$, because otherwise $\lim_{x \to 0} |\psi(x)| = \infty$.

We guess that in this case it is optimal to wait till $X(t)$ reaches or exceeds a value $\bar{x} > c$ and then take out as much as possible, i.e., reduce $X(t)$ to 0. Taking the transaction costs into account this means that we should take out

$$\hat{\zeta}(x) = \frac{x - c}{1 + \lambda} \quad \text{for } x \geq \bar{x}.$$

We therefore propose that $\psi(x)$ has the form

$$\psi(x) = \begin{cases} C_1 x^{\gamma_1} \text{ for } 0 < x < \bar{x} \\ \frac{x-c}{1+\lambda} \text{ for } x \geq \bar{x}. \end{cases}$$

Continuity and differentiability of $\psi(x)$ at $x = \bar{x}$ give the equations

$$C_1\bar{x}^{\gamma_1} = \frac{\bar{x} - c}{1 + \lambda}$$

and

$$C_1\gamma_1\bar{x}^{\gamma_1 - 1} = \frac{1}{1 + \lambda}.$$

Combining these we get

$$\bar{x} = \frac{\gamma_1 c}{\gamma_1 - 1} \quad \text{and} \quad C_1 = \frac{\bar{x} - c}{1 + \lambda}\bar{x}^{-\gamma_1}.$$

With these values of $\bar{x}$ and C_1, we have to verify that ψ satisfies all the requirements of Theorem 6.2. We check some of them:

(ii) $\psi \geq \mathcal{M}\psi$ on $\mathcal{S}$.

Here $\mathcal{M}\psi = \sup\{\psi(x - c - (1+\lambda)\zeta) + \zeta\}; \; 0 \leq \zeta \leq \left\{\frac{x-c}{1+\lambda}\right\}$.

If $x - c - (1+\lambda)\zeta \geq \bar{x}$, then

$$\psi(x - c - (1+\lambda)\zeta) + \zeta = \frac{x - 2c}{1+\lambda} < \frac{x-c}{1+\lambda} = \psi(x)$$

and if $x - c - (1+\lambda)\zeta < \bar{x}$ then

$$h(\zeta) := \psi(x - c - (1+\lambda)\zeta) + \zeta = C_1(x - c - (1+\lambda)\zeta)^{\gamma_1} + \zeta.$$

Since

$$h'\left(\frac{x-c}{1+\lambda}\right) = 1 \text{ and } h''(\zeta) > 0$$

we see that the maximum value of $h(\zeta)$; $0 \leq \zeta \leq \frac{x-c}{1+\lambda}$, is attained at $\zeta = \hat{\zeta}(x) = \frac{x-c}{1+\lambda}$.

Therefore

$$\mathcal{M}\psi(x) = \max\left(\frac{x-2c}{1+\lambda}, \frac{x-c}{1+\lambda}\right) = \frac{x-c}{1+\lambda} \text{ for all } x > c.$$

Hence $\mathcal{M}\psi(x) = \psi(x)$ for $x \geq \bar{x}$.

For $0 < x < \bar{x}$ consider

$$k(x) := C_1 x^{\gamma_1} - \frac{x-c}{1+\lambda}.$$

Since

$$k(\bar{x}) = k'(\bar{x}) = 0 \quad \text{and} \quad k''(x) > 0 \text{ for all } x,$$

we conclude that

$$k(x) > 0 \quad \text{for } 0 < x < \bar{x}.$$

Hence

$$\psi(x) > \mathcal{M}\psi(x) \quad \text{for } 0 < x < \bar{x}.$$

(vi) $A_0\psi(x) \leq 0$ for $x \in \mathcal{S}\backslash\bar{D}$ i.e., for $x > \bar{x}$.

For $x > \bar{x}$, we have

$$A_0\psi(x) = -\rho\frac{x-c}{1+\lambda} + \mu x \cdot \frac{1}{1+\lambda}$$

$$+ \int_{x+\gamma xz < \bar{x}} \left\{C_1(x + \gamma xz)^{\gamma_1} - \frac{x + \gamma xz - c}{1+\lambda}\right\} \nu(dz)$$

$$\leq (1+\lambda)^{-1}[(\mu - \rho)x + (\rho + \|\nu\|)c].$$

Therefore we see that

$$A_0 \psi(x) \le 0 \text{ for all } x > \bar{x}$$
$$\Leftrightarrow (\mu - \rho)x + (\rho + \|\nu\|)c \le 0 \text{ for all } x > \bar{x}$$
$$\Leftrightarrow (\mu - \rho)\bar{x} + (\rho + \|\nu\|)c \le 0$$
$$\Leftrightarrow \bar{x} \ge \frac{(\rho + \|\nu\|)c}{\rho - \mu}$$
$$\Leftrightarrow \frac{\gamma_1 c}{\gamma_1 - 1} \ge \frac{(\rho + \|\nu\|)c}{\rho - \mu}$$
$$\Leftrightarrow \gamma_1 \le \frac{\rho + \|\nu\|}{\mu + \|\nu\|}.$$

Since

$$F\left(\frac{\rho}{\mu}\right) \ge -\rho + \mu\frac{\rho}{\mu} + \frac{1}{2}\sigma^2\frac{\rho}{\mu}\left(\frac{\rho}{\mu} - 1\right) > 0$$

and $F(\gamma_1) = 0$, $\gamma_1 > 1$ we conclude that $\gamma_1 < \frac{\rho}{\mu}$ and hence (vi) holds if $\|\nu\|$ is small enough.

Exercise 6.3

Here $f = g = 0$, $\Gamma(y, \zeta) = (s, 0)$, $K(y, \zeta) = -c + (1 - \lambda)x$ and $\mathcal{S} = \mathbb{R}^2$; $y = (s, x)$. If there are no interventions, the process $Y(t)$ defined by

$$\mathrm{d}Y(t) = \begin{bmatrix} \mathrm{d}t \\ \mathrm{d}X(t) \end{bmatrix} = \begin{bmatrix} 1 \\ \mu \end{bmatrix} \mathrm{d}t + \begin{bmatrix} 0 \\ \sigma \end{bmatrix} \mathrm{d}B(t) + \begin{bmatrix} 0 \\ \int_{\mathbb{R}} \theta z \tilde{N}(\mathrm{d}t, \mathrm{d}z) \end{bmatrix}$$

has the generator

$$A\phi(y) = \frac{\partial\phi}{\partial s} + \mu\frac{\partial\phi}{\partial x} + \frac{1}{2}\sigma^2\frac{\partial^2\phi}{\partial x^2} + \int_{\mathbb{R}} \left\{ \phi(s, x + \theta z) - \phi(s, x) - \theta z\frac{\partial\phi}{\partial x}(s, z) \right\} \nu(\mathrm{d}z) ;$$

$$y = (s, x).$$

The intervention operator $\mathcal{M}$ is given by

$$\mathcal{M}\phi(y) = \sup\{\phi(\Gamma(y, \zeta)) + K(y, \zeta); \zeta \in Z \text{ and } \Gamma(y, \zeta) \in \mathcal{S}\}$$
$$= \phi(s, 0) + (1 - \lambda)x - c.$$

If we try

$$\phi(s,x) = e^{-\rho s}\psi(x),$$

we get that

$$A\phi(s,x) = e^{-\rho s}A_0\psi(x),$$

where

$$A_0\psi(x) = -\rho\psi + \mu\psi'(x) + \tfrac{1}{2}\sigma^2\psi''(x) + \int_{\mathbb{R}} \{\psi(x+\theta z) - \psi(x) - \theta z\psi'(x)\}\,\nu(dz)$$

and

$$\mathcal{M}\phi(s,x) = e^{-\rho s}\mathcal{M}_0\psi(x),$$

where

$$\mathcal{M}_0\psi(x) = \psi(0) + (1-\lambda)x - c.$$

We guess that the continuation region D has the form

$$D = \{(s,x); x < x^*\}$$

for some $x^* > 0$ to be determined. To find a solution ψ_0 of $A_0\psi_0 + f = A_0\psi_0 = 0$, we try

$$\psi_0(x) = e^{rx} \quad (r \text{ constant})$$

and get

$$A_0\psi_0(x) = -\rho e^{rx} + \mu r e^{rx} + \tfrac{1}{2}\sigma^2 r^2 e^{rx}$$

$$+ \int_{\mathbb{R}} \left\{e^{r(x+\theta z)} - e^{rx} - r\theta z e^{rx}\right\}\nu(dz)$$

$$= e^{rx}h(r) = 0,$$

where

$$h(r) = -\rho + \mu r + \tfrac{1}{2}\sigma^2 r^2 + \int_{\mathbb{R}} \{e^{r\theta z} - 1 - r\theta z\}\,\nu(dz).$$

Choose $r_1 > 0$ such that

$$h(r_1) = 0 \quad \text{(see the solution of Exercise 2.1)}.$$

Then we define

$$\psi(x) = \begin{cases} M\,e^{r_1 x}; & x < x^* \\ \psi(0) + (1-\lambda)x - c = M + (1-\lambda)x - c; & x \geq x^* \end{cases} \qquad (11.6.11)$$

for some constant $M = \psi(0) > 0$. If we require continuity and differentiability at $x = x^*$ we get the equations

$$M\,e^{r_1 x^*} = M + (1-\lambda)x^* - c \tag{11.6.12}$$

and

$$M\,r_1 e^{r_1 x^*} = 1 - \lambda. \tag{11.6.13}$$

This gives the following equations for x^* and M:

$$k(x^*) := e^{-r_1 x^*} + r_1 x^* - 1 - \frac{r_1 c}{1-\lambda} = 0, \quad M = \frac{1-\lambda}{r_1}e^{-r_1 x^*} > 0. \tag{11.6.14}$$

Since $k(0) = -\frac{r_1 c}{1-\lambda} < 0$ and $\lim\limits_{x \to \infty} k(x) = \infty$, we see that there exists $x^* > 0$ s.t. $k(x^*) = 0$.

We must verify that with these values of x^* and M the conditions of Theorem 6.2 are satisfied. We consider some of them:
(ii) $\psi(x) \geq \mathcal{M}_0\psi(x)$.

For $x \geq x^*$ we have $\psi(x) = \mathcal{M}_0\psi(x) = M + (1-\lambda)x - c$. For $x < x^*$ we have $\psi(x) = M\,e^{r_1 x}$ and $\mathcal{M}_0\psi(x) = M + (1-\lambda)x - c$. Define

$$F(x) = M\,e^{r_1 x} - (M + (1-\lambda)x - c); \quad x \leq x^*.$$

See Fig. 11.6. We have

$$F(x^*) = F'(x^*) = 0 \text{ and } F''(x) = M\,r_1^2 e^{r_1 x} > 0.$$

Hence $F'(x) < 0$ and so $F(x) > 0$ for $x < x^*$. Therefore

$$\psi(x) \geq \mathcal{M}_0\psi(x) \text{ for all } x.$$

(vi) $A_0\psi \leq 0$ for $x > x^*$:

For $x > x^*$ we have

$$A_0\psi(x) = -\rho[M + (1-\lambda)x - c] + \mu(1-\lambda)$$

$$+ \int_{x+\theta z < x^*} \left\{Me^{r_1(x+\theta z)} - (M + (1-\lambda)(x+\theta z) - c)\right\}\nu(dz)$$

$$\leq -\rho(1-\lambda)x + \rho(c - M) + \mu(1-\lambda) + c\|\nu\|.$$

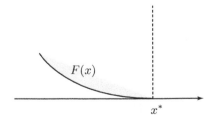

Fig. 11.6. The function F

So

$$A_0\psi(x) \leq 0 \text{ for all } x > x^* \Leftrightarrow x \geq \frac{\mu}{\rho} + \frac{c - M}{1 - \lambda} + \frac{c\|\nu\|}{\rho(1 - \lambda)} \quad \text{for all } x > x^*$$

$$\Leftrightarrow x^* \geq \frac{\mu}{\rho} + \frac{c - M}{1 - \lambda} + \frac{c\|\nu\|}{\rho(1 - \lambda)}$$

$$\Leftrightarrow x^* \geq \frac{\mu}{\rho} + \frac{c}{1 - \lambda} - \frac{1}{r_1}e^{-r_1 x^*} + \frac{c\|\nu\|}{\rho(1 - \lambda)}$$

$$\Leftrightarrow e^{-r_1 x^*} + r_1 x^* - \frac{c}{1 - \lambda} \geq \frac{\mu}{\rho} + \frac{c\|\nu\|}{\rho(1 - \lambda)}$$

$$\Leftrightarrow 1 \geq \frac{\mu}{\rho} + \frac{c\|\nu\|}{\rho(1 - \lambda)}$$

$$\Leftrightarrow \mu + \frac{c\|\nu\|}{1 - \lambda} \leq \rho.$$

So we need to assume that $\mu + \frac{c\|\nu\|}{1-\lambda} \leq \rho$ for (vi) to hold.

Conclusion

Let

$$\psi(s, x) = e^{-\rho s}\psi(x)$$

where ψ is given by (11.6.11). Assume that

$$\mu + \frac{c\|\nu\|}{1 - \lambda} \leq \rho.$$

Then

$$\phi(s, x) = \sup_v J^{(v)}(s, x)$$

and the optimal strategy is to cut the forest every time the biomass reaches the value x^* (see Fig. 11.7).

11.7 Exercises of Chapter 7

Exercise 7.1

As in Exercise 6.3, we have

$$f = g = 0,$$
$$\Gamma(y, \zeta) = (s, 0); \ y = (s, x)$$
$$K(y, \zeta) = (1 - \lambda)x - c$$
$$S = [0, \infty) \times \mathbb{R}$$

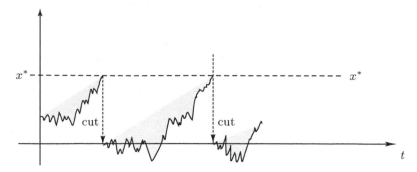

Fig. 11.7. The optimal forest management of Exercise 6.3

If there is no intervention, then $\phi_0 \equiv 0$ and

$$\mathcal{M}\phi_0 = \sup\{(1-\lambda)\zeta - c;\ \zeta = x\} = (1-\lambda)x - c.$$

Hence

$$\phi_1(y) = \sup_{\tau \le \tau_S} E^y[\mathcal{M}\phi_0(Y(\tau))] = \sup_{\tau \le \tau_S} E^y\left[e^{-\rho(s+\tau)}((1-\lambda)X(\tau) - c)\right].$$
(11.7.1)

This is an optimal stopping problem that can be solved by exploiting the three basic variational inequalities. We assume that the continuation region

$$D_1 = \{\phi_1 > \mathcal{M}\phi_0\}$$

is of the form

$$D_1 = \{(s,x);\ x < x_1\}\ \text{for some}\ x_1 > 0$$

and that the value function has the form $\phi_1(s,x) = e^{-\rho s}\psi_1(x)$ for some function ψ_1. On D_1, ψ_1 is the solution of

$$-\rho\psi_1(x) + \mu\psi_1'(x) + \tfrac{1}{2}\sigma^2\psi_1''(x) + \int_{\mathbb{R}}\{\psi_1(x + \theta z) - \psi_1(x) - \theta z\psi_1'(x)\}\nu(dz) = 0.$$
(11.7.2)

A solution of (11.7.2) is

$$\psi_1(x) = Ae^{\gamma_1 x} + Be^{\gamma_2 x}$$

where $\gamma_2 < 0$ and $\gamma_1 > 1$, A and B arbitrary constants to be determined.
We choose $B = 0$ and put $A_1 = A > 0$. We get

$$\psi_1(x) = \begin{cases} A_1 e^{\gamma_1 x} & x < x_1 \\ (1-\lambda)x - c & x \ge x_1. \end{cases}$$

We impose the continuity and differentiability conditions of ψ_1 at $x = x_1$.

(i) Continuity: $A_1 e^{\gamma_1 x_1} = (1 - \lambda)x_1 - c$.
(ii) Differentiability: $A_1 \gamma_1 e^{\gamma_1 x_1} = 1 - \lambda$.

We get $A_1 = \frac{(1-\lambda)}{\gamma_1} e^{-\gamma_1 x_1}$ and $x_1 = \frac{1}{\gamma_1} + \frac{c}{1-\lambda}$.
    As a second step, we evaluate

$$\phi_2(y) = \sup_\tau E^y [\mathcal{M}\phi_1(Y(\tau))].$$

We suppose $\phi_2(s, x) = e^{-\rho s} \psi_2(x)$ and consider

$$\mathcal{M}\psi_1(x) = \sup \{\psi_1(0) + (1 - \lambda)\zeta - c; \zeta \le x\} = \psi_1(0) + (1 - \lambda)x - c$$
$$= (1 - \lambda)x + A_1 - c.$$

Hence

$$\phi_2(y) = \sup_{\tau \le \tau_S} E^y \left[e^{-\rho(s+\tau)} \left((1 - \lambda)X(\tau) - (c - A_1)\right) \right]. \tag{11.7.3}$$

By the same argument as before, we get $\Phi_2(s, x) = e^{-\rho s} \psi_2(x)$, where

$$\psi_2(x) = \begin{cases} A_2 e^{\gamma_1 x} & x < x_2 \\ (1 - \lambda)x + A_1 - c & x \ge x_2 \end{cases}$$

where $x_2 = \frac{1}{\gamma_1} + \frac{c - A_1}{1 - \lambda}$ and $A_2 = \frac{1-\lambda}{\gamma_1} e^{-\gamma_1 x_2}$. Note that $x_2 < x_1$ and $A_2 > A_1$.
    Since $\mathcal{M}\phi_0$ and $\mathcal{M}\phi_1$ have linear growth, the conditions of Theorem 7.2 are satisfied. Hence ϕ_1 and ϕ_2 are the solutions for our impulse control problems when respectively one intervention and two interventions are allowed. The impulses are given by $\zeta_1 = \zeta_2 = x$ and $\tau_1 = \inf \{t : X(t) \ge x_2\}$ and $\tau_2 = \inf \{t > \tau_1 : X(t) \ge x_1\}$.

Exercise 7.2

Here we have (see the notation of Chap. 6)

$$f = g \equiv 0$$
$$K(x, \zeta) = \zeta$$
$$\Gamma(x, \zeta) = x - (1 + \lambda)\zeta - c$$
$$S = \{(s, x); \ x > 0\}.$$

We put $y = (s, x)$ and suppose $\phi_0(s, x) = e^{-\rho s} \psi_0(x)$. Since $f = g = 0$ we have

$$\phi_0(y) = 0$$

and $\mathcal{M}\psi_0(y) = \sup \left\{\zeta : 0 \le \zeta \le \frac{x-c}{1+\lambda}\right\} = \left(\frac{x-c}{1+\lambda}\right)^+$. As a second step, we consider

$$\phi_1(s,x) = \sup_{\tau \le \tau_S} E^x[\mathcal{M}\phi_0(X(\tau))] = \sup_{\tau \le \tau_S} E^x\left[e^{-\rho(\tau+s)}\frac{(X(\tau+s)-c)^+}{1+\lambda}\right].$$

(11.7.4)

We distinguish between three cases

(a) $\mu > \rho$

Then

$$\phi_1(s,x) \ge \frac{xe^{(\mu-\rho)(t+s)} - ce^{-\rho(t+s)}}{1+\lambda}.$$

Hence if $t \to +\infty$

$$\phi_1(s,x) \to +\infty.$$

We obtain $\mathcal{M}\phi_1(s,x) = +\infty$ and clearly $\phi_n = +\infty$ for all n. In this case, the optimal stopping time does not exist.

(b) $\mu < \rho$

In this case we try to put $\phi_1(s,x) = e^{-\rho s}\psi_1(x)$ and solve the optimal stopping problem (11.7.4) by using Theorem 2.2.

We guess that the continuation region is of the form $D = \{0 < x < x_1^*\}$ and solve

$$L\psi_1(x) := -\rho\psi_1(x) + \mu x\psi_1'(x) + \frac{\sigma^2 x^2}{2}\psi_1''(x)$$

$$+ \int_{\mathbb{R}}\{\psi_1(x+\theta xz) - \psi_1(x) - \theta xz\psi_1'(x)\}\nu(dx) = 0. \quad (11.7.5)$$

A solution of (11.7.5) is

$$\psi_1(x) = c_1 x^{\gamma_1} + c_2 x^{\gamma_2}$$

where $\gamma_2 < 0$ and $\gamma_1 > 1$ are solutions of the equation

$$k(\gamma) := -\rho + \mu\gamma + \frac{1}{2}\sigma^2\gamma(\gamma-1) + \int_{\mathbb{R}}\{(1+\theta z)^\gamma - 1 - \gamma\theta z\}\nu(dz) = 0$$

and c_1, c_2 are arbitrary constants. Since $\gamma_2 < 0$, we put $c_2 = 0$. We obtain

$$\psi_1(x) = \begin{cases} c_1 x^{\gamma_1} & 0 < x < x_1^* \\ \frac{x-c}{1+\lambda} & x \ge x_1^*. \end{cases}$$

By imposing the condition of continuity and differentiability, we can compute c_1 and x_1^*. The result is:

1. $x_1^* = \frac{\gamma_1 c}{\gamma_1 - 1}$
2. $c_1 = \frac{1}{\gamma_1(1+\lambda)}\left(\frac{\gamma_1 c}{\gamma_1 - 1}\right)^{1-\gamma_1}$

Note that $\gamma_1 > 1$ and $x_1^* > c$. We check some of the conditions of Theorem 2.2:

(ii) $\psi_1(x) \geq \mathcal{M}\psi_0(x)$ for all x:

We know that $\phi_1(x) = \mathcal{M}\phi_0(x)$ for $x > x_1^*$. Consider $h_1(x) := \psi_1(x) - \mathcal{M}\psi_0(x)$. We have

$$h_1(x_1^*) = 0,$$
$$h_1'(x_1^*) = 0,$$
$$h_1''(x_1^*) = c_1\gamma_1(\gamma_1 - 1)(x_1^*)^{\gamma_1 - 2} > 0.$$

Hence x_1^* is a minimum for h_1 and $\psi_1(x) \geq \mathcal{M}\psi_0(x)$ for every $0 < x < x_1^*$.

(vi) $L\psi_1 \leq 0$ for all $x > 0$:

Clearly $L\psi_1 = 0$ for $0 < x < x_1^*$.
If $x > x_1^*$ then

$$L\psi_1(x) = ((\mu - \rho)x + c\rho)\frac{1}{1+\lambda}$$
$$+ \int_{x+\theta xz < x_1^*} \left\{ c_1(x + \theta xz)^{\gamma_1} - \frac{x + \theta xz - c}{1 + \lambda} \right\} \nu(dz) \leq 0$$

if

$$x \geq \frac{c(\rho + \|\nu\|)}{\rho - \mu}.$$

Define

$$F(\gamma) := x^{-\gamma}L(x^\gamma) = -\rho + \mu\gamma + \frac{1}{2}\gamma(\gamma - 1)$$
$$+ \int_{\mathbb{R}} \{(1 + \theta z)^\gamma - 1 - \gamma\theta z\} \nu(dz).$$

Then we know that $F(\gamma_2) = F(\gamma_1) = 0$ and $F'(\gamma) > 0$ for $\gamma \geq \gamma_1$. Since $F\left(\frac{\rho}{\mu}\right) > 0$ we have that $\frac{\rho}{\mu} > \gamma_1$, which implies that $x_1^* = \frac{\gamma_1 c}{\gamma_1 - 1} > \frac{\rho c}{\rho - \mu}$.
Hence $L\psi_1(x) \leq 0$ for all $x \geq x_1^*$ if $\|\nu\|$ is small enough.

We conclude that under these conditions $\phi_1(s, x) = e^{-\rho s}\psi_1(x)$ actually solves the optimal stopping problem (11.7.4).

Next we consider

$$\mathcal{M}\psi_1(x) = \sup\left\{ \psi_1(x - (1 + \lambda)\zeta - c) + \zeta; 0 \leq \zeta \leq \frac{x - c}{1 + \lambda} \right\} = \frac{(x - c)^+}{1 + \lambda}$$

and repeat the same procedure to find ψ_2.

By induction, we obtain $\mathcal{M}\psi_n = \mathcal{M}\psi_{n-1} = \mathcal{M}\psi_1 = \mathcal{M}\psi_0$. Consequently, we also have

$$\Phi = \Phi_1$$

and $\Phi(s,x) = \Phi_n(s,x)$ for every n. Moreover, we achieve the optimal result with just one intervention.

(c) $\mu = \rho$

This case is left to the reader.

11.8 Exercises of Chapter 8

Exercise 8.1

In this case we have

$$\Gamma_1(\zeta, x_1, x_2) = x_1 - \zeta - c - \lambda|\zeta|$$
$$\Gamma_2(\zeta, x_1, x_2) = x_2 + \zeta$$
$$K(\zeta, x_1, x_2) = c + \lambda|\zeta|$$
$$g = 0$$
$$f(s, x_1, x_2, u) = \frac{e^{-\rho s}}{\gamma} u^\gamma.$$

The generator is given by

$$L^u = \frac{\partial}{\partial t} + (rx_1 - u)\frac{\partial}{\partial x_1} + \alpha x_2 \frac{\partial}{\partial x_2} + \frac{\sigma^2}{2}x_2^2 \frac{\partial^2}{\partial x_2^2}.$$

Let $\phi(s, x_1, x_2)$ be the value function of the optimal consumption problem

$$\sup_{w \in \mathcal{W}} E^y \left[\int_0^\infty e^{-\rho(s+t)} \frac{u(t)^\gamma}{\gamma} dt \right]; \quad y = (s, x_1, x_2)$$

with $c, \lambda > 0$ and $\phi_0(s, x_1, x_2)$ the corresponding value function in the case when there are no transaction costs, i.e.

$$c = \lambda = 0.$$

In order to prove that

$$\Phi(s, x_1, x_2) \leq Ke^{-\rho s}(x_1 + x_2)^\gamma = \Phi_0(s, x_1, x_2)$$

we check the hypotheses of Theorem 8.1a):

(vi) $L^u \phi_0 + f \leq 0$

Since ϕ_0 is the value function in the absence of transaction costs, we have

$$\sup_{u \geq 0} \{L^u \phi_0 + f\} = 0 \quad \text{in } \mathbb{R}^3.$$

Note that

$$\mathcal{M}\phi_0(s, x_1, x_2) = \sup_{\zeta \in \mathbb{R} \backslash \{0\}} \phi_0(s, x_1 - \zeta - \lambda|\zeta| - c, x_2 + \zeta)$$

$$= \sup_{\zeta \in \mathbb{R} \backslash \{0\}} Ke^{-\rho s}(x_1 + x_2 - c - \lambda|\zeta|)^\gamma$$

$$= Ke^{-\rho s}(x_1 + x_2 - c)^\gamma.$$

Therefore

$$D = \{\phi_0 > \mathcal{M}\phi_0\} = \mathbb{R}^3.$$

Hence we can conclude that

$$e^{-\rho s}(x_1 + x_2)^\gamma \geq \Phi(s, x_1, x_2).$$

Exercise 8.2

The HJBQVI's for this problem can be formulated in one equation as follows:

$$\min\left(\inf_{u \in \mathbb{R}} \left\{ \frac{\partial\varphi}{\partial s} + u\frac{\partial\varphi}{\partial x} + \frac{1}{2}\sigma^2\frac{\partial^2\varphi}{\partial x^2} + x^2 + u^2 \right\}, \varphi - \mathcal{M}\varphi\right) = 0,$$

where

$$\mathcal{M}\varphi(s, x) = \inf_{\zeta \in \mathbb{R}} \{\varphi(s, x + \zeta) + c\}.$$

Since φ is a candidate for the value function Φ it is reasonable to guess that, for each s, $\varphi(s, z)$ is minimal for $z = 0$. Hence

$$\mathcal{M}\varphi(s, x) = \varphi(s, 0) + c, \text{ attained for } \zeta = \zeta^*(x) = -x.$$

The minimum of the function

$$k(u) := \frac{\partial\varphi}{\partial s} + u\frac{\partial\varphi}{\partial x} + \frac{1}{2}\sigma^2\frac{\partial^2\varphi}{\partial x^2} + x^2 + u^2 \; ; \quad u \in \mathbb{R}$$

is attained when

$$u = \hat{u}(s, x) := -\frac{1}{2}\frac{\partial\varphi}{\partial x}(s, x)$$

and this gives the minimum value

$$k_{\min} = k(\hat{u}(s, x)) = \frac{\partial\varphi}{\partial s} - \frac{1}{4}\left(\frac{\partial\varphi}{\partial x}\right)^2 + \frac{1}{2}\sigma^2\frac{\partial^2\varphi}{\partial x^2} + x^2.$$

Hence the HJBQVI gets the form

$$\min\left(\frac{\partial\varphi}{\partial s} - \frac{1}{4}\left(\frac{\partial\varphi}{\partial x}\right)^2 + \frac{1}{2}\sigma^2\frac{\partial^2\varphi}{\partial x^2} + x^2, \varphi(s, 0) + c\right) = 0.$$

If we guess that the continuation region

$$D := \{(s,x);\ \varphi(s,x) < \varphi(s,0) + c\}$$

has the form

$$D = \{(s,x);\ |x| < x^*(s),\ 0 \le s \le T\}$$

for some function $x^*(s)$, then the HJBQVI can be split into the two equations

$$\frac{\partial \varphi}{\partial s} = -\frac{1}{2}\sigma^2 \frac{\partial^2 \varphi}{\partial x^2} + \frac{1}{4}\left(\frac{\partial \varphi}{\partial x}\right)^2 + \frac{1}{2}\sigma^2 \frac{\partial^2 \varphi}{\partial x^2} - x^2 \text{ for} |x| < x^*(s) \qquad (11.8.1)$$

$$\varphi(s,x) = \varphi(s,0) + c \text{ for } |x| \ge x^*(s). \qquad (11.8.2)$$

Equation (11.8.1) is known as a *Burgers* equation, which can be linearized by the transformation

$$w(s,x) := \exp(-\frac{1}{2\sigma^2}\varphi(s,x)). \qquad (11.8.3)$$

This transforms (11.8.1) into the equation

$$\frac{\partial w}{\partial s} = -\frac{1}{2}\sigma^2 \frac{\partial^2 w}{\partial x^2} + \frac{x^2}{2\sigma^2}w\ ;\quad |x| < x^*(s) \qquad (11.8.4)$$

with boundary values

$$w(s,x) = b(s) := w(s,0)\exp(-\frac{c}{2\sigma^2});\ x = \pm x^*(s),\ s < T \qquad (11.8.5)$$

$$w(T,x) = 1. \qquad (11.8.6)$$

By the Feynman–Kac formula, the solution of (11.8.4)–(11.8.6) can be written

$$w(s,x) = E_{\hat{P}}\left(\exp(-\frac{1}{2}\int_0^{\hat{\tau}_D - s} \hat{B}^2(t)dt)\ b(\hat{B}(\hat{\tau}_D - s))\right);\ (s,x) \in D \quad (11.8.7)$$

where $\hat{B}(t) = \hat{B}(t,\hat{\omega})$ is an auxiliary Brownian motion with law $\hat{P}$ starting at x and

$$\hat{\tau}_D = \hat{\tau}_D(\hat{\omega}) = \inf\left\{t > 0;\ |\hat{B}(t)| \ge \frac{1}{\sigma}x^*(t)\right\} \wedge T. \qquad (11.8.8)$$

The high contact condition for determination of the curve $x^*(t)$ is then

$$\left(\frac{\partial w}{\partial s}, \frac{\partial w}{\partial x}\right)(s, x^*(s)) = \exp(-\frac{c}{2c^2})\left(\frac{\partial w}{\partial s}(s,0), 0\right). \qquad (11.8.9)$$

The suggested shape of $x^*(t)$ is shown in Fig. 11.8.

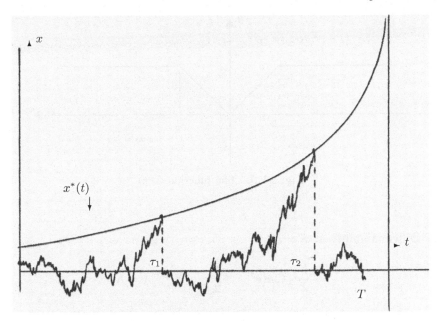

Fig. 11.8. The suggested optimal combined control of Exercise 8.2

11.9 Exercises of Chapter 9

Exercise 9.1

Because of the symmetry of h, we assume that the continuation region is of the form
$$D = \{(s, x) : -x^* < x < x^*\}$$
with $x^* > 0$.

We assume that the value function $\phi(s, x) = e^{-\rho s} \psi(x)$. On D, ϕ is the solution of
$$L\phi(s, x) = 0, \qquad (11.9.1)$$
where $L = \frac{\partial}{\partial s} + \frac{1}{2} \frac{\partial^2}{\partial x^2}$. We obtain
$$L_0 \psi(x) := -\rho \psi(x) + \frac{1}{2} \psi''(x) = 0. \qquad (11.9.2)$$
The general solution of (11.9.2) is
$$\psi(x) = c_1 e^{\sqrt{2\rho}\, x} + c_2 e^{-\sqrt{2\rho}\, x}.$$

We must have $\psi(x) = \psi(-x)$, hence $c_1 = c_2$. We put $c_1 = \frac{1}{2} c$ and assume $c > 0$. We impose continuity and differentiability conditions at $x = x^*$ (Fig. 11.9):

(i) Continuity at $x = x^*$
$$\frac{1}{2} c \left(e^{\sqrt{2\rho}\, x^*} + e^{-\sqrt{2\rho}\, x^*} \right) = K x^*$$

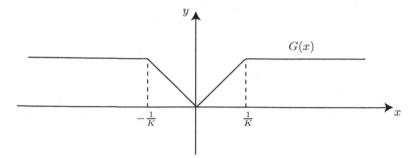

Fig. 11.9. The function $G(x)$

(ii) Differentiability at $x = x^*$

$$\frac{1}{2}c\sqrt{2\rho}\big(e^{x^*\sqrt{2\rho}}\frac{1}{2} - e^{-\sqrt{2\rho}\,x^*}\big) = K.$$

Then x^* is the solution of

$$\frac{1}{x^*\sqrt{2\rho}} = \frac{e^{x^*\sqrt{2\rho}} - e^{-\sqrt{2\rho}\,x^*}}{e^{x^*\sqrt{2\rho}} + e^{-x^*\sqrt{2\rho}}} = \mathrm{tgh}(x^*\sqrt{2\rho})$$

and

$$c = \frac{K}{\sqrt{2\rho}}\frac{1}{+\sinh(x^*\sqrt{2\rho})}.$$

We must check if $x^* < \frac{1}{K}$. If we put $z^* = x^*\sqrt{2\rho}$, then z^* is the solution of

$$\frac{1}{z^*} = \mathrm{tgh}(z^*).$$

We distinguish between two cases:

Case 1.   For $K < \dfrac{1}{x^*} = \dfrac{\sqrt{2\rho}}{z^*}$ we have (Fig. 11.10)

$$\psi(x) = \begin{cases} 1 & |x| > \frac{1}{K} \\ K|x| & x^* < |x| < \frac{1}{K} \\ c\cosh(x\sqrt{2\rho}) & |x| < x^* \end{cases}$$

Since ψ is not C^2 at $x = x^*$ we prove that ψ is a viscosity solution for our optimal stopping problem.

(i)   We first prove that ψ is a viscosity subsolution.

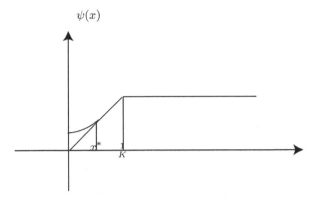

Fig. 11.10. The function $\psi(x)$ in Case 1 for $x \geq 0$

Let u belong to $C^2(\mathbb{R})$ and $u(x) \geq \psi(x)$ for all $x \in \mathbb{R}$ and let $y_0 \in \mathbb{R}$ be such that

$$u(y_0) = \psi(y_0).$$

Then ψ is a viscosity subsolution if and only if

$$\max(L_0 u(y_0), G(y_0) - u(y_0)) \geq 0 \text{ for all such } u, y_0. \qquad (11.9.3)$$

We need to check (11.9.3) only for $y_0 = x^*$. We have

$$u(x^*) = \psi(x^*) = G(x^*)$$

i.e., $(G - u)(x^*) = 0$. Hence $\max(L_0 u(x^*), G(x^*) - u(x^*)) \geq (G - u)(x^*) = 0$.

(ii)   We prove that ψ is a viscosity supersolution.
    Let v belong to $C^2(\mathbb{R})$ and $v(x) \leq \psi(x)$ for every $x \in \mathbb{R}$ and let $y_0 \in \mathbb{R}$ be such that $v(y_0) = \psi(y_0)$. Then ψ is a viscosity supersolution if and only if

$$\max(L_0 v(y_0), G(y_0) - v(y_0)) \leq 0 \text{ for all such } v, y_0.$$

We check it only for $x = x^*$. Then

$$G(x^*) = \psi(x^*) = v(x^*).$$

Since $v \leq \psi$, $x = x^*$ is a maximum point for $H := v - \psi$. We have

$$H(x^*) = 0,$$
$$H'(x^*) = 0,$$
$$H''(x^*) = v''(x^*) - \psi''(x^*_-) \leq 0.$$

Hence $L_0 v(x^*) \leq L_0 \psi(x^*_-) \leq 0$. Therefore ψ is a viscosity supersolution. Since ψ is both a viscosity supersolution and subsolution, ψ is a viscosity solution.

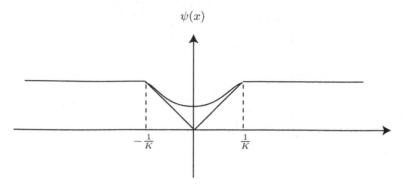

$$\psi(x)$$

Fig. 11.11. The function $\psi(x)$ in Case 2

Case 2.   We consider now the case when $K \geq \frac{\sqrt{2\rho}}{z^*}$.

In this case, the continuation region is given by

$$D = \left\{ -\frac{1}{K} < x < \frac{1}{K} \right\}$$

i.e., $x^* = \frac{1}{K}$.

We have

$$\psi(x) = \begin{cases} 1 & |x| \geq \frac{1}{K} \\ c \cosh(z\sqrt{2\rho}) & |x| < \frac{1}{K} \end{cases}$$

ψ is not C^1 at $|x| = \frac{1}{K}$. See Fig. 11.11. We prove that it is a viscosity solution.

(i) ψ is a viscosity subsolution.

Let u belong to $C^2(\mathbb{R})$ and $u(x) \geq \psi(x)$ for every $x \in \mathbb{R}$. Let y_0 be such that $u(y_0) = \psi(y_0)$. Then ψ is a viscosity subsolution if and only if

$$\max(L_0 u(y_0), G(y_0) - u(y_0)) \geq 0. \tag{11.9.4}$$

For $y_0 = x^*$, $G(x^*) = \psi(x^*) = u(x^*)$. Hence (11.9.4) is trivially satisfied.

(ii) ψ is a viscosity supersolution.

Actually there does not exist any $v \in C^2(\mathbb{R})$ such that $v(\frac{1}{K}) = \psi(\frac{1}{K})$ and $v \leq \psi$. Heuristically, this happens because ψ has an angle at $x = \frac{1}{K}$:

Suppose that there exists $v \in C^2(\mathbb{R})$ such that $v \leq \psi$, $v(\frac{1}{K}) = \psi(\frac{1}{K})$. We consider $H := \psi - v$. We have $H(\frac{1}{K}) = 0$ and $H \geq 0$. Hence we must have

$$H'\left(\left(\frac{1}{K}\right)^-\right) := \lim_{x \to (\frac{1}{K})^-} H'(x) \leq 0 \text{ and } H'\left(\left(\frac{1}{K}\right)^+\right) := \lim_{x \to (\frac{1}{K})^+} H'(x) \geq 0.$$

Therefore

$$\psi'\left(\left(\frac{1}{K}\right)^-\right) - v'\left(\left(\frac{1}{K}\right)^-\right) \le \psi'\left(\left(\frac{1}{K}\right)^+\right) - v'\left(\left(\frac{1}{K}\right)^+\right),$$

which implies that $\psi'\left(\left(\frac{1}{K}\right)^-\right) \le \psi'\left(\left(\frac{1}{K}\right)^+\right)$ since $v \in C^2(\mathbb{R})$.

Since ψ' does not satisfy this inequality, we conclude that there does not exist any $v \in C^2(\mathbb{R})$ such that $\psi \ge v$ and $\psi(\frac{1}{K}) = v(\frac{1}{K})$. ($v$ cannot be even C^1!).

We can conclude that ψ is a viscosity solution for our problem.

Exercise 9.2

We intend to apply (the minimum version of) Theorem 9.8 and note that in this case we have, with $y = (s, x) \in \mathbb{R}^2$,

$$L^u\varphi(y) = \frac{\partial\varphi}{\partial s} + u\frac{\partial\varphi}{\partial x} + \frac{1}{2}\frac{\partial^2\varphi}{\partial x^2} + \int_{\mathbb{R}} \left\{ \varphi(s, x+z) - \varphi(s, x) - z\frac{\partial\varphi}{\partial x}(s, x) \right\} \nu(\mathrm{d}z)$$

$$\Gamma(y, \zeta) = y + (0, \zeta), \quad K(y, \zeta) = ce^{-\rho s}, \quad f(y, u) = e^{-\rho s}(x^2 + \theta u^2),$$

$$g = 0 \text{ and}$$

$$\mathcal{M}\varphi(y) = \inf\left\{\varphi(s, x+\zeta) + ce^{-\rho s}; \ \zeta \in \mathbb{R}\backslash\{0\}\right\} = \varphi(s, 0) + ce^{-\rho s}.$$

and the corresponding HJBQVI becomes

$$\min\left(\inf_{\alpha \in U} \{L^\alpha\varphi(y) + f(y, \alpha)\}, \mathcal{M}\varphi(y) - \varphi(y)\right) = 0. \tag{11.9.5}$$

We first prove that

$$\Phi(x) < \Phi_1(x) = e^{-\rho s}(ax^2 + b)$$

(see the solution of Exercise 3.4). Since clearly

$$\Phi(x) \le \Phi_1(x)$$

it suffices, by Theorem 9.8, to prove that $\Phi_1(x)$ does not satisfy equation (11.9.5) in the viscosity sense. In particular, (11.9.5) implies that

$$e^{-\rho s}(ax^2 + b) = \Phi_1(s, x) \le \mathcal{M}\Phi_1(s, x) = e^{-\rho s}(b + c) ; \ x \in \mathbb{R}.$$

Since $a > 0$ this is impossible.

Next we prove that

$$\Phi(x) < \Phi_2(x)$$

where $\Phi_2(x)$ is the solution of Exercise 6.1. It has the form $\Phi_2(s, x) = e^{-\rho s}\psi_2(x)$, with

$$\psi_2(x) = \begin{cases} \psi_0(x); & |x| \leq \bar{x} \\ \psi_0(\hat{x}) + c; & |x| > \bar{x} \end{cases}$$

where

$$\psi_0(x) = \frac{1}{\rho}x^2 + \frac{b}{\rho^2} - a\cosh(\gamma x)$$

is a solution of

$$-\rho\psi_0(x) + \frac{1}{2}\psi_0''(x) + \int_{\mathbb{R}} \{\psi_0(x+z) - \psi_0(x) - z\psi_0'(x)\}\, \nu(dz) + x^2 = 0.$$

Since clearly $\Phi(s,x) \leq \Phi_2(s,x)$, it suffices to prove that $\Phi_2(s,x)$ does not satisfy (11.9.5) in viscosity sense. In particular, (11.9.5) implies that

$$L^u\Phi_2(s,x) + e^{-\rho s}(x^2 + \theta u^2) \geq 0 \text{ for all } u \in \mathbb{R}.$$

For $|x| < \bar{x}$ this reads

$$u\psi_0'(x) + \theta u^2 \geq 0 \text{ for all } u \in \mathbb{R},\ |x| < \bar{x}. \tag{11.9.6}$$

The function

$$h(u) := \left(\frac{2x}{\rho} - a\gamma\sinh(\gamma x)\right)u + \theta u^2 \,;\ u \in \mathbb{R}$$

is minimal when

$$u = \hat{u} = \frac{1}{2\theta}\left(a\gamma\sinh(\gamma x) - \frac{2x}{\rho}\right)$$

with corresponding minimum value

$$h(\hat{u}) = -\frac{1}{4\theta}\left(a\gamma\sinh(\gamma x) - \frac{2x}{\rho}\right)^2.$$

Hence (11.9.6) cannot possibly hold and we conclude that Φ_2 cannot be a viscosity solution of (11.9.5). Hence $\Phi \neq \Phi_2$ and hence $\Phi(x) < \Phi_2(x)$ for some x.

11.10 Exercises of Chapter 10

Exercise 10.1

(a) Here $A\phi(x) = (\alpha x + u)\phi'(x) + \frac{1}{2}\sigma^2\phi''(x)$ and hence

$$A^*\psi(x) = -\frac{d}{dx}((\alpha x + u)\psi(x)) + \frac{1}{2}\sigma^2\psi''(x)$$

$$= -\alpha\psi(x) - (\alpha x + u)\psi'(x) + \frac{1}{2}\sigma^2\psi''(x). \tag{11.10.1}$$

Therefore the corresponding complete observation controlled SPDE is

$$\begin{cases} dy(t,x) &= \left[-\alpha y(t,x) - (\alpha x + u(t))\frac{\partial y}{\partial x}(t,x) + \frac{1}{2}\sigma^2\frac{\partial^2}{\partial x^2}y(t;x) \right] dt \\ & \quad + h(x)y(t,x)d\zeta(t) \; ; \; t > 0 \\ y(0,x) &= F(x) \end{cases}$$

(11.10.2)

with performance criterion

$$J(u) = E_Q\left[\int_0^T \left(\int_{\mathbb{R}} (x^2 + u^2(t))y(t,x)dx \right) dt \right].$$

(11.10.3)

(b) Here $A\phi(x) = \alpha x u \phi'(x) + \frac{1}{2}\beta^2 x^2 u^2 \phi''(x)$ and hence

$$A^*\psi(x) = -\frac{d}{dx}(\alpha ux\psi(x)) + \frac{1}{2}\frac{d^2}{dx^2}(\beta^2 u^2 x^2\psi(x))$$

$$= (\beta^2 u^2 - \alpha u)\psi(x) + (2\beta^2 u^2 x - \alpha ux)\psi'(x) + \frac{1}{2}\beta^2 u^2 x^2 \psi''(x).$$

(11.10.4)

Hence the corresponding controlled complete observation SPDE is

$$\begin{cases} dy(t,x) &= \left[(\beta^2 u^2(t) - \alpha u(t))y(t,x) + (2\beta^2 u^2(t)x - \alpha u(t)x)\frac{\partial}{\partial x}y(t,x) \right. \\ & \quad \left. + \frac{1}{2}\beta^2 u^2(t)x^2\frac{\partial^2}{\partial x^2}y(t,x) \right] dt + h(x)y(t,x)d\zeta(t) \\ y(0,x) &= F(x), \end{cases}$$

(11.10.5)

with performance criterion

$$J(u) = E_Q\left[\int_{\mathbb{R}} x^\gamma y(T,x)dx \right].$$

(11.10.6)

References

[A]   D. Applebaum: Lévy Processes and Stochastic Calculus. Cambridge University Press 2003.

[Aa]   K. Aase: Optimum portfolio diversification in a general continuous-time model. Stoch. Proc. Appl. 18 (1984), 81–98.

[Am]   A.L. Amadori: Nonlinear integro-differential operators arising in option pricing: a viscosity solutions approach. J. Diff. Integral Eqn. 13 (2003), 787–811.

[AK]   L. H. R. Alvarez and J. Keppo: The impact of delivery lags on irreversible investment under uncertainty. Eur. J. Oper. Res. 136 (2002), 173–180.

[AKL]   A.L. Amadori, K.H. Karlsen and C. La Chioma: Non-linear degenerate integro-partial differential evolution equations related to geometric Lévy processes and applications to backward stochastic differential equations. Stoch. Stoch. Rep. 76(2) (2004), 147–177(31).

[AKy]   L. Alili and A.Kyprianou: Some remarks on first passage of Lévy processes, the American put and pasting principles. Ann. Appl. Probab. (2004).

[AMS]   M. Akian, J.L. Menaldi and A. Sulem: On an investment-consumption model with transaction costs. SIAM J. Control Opt. 34 (1996), 329–364.

[AO]   R. Anderson and S. Orey: Small random perturbations of dynamical systems with reflecting boundary. Nagoya Math. J. 60 (1976), 189-216.

[AT]   O. Alvarez and A. Tourin: Viscosity solutions of nonlinear integro-differential equations. Ann. Inst. Henri Poincaré 13 (1996), 293–317.

[B]   G. Barles: Solutions de Viscosité des Équations de Hamilton-Jacobi. Math. Appl. 17. Springer, Berlin Heidelberg New York 1994.

[BG]   R.M. Blumenthal and R. K. Getoor: Markov Processes and Potential Theory. Academic, New York 1968.

[B-N]   O. Barndorff-Nielsen: Processes of normal inverse Gaussian type. Finance Stoch. 1 (1998), 41–68.

[Be]   J. Bertoin: Lévy Processes. Cambridge University Press 1996.

[BEK]   J.S. Baras, R.J. Elliott and M. Kohlmann: The partially observed stochastic minimum principle. SIAM J. Control Opt. 27 (1989), 1279–1292.

[Ben1]   A. Bensoussan: Maximum principle and dynamic programming approaches of the optimal control of partially observed diffusions. Stochastics 9 (1983), 169–222.

[Ben2]   A. Bensoussan: Stochastic maximum principle for systems with partial information and application to the separation principle. In M. Davis and R. Elliott (eds.): Applied Stochastic Analysis. Gordon Breach, New York 1991, pp. 157–172.

[Ben3]   A. Bensoussan: Stochastic Control of Partially Observable Systems. Cambridge University Press 1992.

[Bi]   J.-M. Bismut: Conjugate convex functions in optimal stochastic control. J. Math. Anal. Appl. 44 (1973), 384–404.

[BCa]   M. Bardi and I. Capuzzo-Dolcetta: Optimal Control and Viscosity Solutions of Hamilton-Jacobi-Bellman Equations. Birkhäuser, Basel 1997.

[BCe]   B. Bassan and C. Ceci: Optimal stopping problems with discontinuous reward: regularity of the value function and viscosity solutions. Stoch. Stoch. Rep. 72 (2002), 55–71.

[BDLØP]   F.E. Benth, G. Di Nunno, A. Løkka, B. Øksendal and F. Proske: Explicit representation of the minimal variance portfolio in markets driven by Lévy processes. Math. Finance 13 (2003), 55–72.

[BK]   F.E. Benth and K.H. Karlsen: Portfolio Optimization in Lévy Markets. World Scientific (in preparation).

[BKR1]   F.E. Benth, K. Karlsen and K. Reikvam: Optimal portfolio management rules in a non-Gaussian market with durability and intertemporal substitution. Finance Stoch. 4 (2001), 447–467.

[BKR2]   F.E. Benth, K. Karlsen and K. Reikvam: Optimal portfolio selection with consumption and nonlinear integro-differential equations with gradient constraint: a viscosity solution approach. Finance Stoch. 5 (2001), 275–303.

[BKR3]   F.E. Benth, K.H. Karlsen and K.Reikvam: A note on portfolio management under non-gaussian logreturns. Int. J. Appl. Theor. Finance 4 (2001), 711–732.

[BKR4]   F.E. Benth, K.H. Karlsen and K. Reikvam: Portfolio optimization in a Lévy market with intertemporal substitution and transaction costs. Stoch. Stoch. Rep. 74 (2002), 517–569.

[BKR5]   F.E. Benth, K.H. Karlsen and K. Reikvam: A semilinear Black & Scholes partial differential equation for valuing American options. Finance Stoch. 7 (2003), 277–298.

[BKR6]   F.E. Benth, K.H. Karlsen and K. Reikvam: Mertons portfolio optimization problem in a Black and Scholes market with non-gaussian stochastic volatility of Ornstein-Uhlenbeck type. Math. Finance 13 (2003), 215–244.

[BL]   A. Bensoussan and J.L. Lions: Impulse Control and Quasi-Variational Inequalities. Gauthiers-Villars, Paris 1984.

[BØ1]   K.A. Brekke and B. Øksendal: Optimal switching in an economic activity under uncertainty. SIAM J. Control Opt. 32 (1994), 1021–1036.

[BØ2]   K.A. Brekke and B. Øksendal: A verification theorem for combined stochastic control and impulse control. In L. Decreusefond et al. (eds.): Stochastic Analysis and Related Topics, Vol. 6, Birkhäuser, Basel 1998, 211–220.

[BR]   F.E. Benth and K. Reikvam: On a connection between singular control and optimal stopping. Appl. Math. Opt. 49 (2004), 27–41.

[BS]   A. Bar-Ilan and A. Sulem: Explicit solution of inventory problems with delivery lags. Math. Oper. Res. 20 (1995), 709–720.

[BT]   D.P. Bertsekas and J.N.Tsitsiklis: An analysis of stochastic shortest path problems. Math. Oper. Res. 16 (1991), 580-595.

[C]   T. Chan: Pricing contingent claims on stocks driven by Lévy processes. Ann. Appl. Probab. 9 (1999), 504-528.

[CE]   C.D. Charalambous and R.E. Elliott: Classes of nonlinear partially observable stochastic optimal control problems with explicit optimal control laws. SIAM J. Control Opt. 36 (1998), 542-578.

[CB]   C. Ceci and B. Bassan: Mixed optimal stopping and stochastic control problems with semi-continuous final reward for diffusion processes. Stoch. Stoch. Rep. 76, (2004), 323-347.

[CEM]   M. Chaleyat-Maurel, N. El Karoui and B. Marchal: Réflexion discontinue et systèmes stochastiques. Ann. Probab. 8 (1980), 1049-1067.

[CIL]   M.G. Crandall, H. Ishii and P.-L. Lions: User's guide to viscosity solutions of second order partial differential equations. Bull. Am. Math. Soc. 27 (1992), 1-67.

[CMS]   J. Ph. Chancelier, M. Messaoud and A. Sulem: A policy iteration algorithm for fixed point problems with nonexpansive operators. Mathematical Methods of Operations Research, 65 (2006).

[CØS]   J.-Ph. Chancelier, B. Øksendal and A. Sulem: Combined stochastic control and optimal stopping, and application to numerical approximation of combined stochastic and impulse control, Stochastic Financial Mathematics, In A.N. Shiryaev (ed.), Steklov Mathematical Institute, Moscow, vol. 237, 149-173, 2002.

[CT]   R. Cont and P. Tankov: Financial Modelling with Jump Processes. Chapman & Hall/CRC, London/Boca Raton 2003.

[D1]   M.H.A. Davis: Linear Estimation and Stochastic Control. Chapman & Hall, London 1977.

[D2]   M.H.A. Davis: Lectures on Stochastic Control and Nonlinear Filtering. Tata Institute of Fundamental Research, Bombay 1984.

[DM]   M.H.A. Davis and I. Markus: An introduction to nonlinear filtering. In M. Hazewinkel and J.C. Willems (eds.): Stochastic Systems; The Mathematics of Filtering and Identification and Applications. Reidel, Dordrecht 1981, 53-75.

[DN]   M.H.A. Davis and A. Norman: Portfolio selection with transaction costs. Math. Oper. Res. 15 (1990), 676-713.

[DS]   F. Delbaen and W. Schachermayer: A general version of the fundamental theorem of asset pricing. Math. Ann. 300 (1994), 463-520.

[DZ]   K. Duckworth and M. Zervos: A model for investment decisions with switching costs. Ann. Appl. Probab. 11 (2001), 239-260.

[E]   H.M. Eikseth: Optimization of dividends with transaction costs. Manuscript 2001.

[Eb]   E. Eberlein: Application of generalized hyperbolic Lévy motion to finance. In O.E. Barndorff-Nielsen (ed.): Lévy Processes. Birkhäuser, Basel 2001, 319-336.

[EK]   E. Eberlein and U. Keller: Hyperbolic distributions in finance. Bernouilli 1 (1995), 281-299.

[F]   N.C. Framstad: Combined Stochastic Control for Jump Diffusions with Applications to Economics. Canadian Scientist Thesis, University of Oslo 1997.

[FØS1]   N.C. Framstad, B. Øksendal and A. Sulem: Optimal consumption and portfolio in a jump diffusion market. In A. Shiryaev and A. Sulem (eds.): Math. Finance INRIA, Paris 1998, 8–20.

[FØS2]   N.C. Framstad, B. Øksendal and A. Sulem: Optimal consumption and portfolio in a jump diffusion market with proportional transaction costs. J. Math. Econ. 35 (2001), 233–257.

[FØS3]   N.C. Framstad, B. Øksendal and A. Sulem: Sufficient stochastic maximum principle for optimal control of jump diffusions and applications to finance. J. Opt. Theor. Appl. 121 (2004), 77–98.

[Fre]   M. Freidlin: Functional Integration and Partial Differential Equations. Princeton University Press, 1985.

[FS]   W. Fleming and M. Soner: Controlled Markov Processes and Viscosity Solutions. Springer, Berlin Heidelberg New York 1993.

[GK]   I. Gyöngy and N.V. Krylov: Stochastic partial differential equations with unbounded coefficients I, II. Stochastics 1990.

[GS]   I.I. Gihman and A.V. Skorohod: Controlled Stochastic Processes. Springer, Berlin Heidelberg New York 1979.

[H1]   U.G. Haussmann: A Stochastic Maximum Principle for Optimal Control of Diffusions. Longman, London 1986.

[H2]   U.G. Haussmann: The maximum principle for optimal control of diffusions with partial information. SIAM J. Control Opt. 25 (1987), 341–361.

[HST]   M.J. Harrison, T. Selke and A. Taylor: Impulse control of a Brownian motion. Math. Oper. Res. 8 (1983), 454–466.

[I]   K. Itô: Spectral type of the shift transformation of differential processes with stationary increments. Trans. Am. Math. Soc. 81 (1956), 253–263.

[Is1]   H. Ishii: Viscosity solutions of nonlinear second order elliptic PDEs associated with impulse control problems. Funkciala Ekvacioj. 36 (1993), 132–141.

[Is2]   H. Ishii: On the equivalence of two notions of weak solutions, viscosity solutions and distribution solutions. Funkciala Ekvacioj. 38 (1995), 101–120.

[Isk]   Y. Iskikawa: Optimal control problem associated with jump processes. Appl. Math. Opt. 50 (2004), 21–65.

[J-P]   M. Jeanblanc-Picqué: Impulse control method and exchange rate. Math. Finance 3 (1993), 161–177.

[JK1]   E.R. Jakobsen and K.H. Karlsen: Continuous dependence estimates for viscosity solutions of integro-PDEs.. J. Differential Equations (2005), 278–318.

[JK2]   E.R. Jakobsen and K.H. Karlsen: A maximum principle for semicontinuous functions applicable to integro-partial differential equations. Nonlinear Diff. Eqn. Appl. 13 (2006), 1–29.

[JS]   J. Jacod and A. Shiryaev: Limit Theorems for Stochastic Processes. Springer, Berlin Heidelberg New York 1987.

[J-PS]   M. Jeanblanc-Picqué and A.N. Shiryaev: Optimization of the flow of dividends. Russian Math. Surv. 50 (1995), 257–277.

[K]   N.V. Krylov: Controlled Diffusion Processes. Springer, Berlin Heidelberg New York 1980.

[Ka]   O. Kallenberg: Foundations of Modern Probability. Second Edition. Springer-Verlag 2002.

[Kalli]   G. Kallianpur: Stochastic Filtering Theory. Springer, Berlin Heidelberg New York 1980.

[Ko1]   R. Korn: Portfolio optimization with strictly positive transaction costs and impulse control. Finance Stoch. 2 (1998), 85–114.

[Ko2]   R. Korn: Optimal Portfolios: Stochastic Models for Optimal Investment and Risk Management in Continuous Time. World Scientific, Singapore 1997.

[Ku]   H.J. Kushner: Necessary conditions for continuous parameter stochastic optimization problems. SIAM J. Control. 10 (1972), 550-565.

[KD]   H.J. Kushner and P. Dupuis: Numerical Methods for Stochastic Control Problems in Continuous Time. Springer-Verlag 1992.

[KS]   I. Karatzas and S. Shreve: Brownian Motion and Stochastic Calculus. 2nd Edition. Springer, Berlin Heidelberg New York 1991.

[Ky]   A. Kyprianou: Introductory Lectures on Fluctuations of Lévy Processes with Applications. Springer, Berlin Heidelberg New York 2006.

[La]   B. Larssen: The partially observed stochastic linear quadratic regulator: A direct approach. Preprint, University of Oslo 29/2001.

[Le]   D. Lefèvre: An introduction to utility maximization with partial observation. Finance 23 (2002), 93–126.

[LL]   P.-L. Lions and J.-M. Lasry: Stochastic control under partial information and applications to finance. CEREMADE (UMR 7534), Number 9912, 10/03/1999.

[LST]   B. Lapeyre, A. Sulem and D. Talay: Understanding Numerical Analysis for Financial Models. Cambridge University Press (in press).

[LS]   S. Leventhal and A.V. Skorohod: A necessary and sufficient condition for absence of arbitrage with tame portfolios. Ann. Appl. Probab. 5 (1995), 906–925.

[L]   A. Løkka: Martingale representation and functionals of Lévy processes. Stoch. Anal. Appl. 22 (2004), 867–892.

[LZ]   R.R. Lumley and M. Zervos: A model for investment in the natural resource industry with switching costs. Math. Oper. Res. 26 (2001), 637–653.

[M]   R. Merton: Optimal consumption and portfolio rules in a continuous time model. J. Econ. Theor. 3 (1971), 373–413.

[Ma]   C. Makasu: On some optimal stopping and stochastic control problems with jump diffusions. Ph.D. Thesis, University of Zimbabwe 2002.

[Me]   J.-L. Menaldi: Optimal impulse control problems for degenerate diffusions with jumps. Acta Appl. Math. 8 (1987), 165–198.

[MR]   J.-L. Menaldi and M. Robin: On a singular control problem for diffusions with jumps, IEEE Trans. Automatic Control, AC-29 (1984), 991–1004.

[MS]   M. Mnif and A. Sulem: Optimal risk control and divident pay-outs under excess of loss reinsurance. Res. Rep. RR-5010, November 2003, Inria-Rocquencourt.

[Mo]   E. Mordecki: Optimal stopping and perpatual options for Lévy processes. Finance Stoch. 6 (2002), 473–493.

[MØ]   G. Mundaca and B. Øksendal: Optimal stochastic intervention control with application to the exchange rate. J. Math. Econ. 29 (1998), 225–243.

[Mort]   R.E. Mortensen: Stochastic optimal control with noisy observations. Int. J. Control 4 (1966), 455–464.

[MP]   T. Meyer-Brandis and F. Proske: Explicit solution of a nonlinear filtering problem for Lévy processes with application to finance. Appl. Math. Opt. 50 (2004), 119–134.

[Ø1]   B. Øksendal: Stochastic Differential Equations. 6th Edition. Springer, Berlin Heidelberg New York 2003.

[Ø2]   B. Øksendal: Stochastic control problems where small intervention costs have big effects. Appl. Math. Opt. 40 (1999), 355–375.

[Ø3]   B. Øksendal: An Introduction to Malliavin Calculus with Applications to Economics. NHH Lecture Notes 1996.

[Ø4]   B. Øksendal: Optimal stopping with delayed information. Stoch. Dynam. 5 (2005), 271–280.

[Ø5]   B. Øksendal: Optimal control of stochastic partial differential equations. Stoch. Anal. Appl. 23 (2005), 165–179.

[ØR]   B. Øksendal and K. Reikvam: Viscosity solutions of optimal stopping problems. Stoch. Stoch. Rep. 62 (1998), 285–301.

[ØPZ]   B. Øksendal, F. Proske and T. Zhang: Backward stochastic partial differential equations with jumps and application to optimal control of random jump fields. Stochastics 77 (2005), 381–399.

[ØS]   B. Øksendal and A. Sulem: Optimal consumption and portfolio with both fixed and proportional transaction costs. SIAM J. Control Opt. 40 (2002), 1765–1790.

[ØUZ]   B. Øksendal, J. Ubøe and T. Zhang: Nonrobustness of some impulse control problems with respect to intervention costs. Stoch. Anal. Appl. 20 (2002), 999–1026.

[Pa1]   E. Pardoux: Stochastic partial differential equations and filtering of diffusion processes. Stochastics 3 (1979), 127–167.

[Pa2]   E. Pardoux: Filtrage non linéaire et équations aux dérivées partielles stochastiques associées. Ecole d'Eté de Probabilités de Saint-Flour 1989.

[P]   P. Protter: Stochastic Integration and Differential Equations. 2nd Edition. Springer, Berlin Heidelberg New York 2003.

[Ph]   H. Pham: Optimal stopping of controlled jump diffusion processes: a viscosity solution approach. J. Math. Syst. Estimation Control 8 (1998), 1–27.

[Pu]   M.L. Puterman: Markov Decision Processes: Discrete Stochastics Dynamic Programming. Probability and Mathematical Statistics: applied probability and statistics section. Wiley 1994.

[PBGM]   L.S. Pontryagin, V.G. Boltyanskii, R.V. Gamkrelidze and E.F. Mishchenko: The Mathematical Theory of Optimal Processes. Wiley, New York 1962.

[R]   R.T. Rockafellar: Convex Analysis. Princeton University Press 1970.

[S]   K. Sato: Lévy Processes and Infinitely Divisible Distributions. Cambridge University Press 1999.

[ST]   C. Schwab and R.A. Todor: Convergence rates for sparse chaos approximations of elliptic problems with stochastic coefficients. IMA J. Numer. Anal. (to appear).

[Sh]   A. Shiryaev: Optimal Stopping Rules. Springer, Berlin Heidelberg New York 1978.

[Sc]   W. Shoutens: Lévy Processes in Finance. Wiley, New York 2003.

[SeSy]   A. Seierstad and K. Sydsæter: Optimal Control Theory with Economic Applications. North-Holland, Amsterdam 1987.

[SS]   S.E. Shreve and H.M. Soner: Optimal investment and consumption with transaction costs. Ann. Appl. Probab. 4 (1994), 609–692.

[S1]   A. Sulem: A solvable one-dimensional model of a diffusion inventory system. Math. Oper. Res. 11 (1986), 125–133.

[S2]   A. Sulem: Explicit solution of a two-dimensional deterministic inventory problem. Math. Oper. Res. 11 (1986), 134–146.

[V]   H. Varner: Some Impulse Control Problems with Applications to Economics. Canadian Scientist Thesis, University of Oslo, 1997.

[W]   Y. Willassen: The stochastic rotation problem: a generalization of Faustmann's formula to stochastic forest growth. J. Econ. Dynam. Control 22 (1998), 573–596.

[YZ]   J. Yong and X.Y. Zhou: Stochastic Controls. Springer, Berlin Heidelberg New York 1999.

Notation and Symbols

$\mathbb{R}^n$ — n-dimensional Euclidean space

$\mathbb{R}^+$ — the nonnegative real numbers

$\mathbb{R}^{n \times m}$ — the $n \times m$ matrices (real entries)

$\mathbb{Z}$ — the integers

$\mathbb{N}$ — the natural numbers

$\mathbf{B}_0$ — the family of Borel sets $U \subset \mathbb{R}$ whose closure $\bar{U}$ does not contain 0

$\mathbb{R}^n \simeq \mathbb{R}^{n \times 1}$ — i.e., vectors in $\mathbb{R}^n$ are regarded as $n \times 1$ matrices

I_n — the $n \times n$ identity matrix

A^T — the transposed of the matrix A

$\mathcal{P}(\mathbb{R}^k)$ — set of functions $f : \mathbb{R}^k \to \mathbb{R}$ of at must polynomial growth, i.e., there exists constants C, m such that: $|f(y)| \leq C(1 + |y|^m)$ for all $y \in \mathbb{R}^k$

$C(U, V)$ — the continuous functions from U into V

$C(U)$ — the same as $C(U, \mathbb{R})$

$C_0(U)$ — the functions in $C(U)$ with compact support

$C^k = C^k(U)$ — the functions in $C(U, \mathbb{R})$ with continuous derivatives up to order k

$C_0^k = C_0^k(U)$ — the functions in $C^k(U)$ with compact support in U

$C^{k+\alpha}$ — the functions in C^k whose kth derivatives are Lipschitz continuous with exponent α

$C^{1,2}(\mathbb{R} \times \mathbb{R}^n)$ — the functions $f(t, x); \mathbb{R} \times \mathbb{R}^n \to \mathbb{R}$ which are C^1 w.r.t. $t \in \mathbb{R}$ and C^2 w.r.t. $x \in \mathbb{R}^n$

$C_b(U)$ — the bounded continuous functions on U

$|x|^2 = x^2$ — $\sum_{n=1}^{n} x_i^2$ if $x = (x_1, \ldots, x_n)$

$x \cdot y$ — the dot product $\sum_{n=1}^{n} x_i y_i$ if $x = (x_1, \ldots, x_n), y = (y_1, \ldots, y_n)$

x^+ — $\max(x, 0)$ if $x \in \mathbb{R}$

x^- — $\max(-x, 0)$ if $x \in \mathbb{R}$

$\text{sign } x$	$\begin{cases} 1 & \text{if } x \geq 0 \\ -1 & \text{if } x > 0 \end{cases}$
$\sinh(x)$	hyperbolic sine of x $\left(= \frac{e^x - e^{-x}}{2}\right)$
$\cosh(x)$	hyperbolic cosine of x $\left(= \frac{e^x + e^{-x}}{2}\right)$
$\text{tgh}(x)$	$\frac{\sinh(x)}{\cosh(x)}$
$s \wedge t$	the minimum of s and t $(= \min(s,t))$
$s \vee t$	the maximum of s and t $(= \max(s,t))$
δ_x	the unit point mass at x
$\text{Argmax}_{u \in U} f(u)$	$\{u^* \in U; f(u^*) \geq f(u), \forall u \in U\}$
$:=$	equal to by definition
$\overline{\lim}, \underline{\lim}$	the same as $\liminf, \limsup$
$\text{supp } f$	the support of the function f
∇f	the same as $Df = \left[\frac{\partial f}{\partial x_i}\right]_{i=1}^n$
∂G	the boundary of the set G
$\bar{G}$	the closure of the set G
G^0	the interior of the set G
χ_G	the indicator function of the set G; $\chi_G(x) = 1$ if $x \in G$, $\quad \chi_G(x) = 0$ if $x \notin G$
$(\Omega, \mathcal{F}, (\mathcal{F}_t)_{t \geq 0}, P)$	filtered probability space
$\Delta \eta_t$	the jump of η_t defined by $\Delta \eta_t = \eta_t - \eta_t^-$
P	the probability law of η_t
$N(t, U)$	see (1.1.2)
$\nu(U)$	$E[N(1, U)]$ see (1.1.3)
$\|\nu\|$	the norm (total mass) of the measure ν, i.e., $\nu(\mathbb{R})$
$\tilde{N}(dt, dz)$	see (1.1.7)
$B(t)$	Brownian motion
$P \ll Q$	the measure P is absolutely continuous w.r.t. the measure Q
$P \sim Q$	P is equivalent to Q, i.e., $P \ll Q$ and $Q \ll P$
E_Q	the expectation w.r.t. the measure Q
E	the expectation w.r.t. a measure which is clear from the context (usually P)
$E[Y] = E^\mu[Y] = \int Y \, d\mu$	the expectation of the random variable Y w.r.t. the measure μ
$[X, Y]$	quadratic covariation of X and Y, see Definition 1.28
$\mathcal{T}$	set of all stopping times $\leq \tau_S$ see (2.1.1)
τ_G	the first exit time from the set G of a process X_t: $\tau_G = \inf\{t > 0; X_t \notin G\}$
$\Delta_N Y(t)$	the jump of Y caused by the jump of N, see (5.2.2)
$\check{Y}(t^-)$	$Y(t^-) + \Delta_N Y(t)$ (see (6.1.5))
$\Delta_\xi Y(t)$	the jump of Y caused by the singular control ξ
$\Delta_\xi \phi$	see (5.2.3)

$\xi^c(t)$	continuous part of $\xi(t)$, i.e., the process obtained by removing the jumps of $\xi(t)$
π/K	the restriction of the measure π to the set K
$A = A_Y$	the generator of jump diffusion Y
$\mathcal{M}$	intervention operator, see Definition 6.1
VI	variational inequality
QVI	quasivariational inequality
HJB	Hamilton–Jacobi–Bellman equation
HJBVI	Hamilton–Jacobi–Bellman variational inequality
HJBQVI	Hamilton–Jacobi–Bellman quasivariational inequality
SDE	Stochastic differential equation
càdlàg	right continuous with left limits
càglàd	left continuous with right limits
i.i.d.	independent identically distributed
iff	if and only if
a.a., a.e., a.s.	almost all, almost everywhere, almost surely
w.r.t.	with respect to
s.t.	such that

Index

adjoint equation, 54, 164
adjoint operator, 164
adjoint processes, 164
admissible, 21, 45, 48, 53, 58, 78, 80
admissible combined controls, 123
admissible controls, 163
admissible impulse controls, 92
approximation theorem, 28
arbitrage, 20, 21
Arrow condition, 168
average maximum condition, 175

backward stochastic differential
    equation, 54
backward stochastic partial differential
    equation, 164
bankruptcy time, 27, 45

càdlàg, 1
càglàd, 5
combined control, 123
combined impulse linear regulator
    problem, 132
combined optimal stopping and
    stochastic control, 65, 67
combined stochastic control and impulse
    control, 123, 124
comparison theorem, 149
compensated Poisson random measure,
    4
compound Poisson process, 3
continuation region, 29, 68, 93
control process, 45, 53
controlled jump diffusion, 45

controls which do not depend on x, 174

delayed effect, 37
delayed information, 36
delayed optimal stopping, 42, 44
delayed stopping times, 36
diagonally dominant matrix, 156
discrete maximum principle, 157
Duncan–Mortensen–Zakai, 179, 180
dynamic programming, 45, 57
dynamic programming principle, 110,
    137, 138
Dynkin formula, 12

equivalent measure, 13

finite difference approximation, 153
first exit time from a ball, 23
first Fundamental Theorem of Asset
    Pricing, 21
fixed transaction cost, 96
full observation control, 176

geometric Lévy martingales, 22
geometric Lévy process, 7, 22, 23
Girsanov theorem, 12, 14–17
graph of the geometric Lévy process, 23

Hamilton–Jacobi–Bellman, 46
Hamiltonian, 53, 163
high contact principle, 33, 68
HJB-variational inequalities, 68
HJBQVI verification theorem, 124
Howard algorithm, 158

Universitext

Nikulin, V. V.; Shafarevich, I. R.: Geometries and Groups

Oden, J. J.; Reddy, J. N.: Variational Methods in Theoretical Mechanics

Øksendal, B.: Stochastic Differential Equations

Øksendal, B.; Sulem, A.: Applied Stochastic Control of Jump Diffusions

Orlik, P.; Welker, V.: Algebraic Combinatorics

Poizat, B.: A Course in Model Theory

Polster, B.: A Geometrical Picture Book

Porter, J. R.; Woods, R. G.: Extensions and Absolutes of Hausdorff Spaces

Procesi, C.: Lie Groups

Radjavi, H.; Rosenthal, P.: Simultaneous Triangularization

Ramsay, A.; Richtmeyer, R. D.: Introduction to Hyperbolic Geometry

Rautenberg, W.: A concise Introduction to Mathematical Logic

Rees, E. G.: Notes on Geometry

Reisel, R. B.: Elementary Theory of Metric Spaces

Rey, W. J. J.: Introduction to Robust and Quasi-Robust Statistical Methods

Ribenboim, P.: Classical Theory of Algebraic Numbers

Rickart, C. E.: Natural Function Algebras

Rotman, J. J.: Galois Theory

Rubel, L. A.: Entire and Meromorphic Functions

Ruiz-Tolosa, J. R.; Castillo E.: From Vectors to Tensors

Runde, V.: A Taste of Topology

Rybakowski, K. P.: The Homotopy Index and Partial Differential Equations

Sagan, H.: Space-Filling Curves

Samelson, H.: Notes on Lie Algebras

Sauvigny, F.: Partial Differential Equations I

Sauvigny, F.: Partial Differential Equations II

Schiff, J. L.: Normal Families

Sengupta, J. K.: Optimal Decisions under Uncertainty

Séroul, R.: Programming for Mathematicians

Seydel, R.: Tools for Computational Finance

Shafarevich, I. R.: Discourses on Algebra

Shapiro, J. H.: Composition Operators and Classical Function Theory

Simonnet, M.: Measures and Probabilities

Smith, K. E.; Kahanpää, L.; Kekäläinen, P.; Traves, W.: An Invitation to Algebraic Geometry

Smith, K. T.: Power Series from a Computational Point of View

Smoryński, C.: Logical Number Theory I. An Introduction

Stichtenoth, H.: Algebraic Function Fields and Codes

Stillwell, J.: Geometry of Surfaces

Stroock, D. W.: An Introduction to the Theory of Large Deviations

Sunder, V. S.: An Invitation to von Neumann Algebras

Tamme, G.: Introduction to Étale Cohomology

Tondeur, P.: Foliations on Riemannian Manifolds

Toth, G.: Finite Mbius Groups, Minimal Immersions of Spheres, and Moduli

Tu, L. W.: An Introduction to Manifolds

Verhulst, F.: Nonlinear Differential Equations and Dynamical Systems

Weintraub, S. H.: Galois Theory

Wong, M. W.: Weyl Transforms

Xambó-Descamps, S.: Block Error-Correcting Codes

Zaanen, A.C.: Continuity, Integration and Fourier Theory

Zhang, F.: Matrix Theory

Zong, C.: Sphere Packings

Zong, C.: Strange Phenomena in Convex and Discrete Geometry

Zorich, V. A.: Mathematical Analysis I

Zorich, V. A.: Mathematical Analysis II

Lightning Source UK Ltd.
Milton Keynes UK
UKOW05f0357280317
297672UK00002B/9/P

9 783540 698258